METHODS IN MOLECULAR BIOLOGY

For further volumes:
http://www.springer.com/series/7651

Plant Virology Protocols

New Approaches to Detect Viruses and Host Responses

Third Edition

Edited by

Ichiro Uyeda and Chikara Masuta

Hokkaido University, Sapporo, Japan

Humana Press

Editors
Ichiro Uyeda
Hokkaido University
Sapporo, Japan

Chikara Masuta
Hokkaido University
Sapporo, Japan

ISSN 1064-3745 ISSN 1940-6029 (electronic)
ISBN 978-1-4939-1742-6 ISBN 978-1-4939-1743-3 (eBook)
DOI 10.1007/978-1-4939-1743-3
Springer New York Heidelberg Dordrecht London

Library of Congress Control Number: 2014949862

Printed on acid-free paper

Humana Press is a brand of Springer
Springer is part of Springer Science+Business Media (www.springer.com)

Preface

Recent developments in genome analyses and cytological technologies have made noble approaches to solving many problems in plant virology. The techniques have been applied to not only diagnostic studies of virus diseases but also ecology and pathogenicity of viruses and host responses to viral infection.

The first six chapters describe techniques relevant to the detection of unknown viruses and disease diagnosis. Shimura et al. describe a simple and practical diagnostic method of macroarray. Diagnostic methods using PCR are presented by MacKenzie et al. for *Potato virus X*, Gabino for grapevine viruses, Sasaya for rice viruses, and Noris and Miozzi for *Tomato yellow leaf curl Sardinia virus*. Atsumi et al. describe the dsRNA extraction method for isolating viruses of unknown sequences.

The next nine chapters deal with utilization of meta-genome sequencing and global gene expression analyses for the search and identification of viruses, as well as the elucidation of host responses to viral infection. They also describe analyses of huge sequence data. The methods have potential to detect undescribed viruses and network of host responses not yet known. Kondo et al. look for non-retroviral RNA viruslike sequence in plant genome. Nagano et al. describe sequencing of RNAs by next-generation sequencers to detect viruses in wild plants. Wang et al. describe the cloning and profile analysis of satsiRNAs from satRNAs of *Cucumber mosaic virus*. Miozzi and Pantaleo describe analyses of the huge amount of data collected from viral short interfering RNA sequencing. Motooka et al. detected many viral sequences of plant and algal and protozoal viruses in addition to bacteriophage, mammal, and insect-derived viruses in fecal samples from cows by high-throughput sequencing. Adkar-Purushothama et al. describe a deep sequencing of viroid-infected plants and analyses of viroid-specific small RNAs. Ezawa et al. describe extraction of total dsRNAs from fungi, and their cDNAs are sequenced and searched for viral sequence. Matsuura et al. describe the high-throughput SuperSAGE for host and viral gene expression. Ishihara et al. describe the use of global gene expression analyses for elucidating the complex network for resistance to viruses.

The next three chapters describe construction methods of infectious cDNAs. The use of infectious cDNA clones of RNA viruses provides a homogenous viral genome population and makes it possible to examine critical single-point mutations involved in pathogenicity and breaking resistance of host plants. It is also used as a vector for expressing foreign gene genes. Nakahara et al. describe construction of infectious cDNA to *Bean yellow mosaic virus* and *Clover yellow vein virus*, legume-infecting potyviruses. Li and Yoshikawa describe construction of infectious *Apple latent spherical virus* cDNA vectors and its use for virus-induced gene silencing. Nagata et al. describe simplified methods for the construction of *Pepper mild mottle virus* and *Begomovirus*.

The last two chapters contain methods relevant to plant virus control. Voloudakis et al. describe production of dsRNA of a target RNA virus and its use for inducing RNA silencing. Mochizuki and Ohki detect and examine virus localization in the shoot apical meristem tissue in order to obtain virus-free meristem tips.

We would like to thank professor emeritus John M. Walker for his patient guidance and suggestions to complete the book. Lastly, we would like to thank Dr. Kenji Nakahara for helping us edit the manuscripts and all of the authors for their excellent contributions.

Sapporo, Japan

Chikara Masuta
Ichiro Uyeda

Contents

Contributors

CHARITH RAJ ADKAR-PURUSHOTHAMA • *RNA Group, Département de Biochimie, Pavillon de Recherche Appliquée sur le, Université de Sherbrooke, Sherbrooke, QC, Canada*
GO ATSUMI • *Iwate Biotechnology Research Center, Kitakami, Iwate, Japan*
DENNIS H. BAMFORD • *Department of Biosciences and Institute of Biotechnology, University of Helsinki, Helsinki, Finland*
SOTARO CHIBA • *Institute of Plant Science and Resources (IPSR), Okayama University, Kurashiki, Japan*
TATSUHIRO EZAWA • *Graduate School of Agriculture, Hokkaido University, Sapporo, Hokkaido, Japan*
YUAN-YUAN FANG • *State Key Laboratory of Plant Genomics and National Center for Plant Gene Research, Institute of Microbiology, Chinese Academy of Sciences, Beijing, China*
KAZUYOSHI FURUTA • *Hokusan Co. Ltd., Kitahiroshima, Japan*
GIORGIO GAMBINO • *Institute for Sustainable Plant Protection, National Research Council (IPSP-CNR), Grugliasco Unit, Grugliasco, TO, Italy*
HUI-SHAN GUO • *State Key Laboratory of Plant Genomics and National Center for Plant Gene Research, Institute of Microbiology, Chinese Academy of Sciences, Beijing, China*
KATSURO HAGIWARA • *School of Veterinary Medicine, Rakuno Gakuen University, Ebetsu, Japan*
MARIA C. HOLEVA • *Department of Phytopathology, Laboratory of Bacteriology, Benaki Phytopathological Institute, Kifissia, Greece*
MIE N. HONJO • *Center for Ecological Research, Kyoto University, Otsu, Shiga, Japan*
YOJI IKEDA • *Graduate School of Agriculture, Hokkaido University, Sapporo, Hokkaido, Japan*
ALICE KAZUKO INOUE-NAGATA • *Laboratory of Virology and Molecular Biology, Embrapa Vegetables, Brazilian Agricultural Research Corporation, Brasilia, DF, Brazil*
TAKEAKI ISHIHARA • *National Agricultural Research Center, National Agriculture and Food Research Organization, Tsukuba, Ibaraki, Japan*
GÜNTER KAHL • *Biocenter, University of Frankfurt am Main, Frankfurt, Germany*
KAPPEI KOBAYASHI • *Faculty of Agriculture, Ehime University, Matsuyama, Ehime, Japan*
HIDEKI KONDO • *Institute of Plant Science and Resources (IPSR), Okayama University, Kurashiki, Japan*
DETLEV H. KRÜGER • *Institute of Medical Virology, Charité Universitätsmedizin Berlin, Berlin, Germany*
HIROSHI KUDOH • *Center for Ecological Research, Kyoto University, Otsu, Shiga, Japan*
JOANNE R.M. LEE • *CSIRO Plant Industry, Canberra, ACT, Australia*
CHUNJIANG LI • *Faculty of Agriculture, Iwate University, Morioka, Japan*
SHIFANG LI • *State Key Laboratory of Biology of Plant Diseases and Insect Pests, Institute of Plant Protection, Chinese Academy of Agricultural Sciences, Beijing, China*
TYLER D.B. MACKENZIE • *Agricultural Certification Services, Fredericton, NB, Canada*
CHIKARA MASUTA • *Research Faculty of Agriculture, Hokkaido University, Sapporo, Japan*

HIDEO MATSUMURA • *Gene Research Center, Shinshu University, Ueda, Nagano, Japan*
MOTOHIRO MIHARA • *Dynacom Co. Ltd., Chiba-shi, Chiba, Japan*
LAURA MIOZZI • *Institute for Sustainable Plant Protection of the CNR, National Research Council (CNR), Turin, Italy*
TOMOFUMI MOCHIZUKI • *Graduate School of Life and Environmental Sciences, Osaka Prefecture University, Osaka, Japan*
DAISUKE MOTOOKA • *Department of Infection Metagenomics, Research Institute for Microbial Disease, Osaka University, Osaka, Japan*
ATSUSHI J. NAGANO • *Center for Ecological Research, Kyoto University, Otsu, Shiga, Japan*
TATSUYA NAGATA • *Laboratório de Microscopia e Virologia, Departamento de Biologia Celular, Conselho de Ensino, Pesquisa, e Extensão, Universidade de Brasília, Brasilia, DF, Brazil*
KENJI S. NAKAHARA • *Pathogen-Plant Interactions Group, Plant Breeding Science, Research Faculty of Agriculture, Hokkaido University, Sapporo, Japan*
SHOTA NAKAMURA • *Department of Infection Metagenomics, Research Institute for Microbial Disease, Osaka University, Osaka, Japan*
TAKAAKI NAKAYA • *Department of Infectious Diseases, Kyoto Prefectural University of Medicine, Kyoto, Japan*
XIANZHOU NIE • *Potato Research Centre Agriculture and Agri-Food Canada, Fredericton, NB, Canada*
KEI NISHINO • *Pathogen-Plant Interactions Group, Plant Breeding Science, Research Faculty of Agriculture, Hokkaido University, Sapporo, Japan*
EMANUELA NORIS • *Institute for Sustainable Plant Protection (IPSP), National Research Council (CNR), Torino, Italy*
SATHOSHI T. OHKI • *Graduate School of Life and Environmental Sciences, Osaka Prefecture University, Osaka, Japan*
VITANTONIO PANTALEO • *Institute for Sustainable Plant Protection of the CNR, Research Unit of Bari, Bari, Italy*
MINNA M. PORANEN • *Department of Biosciences, University of Helsinki, Helsinki, Finland*
TERUO SANO • *Plant Pathology Laboratory, Faculty of Agriculture and Life Science, Hirosaki University, Hirosaki, Japan*
L. PETER SARIN • *Department of Biosciences and Institute of Biotechnology, University of Helsinki, Helsinki, Finland; Max Planck Institute for Molecular Biomedicine, Münster, Germany*
TAKAHIDE SASAYA • *Kyusyu Okinawa Agricultural Research Center, National Agriculture and Food Research Organization, Koshi, Kumamoto, Japan*
MASANAO SATO • *National Institute for Basic Biology, Okazaki Institute for Integrative Bioscience, National Institutes of Natural Sciences, Okazaki, Japan*
YUKIYO SATO • *Graduate School of Agricultural Science, Tohoku University, Sendai, Japan*
KEN-TARO SEKINE • *Iwate Biotechnology Research Center, Kitakami, Iwate, Japan*
HANAKO SHIMURA • *Research Faculty of Agriculture, Hokkaido University, Sapporo, Japan*
MATHURESH SINGH • *Agricultural Certification Services, Fredericton, NB, Canada*
NEIL A. SMITH • *CSIRO Plant Industry, Canberra, ACT, Australia*
NOBUHIRO SUZUKI • *Institute of Plant Science and Resources (IPSR), Okayama University, Kurashiki, Japan*
HIDEKI TAKAHASHI • *Graduate School of Agricultural Science, Tohoku University, Sendai, Japan*

Francisco Tenllado • *Centro de Investigaciones Biológicas, Consejo Superior de Investigaciones Científicas, Madrid, Spain*
Ryohei Terauchi • *Iwate Biotechnology Research Center, Kitakami, Iwate, Japan*
Ichiro Uyeda • *Pathogen-Plant Interactions Group, Plant Breeding Science, Research Faculty of Agriculture, Hokkaido University, Sapporo, Japan*
Marisol Vargas • *Departamento de Biología Medioambiental, Centro de Investigaciones Biológicas, CSIC, Madrid, Spain*
Andreas E. Voloudakis • *Laboratory of Plant Breeding and Biometry, Agricultural University of Athens, Athens, Greece*
Ming-Bo Wang • *CSIRO Plant Industry, Canberra, ACT, Australia*
Nobuyuki Yoshikawa • *Plant Pathology Laboratory, Faculty of Agriculture, Iwate University, Morioka, Japan*
Zhixiang Zhang • *State Key Laboratory of Biology of Plant Diseases and Insect Pests, Institute of Plant Protection, Chinese Academy of Agricultural Sciences, Beijing, China*
Jian-Hua Zhao • *State Key Laboratory of Plant Genomics and National Center for Plant Gene Research, Institute of Microbiology, Chinese Academy of Sciences, Beijing, China*

Chapter 1

Detection of Plant Viruses in Mixed Infection by a Macroarray-Assisted Method

Hanako Shimura, Kazuyoshi Furuta, and Chikara Masuta

Abstract

The protocol for a simple, sensitive, and specific method using a cDNA macroarray to detect multiple viruses is provided. The method can be used even at the production sites for crops, which need a reliable routine diagnosis for mixed infection of plant viruses. The method consists of three steps: RNA extraction, duplex RT-PCR, and "microtube hybridization" (MTH). Biotinylated cDNA probes are prepared using RT-PCR and used to hybridize a nylon membrane containing target viral cDNAs by MTH. Positive signals can be visualized by colorimetric reaction and judged by eyes. We here demonstrate this method to detect asparagus viruses (*Asparagus virus 1* and *Asparagus virus 2*) from latently infected asparagus plants.

Key words Asparagus virus 1, Asparagus virus 2, Macroarray, Microtube hybridization, Detection of multiple viruses

1 Introduction

Enzyme-linked immunosorbent assay (ELISA) is the most common and practical method for virus detection [1], but it is used to detect a single virus; in a field, multiple viruses often infect plants. RT-PCR is also a sensitive method to detect plant viruses [2], but we often suffer from nonspecific bands especially as the number of samples goes up although we determined the best conditions for RT-PCR. In addition, if there is high genetic variability in viral sequences, RT-PCR is not necessarily the most sensitive and accurate method [3]. Considering the advantage that we can confirm the viral sequences by a hybridization-based method, we here applied a macroarray-based method to analyze gene expression in human cells [4] and diagnosis of plant viruses [5]. In this chapter, the macroarray system with the microtube hybridization (MTH) is shown to detect two asparagus viruses (*see* Fig. 1). For array production, cDNA

Ichiro Uyeda and Chikara Masuta (eds.), *Plant Virology Protocols*, Methods in Molecular Biology, vol. 1236,
DOI 10.1007/978-1-4939-1743-3_1, © Springer Science+Business Media New York 2015

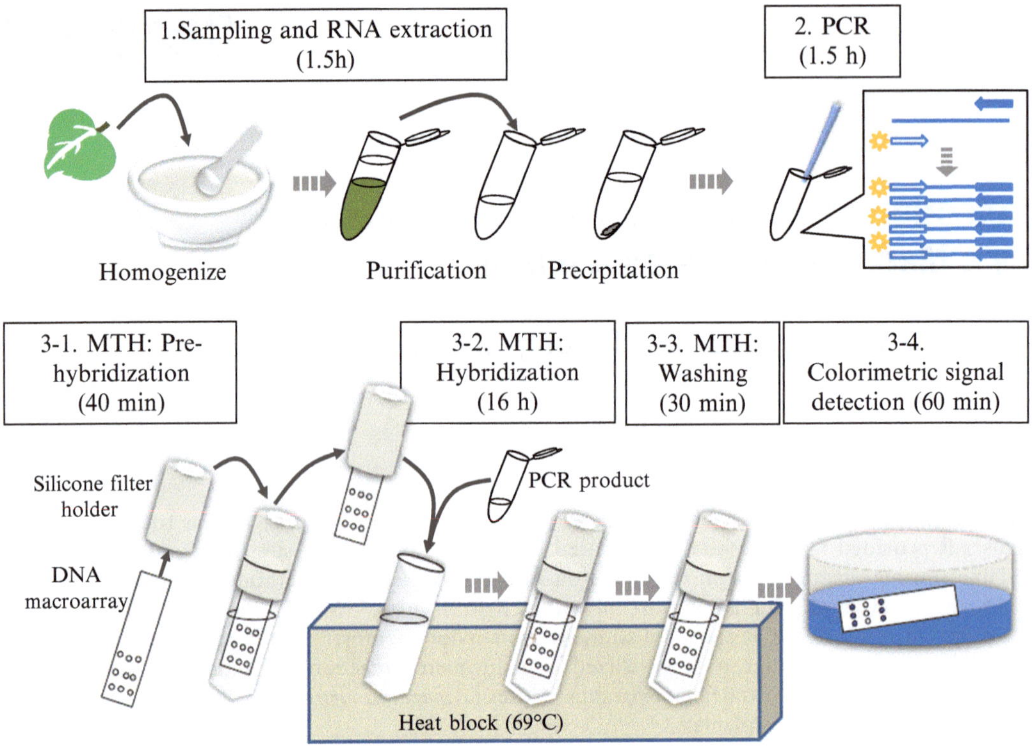

Fig. 1 A schematic outline of the microtube hybridization (MTH) method to detect plant viruses in the macroarray-assisted method

containing the conserved sequence of each virus or plant gene is first amplified by PCR. Next, we spot the cDNA fragments in triplicate on a nylon membrane, using high-throughput liquid handling systems with glass syringes, as reported previously [6]. When the number of target viruses is not so high, we can use a handmade array just as we make a dot blot membrane. For sample preparation, biotinylated cDNA is synthesized with a biotin-labeled primer from RNA sample by conventional RT-PCR in a one-step RT-PCR system (*see* Fig. 2). Hybridization is carried out using MTH, which significantly saves time and labor. With this new method, a vinyl bag is not necessary and instead a tube containing a membrane is simply transferred from one heat block to another (*see* Fig. 3). A cDNA macroarray strip is inserted into the slit of the silicone filter holder, which can be handmade as shown below (*see* Fig. 4). The positive signals on a hybridized membrane can be easily confirmed by colorimetric detection. This method can be used not only by researchers but also by farmers who have a small-scale laboratory.

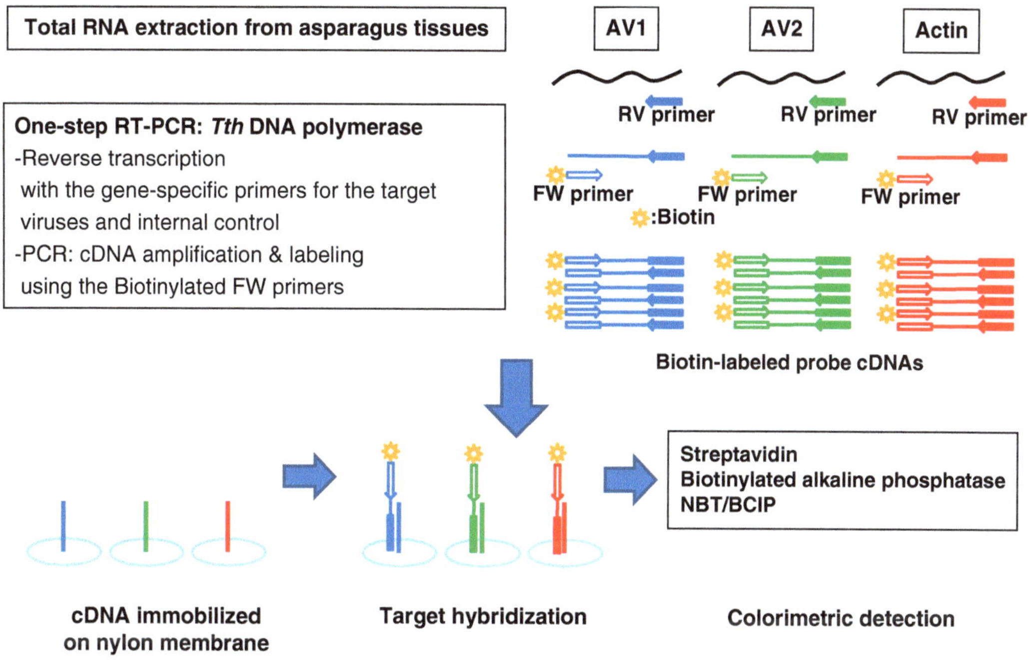

Fig. 2 cDNA preparation for the macroarray hybridization. Viral RNAs in total RNA from asparagus tissues are reverse-transcribed with virus-specific reverse primers to synthesize cDNAs, and then cDNAs are amplified by PCR with pairs of the biotinylated virus-specific forward and reverse primers. cDNA synthesis and amplification are performed simultaneously in one tube by one-step RT-PCR method using *Tth* DNA polymerase. The biotin-labeled cDNA probes are hybridized with cDNAs on macroarray. Because the cDNAs on the macroarray were designed from the region that does not cover the biotinylated forward primer sequence, unreacted biotinylated primers or nonspecific cDNAs generated by RT-PCR theoretically does not mishybridize to the macroarray

2 Material

2.1 Plasmid DNAs of Viral cDNAs and the Actin Gene

1. Cloning vector plasmid: for example, pCR 2.1 (Invitrogen).
2. *Escherichia coli* competent cells: for example, INVαF′ competent cell (Invitrogen).
3. Ampicillin sodium salt is dissolved in water at 100 mg/ml. Store at −30 °C.
4. LB medium or any pre-made medium such as Circlegrow (MP Biomedicals) medium (add 4 capsules/100 ml of water).
5. Any miniprep kit: for example, QIAprep Spin Miniprep Kit (QIAGEN).

2.2 Macroarray Construction

1. Oligonucleotide primers to amplify the cDNA (203 bp) of *Asparagus virus 1* (AV-1) are as follows: AV1-CP-FW7, 5′-TACATGCCAAGGTATGGACG-3′ and AV1-CP-RV4, 5′-TCCGTGTCCTCTTCCTGTGT-3′.

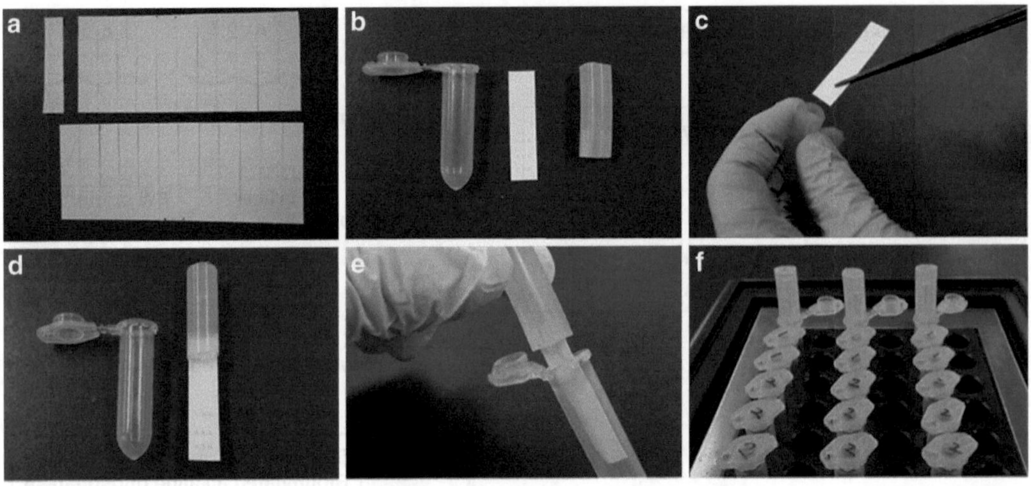

Fig. 3 The MTH method procedures. (**a**) Cut the membrane into pieces suitable for the MTH method. (**b**) 2-ml microtube, a macroarray, and a silicone filter holder. (**c**) Insert the macroarray into the slit of the silicon filter holder. (**d**) Microtube with a macroarray membrane inserted into the silicon filter holder. (**e**) Insert the macroarray into a microtube by the holder as a finger grip. (**f**) All the experimental steps including pre-hybridization, hybridization, and washing are performed in a 2-ml microtube on a heat block. The silicon filter holder serves to shut the microtube instead of the lid, and the membrane is easily transferred to the next tube

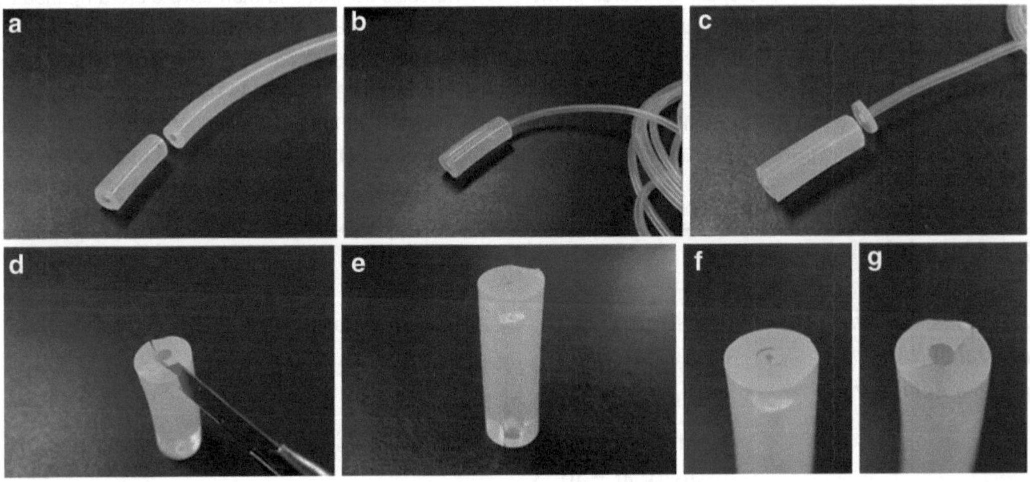

Fig. 4 Procedures to make a silicon filter holder. (**a**) Cut a silicone tube (1 mm internal diameter × 3 mm outside diameter) in approximately 3 cm. (**b**) Insert the silicone tube (1 mm outside diameter) into the tube on one side. (**c**) Cut the silicone tube to make the lid of a filter holder. (**d**) Slit the bottom of the tube to make an insertion site for a membrane. The cutting angle is preferable to be obtuse. (**e**) Example of a silicone filter holder. (**f**) Close-up of the top side. (**g**) Close-up of the bottom side

2. Oligonucleotide primers to amplify the cDNA (133 bp) of *Asparagus virus 2* (AV-2) are as follows: AV2-5-60, 5′-GCTAAGTCCATATGCCCATCT-3′ and AV2-3utr-RV1, 5′-GCATCTCCTTTGGAGGCATCTA-3′.

3. Oligonucleotide primers to amplify the cDNA (224 bp) of asparagus *actin* gene are as follows: AoAct-FW2, 5′-ATACGCCCTTCCTCATGCCA-3′ and AoAct-real-R, 5′-GAAGCTCGTAGTTCTTCTCCAC-3′.
4. Any *Taq* DNA polymerase: for example, KOD plus DNA polymerase (TOYOBO).
5. Any gel extraction kit: for example, QIAquick Gel Extraction Kit (QIAGEN).
6. Xylen Cyanol is dissolved in water at 0.1 mg/ml.
7. Positively charged nylon membrane: for example, Biodyne plus (Pall).

2.3 Plants and Viruses

Both AV-1 and AV-2 will be easily isolated in asparagus fields. We actually found AV-1-infected and AV-2-infected asparagus plants in the field of our university (Hokkaido University). We can isolate AV-2 even in asparagus in the market because most of asparagus plants are latently infected with AV-2.

2.4 Amplification of Biotin-Labeled Probe cDNAs

1. RNA extraction buffer: 25 mM Tris–HCl (pH 7.5), 25 mM $MgCl_2$, 25 ml KCl, and 1 % sodium dodecyl sulfate (SDS).
2. TE-saturated phenol (liquid phenol–TE buffer = 4:1).
3. TE-saturated phenol–chloroform (phenol–chloroform = 1:1).
4. Oligonucleotide primers to amplify the cDNA probe (223 bp) of AV-1 are as follows: AV1-CP-FW6, 5′-TGCGTAACACA ACTGAGAAG-3′ and AV1-CP-RV4 described above. AV1-CP-FW6 is biotinylated at the 5′ end.
5. Oligonucleotide primers to amplify the cDNA probe (151 bp) of AV-2 are as follows: AV2-3utr-FW1, 5′-GCTCAGCACC TAGCCTAAGC-3′ and AV2-3utr-RV1 described above. AV2-3utr-fw1 is biotinylated at the 5′ end.
6. Oligonucleotide primers to amplify the cDNA probe (277 bp) of the asparagus *actin* gene are as follows: AoAct-FW1, 5′-ATTGTGCTGGACTCTGGTGATG-3′ and AoAct-real-R described above. AoAct-FW1 is biotinylated at the 5′ end.
7. Any *Taq* DNA polymerase with reverse transcription activity: for example, *Tth* DNA polymerase (Roche).

2.5 Microtube Hybridization (MTH)

1. Pre-made hybridization solution such as PerfectHyb (TOYOBO).
2. Silicone filter holder: cut a silicone tube (1 mm internal diameter × 3 mm outside diameter) in 3 cm, and put a lid with another silicone tube (1 mm outside diameter) at the top and slit the bottom (*see* **Note 1** and Fig. 4).
3. 2-ml microtube (*see* **Note 2**).

4. Wash buffer-1: 2× SSC and 0.1 % SDS.
5. Wash buffer-2: 0.1× SSC and 0.1 % SDS.
6. Blocking solution: 5 % SDS, 125 mM NaCl and 25 mM phosphate buffer (pH 7.2).
7. Wash buffer-3: 1:10 dilution of the blocking solution.
8. Streptavidin solution: Add 10 μl of 1 mg/ml streptavidin (NEB) to 15 ml of the blocking solution, and mix gently by pipetting.
9. BAP solution: Add 10 μl of biotinylated alkaline phosphatase (Bio-Rad) to 15 ml of the blocking solution, and mix gently by pipetting.
10. Wash buffer-4: 10 mM NaCl, 1 mM $MgCl_2$ and 10 mM Tris–HCl (pH 9.5).
11. NBT/BCIP solution: 1 tablet of NBT/BCIP Ready-to-Use Tablets (Roche) is dissolved in 10 ml of water.
12. Plastic container (*see* **Note 3**).
13. Wrap film.

3 Methods

Asparagus virus 1 (AV-1, *Potyvirus*) and *Asparagus virus 2* (AV-2, *Ilarvirus*) are commonly found in commercial asparagus fields. These viruses cause no obvious symptoms in asparagus plants by single infection, but the double infection with AV-1 and AV-2 significantly causes a reduction in vigor and productivity of asparagus. In addition, it has been shown that asparagus plants doubly infected with AV-1 and AV-2 are more susceptible to *Fusarium* crown rot than those with either AV-1 or AV-2, and thus the mixed infection to asparagus has been thought to be responsible for asparagus decline disease. Because mixed infection of AV-1 and AV-2 causes unfavorable damages in asparagus production, it is important to efficiently and easily detect these viruses in asparagus plants. We here show a simple, efficient macroarray-assisted method to detect mixed infection of AV-1 and AV-2 in asparagus plants (*see* Fig. 5). Our macroarray method is adaptable to other crops with mixed infection such as garlic, edible lily, potatoes, yams, etc.

3.1 Preparation of Plasmid DNA

1. AV-1 cDNA, AV-2 cDNA, and the asparagus *actin* cDNA were cloned in pCR2.1 to create pCR-AV1, pCR-AV2, and pCR-AoAct, respectively. The gene-specific primers for cDNA cloning were designed based on the viral sequences obtained from the Genbank database (AV-2 RNA3, X86352), the previous report by Scott et al. [7] and our unpublished AV-1 sequences.

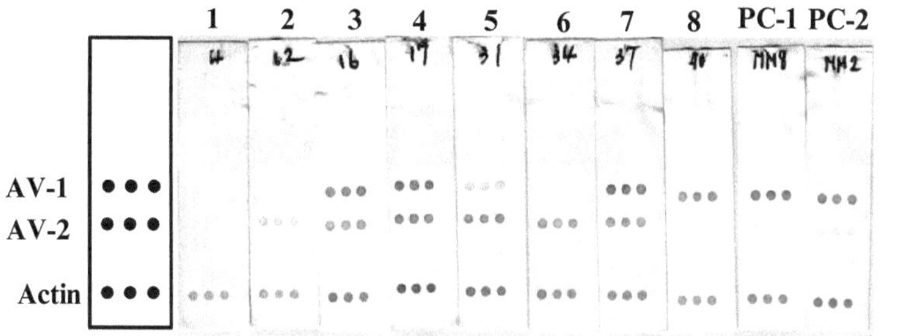

	Sample No.									
	1	2	3	4	5	6	7	8	PC-1	PC-2
AV-1	-	-	+	+	+	-	+	+	+	+
AV-2	-	+	+	+	+	+	+	-	+	+
Actin	+	+	+	+	+	+	+	+	+	+

Fig. 5 Representative results of the macroarray detection of asparagus viruses. Macroarray-assisted detection of *Asparagus virus 1* (AV-1) and *Asparagus virus 2* (AV-2) was conducted using total RNA from leaves of asparagus plants, which were collected from fields. We can clearly diagnose the virus-infected plants as indicated below the array image. Actin, the asparagus *acitn* gene as an internal control. PC-1 and PC-2, positive controls for RNA from the asparagus plants doubly infected with both AV-1 and AV-2

2. *E. coli* INVαF′ is transformed with one of plasmid DNAs. The *E. coli* cells are cultured overnight at 37 °C in Circlegrow medium containing 100 μg/ml ampicillin. Plasmid DNAs are prepared by using a QIAprep Spin Miniprep Kit according to the manufacturer's instructions.

3.2 Primer Construction

1. For amplification of cDNA for macroarray and probe cDNAs, design gene-specific primers to obtain a DNA fragment containing the conserved sequence of AV-1 coat protein gene or AV-2 3′-noncoding region.
2. As a control, design primers to amplify a DNA fragment of the asparagus *actin* gene.

3.3 Amplification of cDNAs for Macroarray

1. The cDNAs used on a macroarray are amplified by PCR with gene-specific primers as described above. The plasmids pCR-AV1, pCR-AV2, and pCR-AoAct are used as PCR templates.
2. PCR reactions (total volume 50 μl) are as follows: 5 μl of 10× reaction buffer, 5 μl of dNTP mix, 2 μl of $MgSO_4$, 2 μl of primer mix (final 0.4 μM each), 1 μl of each plasmid DNA, 1 μl of KOD DNA polymerase, and dH_2O up to 50 μl (*see* **Note 4**).

3. Use the following PCR conditions: one cycle of 94 °C/2 min, 35 cycles of 94 °C/15 s, 56 °C/30 s, 68 °C/30 s.
4. PCR products are purified using a QIAquick Gel Extraction kit according to the manufacturer's instructions.
5. Quantify the purified DNAs spectrophotometrically and dilute to 25 ng/μl (colored by 0.1 mg/ml Xylene Cyanol) to use as cDNAs for array spotting.

3.4 Macroarray Construction

1. The cDNAs (0.2 μl of 25 ng/μl) are spotted in triplicate on a nylon membrane using high-throughput liquid handling systems with glass syringe as reported previously [6] (*see* **Note 5**).
2. To denature cDNAs on the solid phase, incubate the membrane at 120 °C for 30 min in a dry incubator (*see* **Note 6**).
3. For immobilization, irradiate the membrane with UV at 254 nm (120 mJ/cm^2) in a UV cross-linker.
4. Cut the membrane into 38 mm × 9 mm strips suitable for the microtube hybridization method.

3.5 Preparation of Total RNA

1. Total RNA is extracted from fern (cladodes) of asparagus as follows: homogenize 50–100 mg asparagus tissue in 500 μl of TE-saturated phenol and 500 μl of RNA extraction buffer using a mortar and pestle (*see* **Note 7**).
2. Transfer the homogenate to a new 1.5 ml microtube, vortex briefly and then centrifuge for 3 min at 11,000 × *g* at room temperature.
3. Transfer the aqueous phase to a new tube and add an equal volume of TE-saturated phenol–chloroform. Vortex the mixture and then centrifuge for 5 min at 15,000 × *g* at room temperature (*see* **Note 8**).
4. Repeat TE-saturated phenol–chloroform extraction (*see* **Note 9**).
5. Collect the aqueous phase and add 1/10 volume of 3 M sodium acetate (pH 5.2) and 3 volumes of 100 % ethanol, and then turn over the tube (mix gently).
6. Centrifuge for 15 min at 15,000 × *g* at 4 °C. Carefully remove the supernatant and wash the pellet with 500 μl of 70 % ethanol followed by a spin at 15,000 × *g* for 5 min at 4 °C.
7. Discard the supernatant, spin at 15,000 × *g* for 1 min again to remove the residual supernatant.
8. Vacuum-dry or let it stand until it dries (for 5 min at room temperature). The pellet is then resuspended in 50 μl of RNase-free water (*see* **Note 10**).
9. Quantify the nucleic acid spectrophotometrically and check RNA by 1.2 % agarose-gel electrophoresis.

3.6 Amplification of Biotin-Labeled Probe cDNAs (See Note 11)

1. Probe cDNAs are amplified by one-step RT-PCR using biotin-labeled primers (the 5′ end of forward primers are biotinylated).
2. Conduct one-step multiplex RT-PCR reaction for amplification of AV-1 cDNA and the *actin* gene cDNA as follows: 5 μl of 5× RT-PCR buffer, 2.5 μl of $Mn(OAc)_2$, 2.5 μl of 2.5 mM each dNTP mix, 0.5 μl of 20 μM AV1-CP-FW6, 0.5 μl of 20 μM AV1-CP-RV4, 0.5 μl of 20 μM AoAct-FW1, 0.5 μl of 20 μM AoAct-real-R, 0.5 μl of 5 U/μl *Tth* DNA polymerase, 1 μl of template RNA, and RNase-free water up to 25 μl.
3. Conduct one-step RT-PCR reaction for amplification of AV-2 cDNA as follows: 5 μl of 5× RT-PCR buffer, 2.5 μl of $Mn(OAc)_2$, 2.5 μl of 2.5 mM each dNTP mix, 0.5 μl of 20 μM AV2-3utr-FW1, 0.5 μl of 20 μM AV2-3utr-RV1, 0.5 μl of 5 U/μl *Tth* DNA polymerase, 1 μl of template RNA, and RNase-free water up to 25 μl.
4. Use the following one-step RT-PCR: One cycle of 60 °C/30 min, 94 °C/1 min and 35 cycles of 94 °C/30 s, 56 °C/30 s, 72 °C/30 s.
5. Denature RT-PCR products by incubating at 98 °C for 5 min, and then chill on ice immediately. Keep on ice until use.

3.7 Microtube Hybridization

1. All the experimental steps including pre-hybridization, hybridization, and washing are performed in a 2-ml microtube on a heat block at 69 °C.
2. Set a heat block at 69 °C and put 2-ml microtubes filled with 1.7 ml of PerfectHyb solution on a heat block (2 tubes per test sample). Incubate the tubes for at least 20 min.
3. Insert a cDNA macroarray strip (membrane) into the slit of the silicone filter holder. For pre-hybridization, put the membrane with a filter holder in 2-ml microtubes containing PerfectHyb and incubate at 69 °C for at least 40 min.
4. Add 15 μl of denatured probe cDNAs to another 2-ml microtube of PerfectHyb kept at 69 °C (*see* **Note 12**).
5. Transfer the membranes to the 2-ml microtube with probe cDNAs, and incubate at 69 °C overnight (*see* **Note 13**).
6. Prepare 2-ml microtubes with wash buffer-1 and wash buffer-2 (3 tubes per test sample) on a heat block at 69 °C (at least 20 min prior to washing).
7. Wash the membrane as follows. Primary washing: transfer the membrane to the 2-ml microtube with wash buffer-1 three times (incubate for 1 min, 1 min and 10 min). Secondary washing: transfer the membrane to the 2-ml microtube with wash buffer-2 three times (incubate for 1 min, 1 min, and 10 min) (*see* **Note 14**).

3.8 Colorimetric Detection (See Note 15)

1. Rinse the membrane with 15 ml of the blocking solution in a plastic container.
2. Incubate the membrane with 15 ml of another blocking solution at room temperature for 10 min.
3. Incubate the membrane with 15 ml of streptavidin solution at room temperature for 5 min.
4. Rinse the membrane with 15 ml of wash buffer-3 two times.
5. Incubate the membrane with 15 ml of wash buffer-3 for 5 min.
6. Incubate the membrane with 15 ml of BAP solution at room temperature for 5 min.
7. Rinse the membrane with 15 ml of wash buffer-3 two times.
8. Incubate the membrane with 15 ml of wash buffer-3 for 5 min.
9. Rinse the membrane with 15 ml of wash buffer-4 two times.
10. Incubate the membrane with 15 ml of wash buffer-4 for 5 min.
11. Incubate the membrane with 10 ml of NBT/BCIP solution for 1 min.
12. Wrap the membrane with a wrap film, and conduct a colorimetric reaction for 15–60 min (*see* **Note 16**).
13. Wash the membrane with dH_2O three times and dry in the dark.

4 Notes

1. A silicon filter holder is actually a silicon tube. It has a hole through the tube, which can condition inner pressure in the microtube.
2. We can use any microtubes: for example, SafeSeal Micro Tube 2 ml, PP (SARSTED) and Sampling tube LT-0200 (BIO-BIK).
3. Alternatively, a 50 ml conical centrifuging tube can be used.
4. As appropriate, the reaction volume should be scaled up.
5. Instead of the mechanical spotting, we can manually spot DNA on a membrane by micropipette if target viruses are not so many.
6. During the incubation, the membrane should be sandwiched between two pieces of filter paper.
7. As appropriate, the buffer volume should be scaled up.
8. Approximately 400 μl of aqueous phase can be recovered.
9. TE-saturated phenol–chloroform extraction should be repeated until the middle phase almost disappears.

10. Do not dry the pellet completely as this will decrease its solubility.
11. Alternatively, biotin-labeled cRNA amplification kit for microarray analysis (e.g., Illumina TotalPrep RNA Amplification kit, Ambion) can be used, if you want to increase the sensitivity.
12. Pre-hybridization and hybridization can be performed in the same tube without changing the PerfectHyb solution.
13. The holder may be tied by Parafilm so that the microtube can be tightly held in the holder.
14. During washing on a heat block, shaking is not necessary because of the agitation effect in the buffer.
15. We can simultaneously handle at most 12 macroarray strips for 15 ml buffer. With increasing the number of macroarray strips, the volume of buffer should be scaled up.
16. We recommend that the staining incubation should be stopped in 60 min. Over-incubation will result in background staining and false positive signals.

Acknowledgment

We thank Ms. Chieko Hirata for her excellent technical help. This work was supported in part by a Grant-in-Aid for Research and Development Projects for Application in Promoting New Policy of Agriculture, Forestry and Fisheries from the Ministry of Agriculture, Forestry and Fisheries (MAFF) (Japan).

References

1. McLaughlin MR, Barnett OW, Burrows PM, Baum RH (1981) Improved ELISA conditions for detection of plant viruses. J Virol Methods 3:13–25
2. Li R, Hartung JS (2007) Reverse transcription-polymerase chain reaction-based detection of plant viruses. In: Current Protocols in Microbiology Supplement 6, Wiley & Sons Inc:16C .1.1–1.9
3. Kominek P, Glasa M, Kominkova M (2009) Analysis of multiple virus-infected grapevine plant reveals persistence but uneven virus distribution. Acta Virol 53:281–285
4. Sugiyama S, Yamamoto K, Nishimura N, Nakagawa M, Maruta Y, Ando J (2007) Adequate design of customized cDNA macroarray for convenient multiple gene expression analysis. J Biosci Bioeng 103:74–81
5. Sugiyama S, Masuta C, Sekiguchi H, Uehara T, Shimura H, Maruta Y (2008) A simple, sensitive, specific detection of mixed infection of multiple plant viruses using macroarray and microtube hybridization. J Virol Methods 153: 241–244
6. Shieh J, To C, Carramao J, Nishimura N, Maruta Y, Hashimoto Y, Wright D, Wu HC, Azarani A (2002) High-throughput array production using precision glass syringes. Biotechniques 32:1360–1365
7. Scott SW, Zimmerman MT (2009) The nucleotide sequences of the RNA 1 and RNA 2 of asparagus virus 2 show a close relationship to citrus variegation virus. Arch Virol 154: 719–722

Chapter 2

RT-PCR and Real-Time RT-PCR Methods for the Detection of *Potato Virus Y* in Potato Leaves and Tubers

Tyler D.B. MacKenzie, Xianzhou Nie, and Mathuresh Singh

Abstract

Potato virus Y (PVY) is a major threat to potato crops around the world. It is an RNA virus of the family Potyviridae, exhibiting many different strains that cause a range of symptoms in potato. ELISA detection of viral proteins has traditionally been used to quantify virus incidence in a crop or seed lot. ELISA, however, cannot reliably detect the virus directly in dormant tubers, requiring several weeks of sprouting tubers to produce detectable levels of virus. Nor can ELISA fully discriminate between the wide range of strains of the virus. Several techniques for directly detecting the viral RNA have been developed which allow rapid detection of PVY in leaf or tuber tissue, and that can be used to easily distinguish between different strains of the virus. Described in this chapter are several protocols for the extraction of RNA from leaf and tuber tissues, and three detection methods based upon reverse-transcription-PCR (RT-PCR). First described is a traditional two-step protocol with separate reverse transcription of viral RNA into cDNA, then PCR to amplify the viral cDNA fragment. Second described is a one-step RT-PCR protocol combining the cDNA production and PCR in one tube and one step, which greatly reduces material and labor costs for PVY detection. The third protocol is a real-time RT-PCR procedure which not only saves on labor but also allows for more precise quantification of PVY titre. The three protocols are described in detail, and accompanied with a discussion of their relative advantages, costs, and possibilities for cost-saving modifications. While these techniques have primarily been developed for large-scale screening of many samples for determining viral incidence in commercial fields or seed lots, they are also amenable to use in smaller-scale research applications.

Key words *Potato virus Y*, RNA extraction, RT-PCR, Real-time RT-PCR

1 Introduction

Potato virus Y (PVY) is one of the most important potato pathogens, affecting potato growing regions globally, and it represents one of the most economically damaging crop viruses in the world [1]. PVY is a positive sense RNA virus of the genus *Potyvirus*, family Potyviridae. A number of different strains of PVY causing symptoms of varying severity have been identified around the world in recent years [2, 3]. PVY is spread between potato plants and other host plants naturally by aphids, but it can also spread through the

Ichiro Uyeda and Chikara Masuta (eds.), *Plant Virology Protocols*, Methods in Molecular Biology, vol. 1236,
DOI 10.1007/978-1-4939-1743-3_2, © Springer Science+Business Media New York 2015

sale and distribution of seed potatoes from infected plants. Rapid and sensitive molecular technologies have been developed over the past c. 15 years to identify PVY infection in the field or seed and minimize its spread [3, 5], and to aid research into strain diversity [6].

Traditionally, PVY has been detected in the field or in grow-out tests of tubers by observation of mosaic symptoms on the potato foliage, and later by direct immunological detection of the virus (ELISA). There are several drawbacks of these techniques, however, that can be addressed by nucleic acid detection methods. The mosaic symptoms of PVY are difficult to observe on many potato varieties, hampering accurate detection of the virus visually. Also, observation of symptoms can only be done on foliage, and ELISA can only be applied to foliage or tuber sprouts, requiring either fresh tissue during the growing season or significant time and effort in sprouting or grow-out of dormant tubers after harvest [5]. In addition, there are many strains of PVY that vary in the type and severity of symptoms [7], and visual observation and ELISA methods are limited in their ability to differentiate them [6].

Polymerase chain reaction (PCR)-based detection techniques are sensitive enough to be applied to any tissues that could not be assessed by visual observation or ELISA, and also directly to the dormant tuber. Unlike visual or ELISA screening, PCR methods can be used to rapidly screen tubers for PVY immediately after harvest [5]. PCR methods are also ideal for differentiating between different PVY strains, and can be multiplexed to detect multiple strains within the same sample or plant [8]. As PVY is an RNA virus, the particular variant of PCR applied is reverse-transcription-PCR (RT-PCR) in which the extracted RNA of a tissue sample is subjected to reverse transcriptase to generate a complementary DNA (cDNA) strand from the viral genome that is then amplified by PCR. The amplification step can either be traditional end-point PCR cycled to near saturation in a thermocycler then detected by gel electrophoresis, or real-time RT-PCR, in which a fluorescent probe reports the changing concentration of amplified viral sequence during thermocycling, allowing more accurate quantification of initial viral RNA levels [9].

2 Materials

2.1 Plant Material Preparation and Viral RNA Extraction

1. Potato (*Solanum tuberosum* L.) leaves or tubers to be screened for PVY infection and similar potato material known to be infected with PVY to use as positive control.
2. RNA extraction buffer: 100 mM Tris–HCl pH 7.5, 2.5 mM $MgCl_2$ in DNase/RNase-free water (Ultrapure water, Invitrogen); prepared fresh daily from autoclaved stock solutions.
3. TriReagent (Molecular Research Center, Inc.)

4. Chloroform, HPLC grade (VWR).
5. Isopropanol, ACS grade (VWR).
6. 75 % ethanol: 75 % anhydrous ethanol (Commercial Alcohols) in Ultrapure water.
7. RNaseOUT 40 U/μl (Invitrogen).

2.2 Two-Step Reverse-Transcription of Viral RNA

1. Moloney-Murine Leukemia Virus (M-MLV) reverse transcriptase 200 U/μl (Cat. No. 28025-013, Invitrogen), with supplied 5× First Strand Buffer and 0.1 M dithiothreitol stocks.
2. Hexanucleotide mix (ref: 11 277 081 001, Roche Diagnostics GmbH) diluted 1:4.16 with Ultrapure water.
3. 5 mM dNTP mix (100 mM stocks of dATP, dCTP, dGTP, dTTP (Promega) combined at 5 mM each).
4. RNasin 40 U/μl (Promega).

2.3 Two-Step PCR Amplification and Gel Electrophoresis

1. Amplitaq DNA polymerase 5 U/μl (N808-0172, Applied Biosystems) with supplied 10× GeneAmp PCR Buffer II and $MgCl_2$ stocks.
2. 5 mM dNTP mix (see above).
3. Oligonucleotide primers "Potato Virus Y 1" sense: 5′-ACGTC CAAAATGAGAATGCC-3′ and "Potato Virus Y 1" antisense: 5′-TGGTGTTCGTGATGTGACCT-3′ [8]. Under the conditions of the described PCR reaction, these primers have estimated T_m of 55 °C and 59 °C respectively. The primers amplify a 480 bp product at a position on the PVY genome between bases 8721 and 9200 (reference sequence $PVY^{N:O}$-Mb112, accession number AY745491). The RNA sequences complementing these primers are within the coding region for the viral coat protein and are highly conserved across most known PVY strains. Concentration 16 μM.
4. PCR thermocycler.
5. Ultrapure agarose (Invitrogen).
6. 1× TAE: 40 mM Tris base, 20 mM acetic acid, 1 mM EDTA.
7. Loading dye: 0.25 % bromophenol blue, 40 % glycerol in water.
8. Ethidium bromide (Invitrogen) diluted to 8.3 μg/ml in distilled water.

2.4 One-Step Reverse-Transcription/PCR Amplification of Viral RNA

1. HotStarTaq DNA polymerase 5 U/μl (Qiagen) with supplied 10× PCR Buffer including 15 mM $MgCl_2$.
2. 5 mM dNTP mix (see above).
3. Sense and antisense primers, thermocycler and gel electrophoresis materials identical to those described in Subheading 2.3 above.

4. Moloney-Murine Leukemia Virus (M-MLV) reverse transcriptase 200 U/μl (Cat. No. 28025-013, Invitrogen).
5. RNaseOUT 40 U/μl (Invitrogen).

2.5 Real-Time Reverse-Transcription/PCR Amplification of Viral RNA

1. ABI Master Mix (includes Taq polymerase, PCR buffer, and dNTPs) (Applied Biosystems, Part no. #4318157).
2. Oligonucleotide primers PVY-1 FP (sense primer, 5′- CCA ATC GTT GAG AAT GCA AAA C-3′) and PVY-1 RP (antisense primer, 5′- ATA TAC GCT TCT GCA ACA TCT GAG A-3′) [5]. These primers have estimated T_m of 62 °C and 61 °C respectively. The primers amplify a 74 bp product at a position on the PVY genome between nt. 9009 and 9082 (reference sequence $PVY^{N:O}$-Mb112, accession number AY745491) in the same coat protein coding region as the traditional RT-PCR protocol described above, and are also well conserved across most known PVY strains. Concentration 50 μM.
3. Oligonucleotide Taqman probe 5′- TTA GGC AAA TCA TGG CAC AT -3′ using 5′ reporter dye FAM (6-carboxyfluorescein) and 3′ MGB-TAMRA (Minor Groove Binder with Tetramethyl-rhodamine quencher). The probe binds at a position on the PVY genome between nt. 9037 and 9057 (reference sequence $PVY^{N:O}$-Mb112, accession number AY745491). Concentration 10 μM.
4. Sample cDNA produced from reverse-transcription protocol above.

2.6 General Laboratory Equipment

1. 1.5 ml microcentrifuge tubes.
2. 10 μl, 200 μl and 1,000 μl filtered pipette tips.
3. 200 μl PCR tubes, strip-tubes, or 96-well plates.
4. DNase/RNase-free water (Ultrapure water, Invitrogen).

3 Methods

PCR-based protocols are rapid and sensitive methods for detection of PVY in any part of the potato plant. The protocols described here originated with Singh [4], though considerable modifications to those protocols has occurred more recently [5, 10] to reduce labor and make them more compatible with alternatives such as one-step RT-PCR and real-time RT-PCR. These protocols have been developed primarily for screening of large numbers of samples to estimate viral incidence in a field or seed lot, for example, but can also be applied to smaller-scale research projects. For screening large numbers of samples, several time- and cost-saving modifications can be made to the protocols, and these are described in **Note 2**. In a typical screening run of 96 samples, to fill the most common format of PCR thermocyclers, leaf/tuber preparation and RNA extraction requires about 6 h of lab time. The traditional two-step RT-PCR and gel electrophoresis can be performed in about 8 h, while the one-step RT-PCR and gel electrophoresis or the real-time RT-PCR procedure without gel can each reduce overall protocol time by about 2 h. These protocols can be paused at several stages to divide this time comfortably into two or more working days, and these stages where pausing is appropriate are noted within the protocols below.

The cDNA and PCR protocols described here are suitable to apply to fresh foliage (leaves) of the potato plant as well as developing, dormant or sprouted tubers. Due to the different physical qualities and typically differing concentration of virus in these tissues, the preparation and RNA extraction varies slightly depending on the plant material used. After RNA extraction, however, cDNA, PCR, and real-time RT-PCR methods are the same, regardless of the tissue source of the RNA. Below, we describe the tissue preparation and RNA extraction techniques, highlighting the different procedures for leaf versus tuber tissues. Also described are three alternative types of PCR-based detection: (1) a traditional two-step RT-PCR protocol, (2) a rapid and less costly one-step RT-PCR protocol, and (3) a rapid and quantitative real-time RT-PCR protocol. The traditional two-step RT-PCR is similar to typical government regulatory methods in North America for the certification of seed potatoes, and allows multiplex detections of different viruses or PVY strains within the same sample. The one-step RT-PCR protocol has the advantage of using fewer costly PCR reagents, and combining the cDNA and PCR reactions into a single step, greatly reducing labor and potential for contamination. By combining cDNA and PCR, however, the one-step protocol does not yield a separate volume of cDNA as in the two-step and real-time protocols which can be used in multiple PCR reactions for different targets. The comparison of the two-step and one-step RT-PCR protocols detecting PVY in leaf, tuber and composite samples (*see* **Note 2**) is shown in Fig. 1. The two-step protocol has also been modified to use primers specific to different strains of PVY, and to detect these different strains simultaneously within single multiplex RT-PCR reactions (Fig. 2).

The real-time RT-PCR protocol is the most expensive of the three protocols described here because of the more specialized PCR reagents, but it saves on labor costs by eliminating gel electrophoresis, and moreover, it allows for a more quantitative and potentially more sensitive assessment of viral RNA titre than the other protocols [9]. Typical output from a real-time RT-PCR instrument is shown in Fig. 3. The relative performance of traditional two-step RT-PCR and real-time RT PCR protocols and how they compare to the more common ELISA assay has been studied and reported by Singh et al. [5]. A detailed costing of reagents, equipment, and labor for each protocol is shown in **Note 3**.

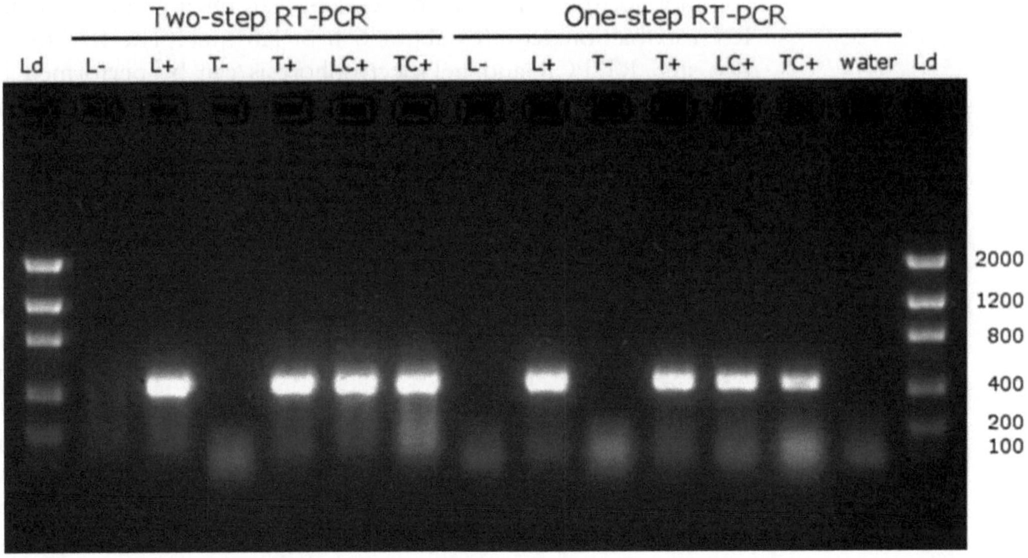

Fig. 1 Comparison of PCR products from two-step RT-PCR and one-step RT-PCR on PVY-free and infected leaf tissue (L−, L+), PVY-free and infected tuber tissue (T−, T+), and composite samples with four PVY free plants and one PVY infected plant (LC + leaves, TC + tubers). Lane marked "water" is a negative control with no added RNA template. "Ld" is a Low Mass Ladder (Invitrogen) with band sizes indicated to right of image (in bp). PVY amplicon is 480 bp

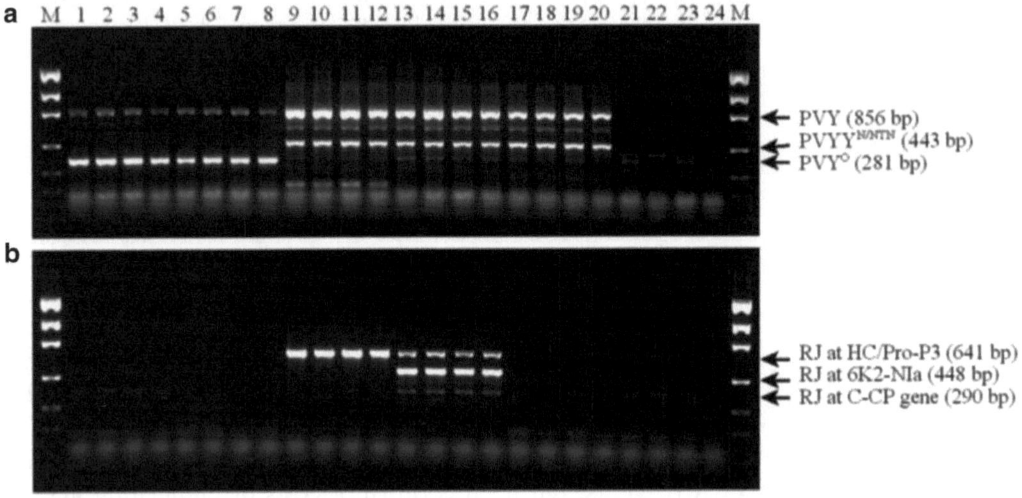

Fig. 2 Two-step multiplex reverse transcription-polymerase chain reaction (RT-PCR) for detection and differentiation of *Potato virus Y* (PVY) strains. (**a**) P1 gene-based RT-PCR assay for discrimination of PVY^{O} from PVY$^{N/N:O/NTN}$ as described by Nie and Singh [9]. (**b**) Recombinant joint (RJ)-based RT-PCR assay results for detection of common PVY recombinant events as described in Nie and Singh [12]. *Lanes 1–4*: field samples of PVY infected plants; *lanes 5–8*: PVY^{O} (isolate PVY^{O}-RB); *9–12*: PVY$^{N:O}$ (isolate PVY$^{N:O}$-Mb58); *lanes 13–16*: PVY^{NTN} (isolate PVY^{NTN}-SI); *lanes 17–20*: PVY^{N} (isolate PVY^{N}-Jg); *lanes 21–24*: healthy potato plant samples; M = DNA ladder (from *top* to *bottom*: 2,000, 1,200, 800, 400, 200, and 100 bp). This figure is adapted from Nie et al. [13]

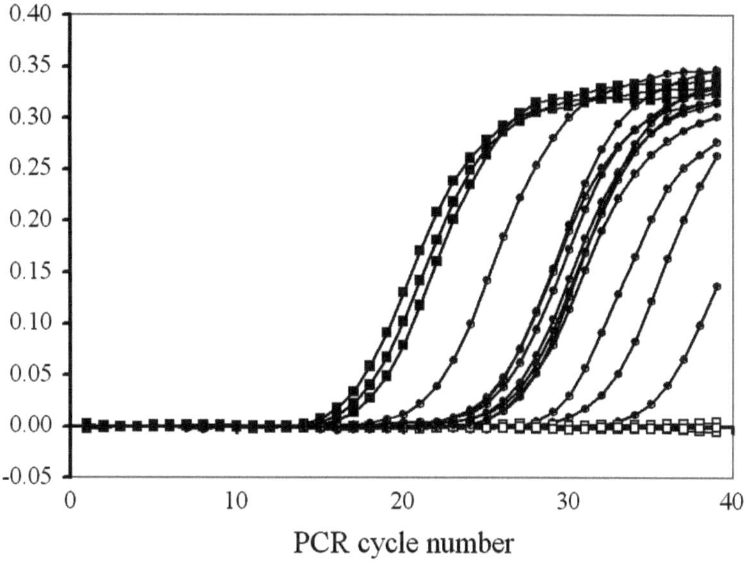

Fig. 3 Typical real-time RT-PCR results for simultaneously processing tuber samples containing viral RNA. Included are three negative control samples (*open squares*) that remain at baseline fluorescence through the entire PCR routine, and three positive control samples containing virus (*closed squares*). Eleven tuber samples representing different potato varieties and different states of infection (*closed circles*) show a range of response in the assay indicative of differences in initial concentration of viral template. The specific difference in initial concentration of viral template can be quantified using these profiles [9]. Fluorescence of the TaqMan probe, responding to the amplified PCR product, is normalized by the internal control dye

3.1 Plant Material Preparation and Viral RNA Extraction

1. For extraction of RNA from foliage, young but fully expanded entire leaves are picked and can be stored for up to 3 days at 4 °C prior to the RNA extraction. Tubers, however, can be stored for many months in the dark under industry-standard cold storage conditions prior to extraction. Typically, at least two leaves or two tubers from each plant are sampled, cleaned of soil or other contaminants, and combined during extraction to identify the PVY-status of the plant (*see* **Note 4**).
2. RNA from leaves can be extracted directly without further leaf preparation. Tubers, however, must be washed and dried, then sampled by slicing small sections of skin from the eye- and stem-ends of the tuber. Slices should be approximately 4 cm^2 each, and as thin as possible to avoid excessive sampling of the unwanted starchy interior of the tuber.
3. Leaves or tuber slices are ground on a Pollähne press and 5–7 drops of the exudate ("sap") are collected into a 1.5 ml microcentrifuge tube pre-filled with 150 μl cold RNA extraction buffer. Between samples, the press rollers must be washed

sufficiently to avoid carryover contamination of subsequent samples (*see* **Note 5**). Exuded sap should be kept cold (on ice or at 4 °C) and extracted within 90 min or frozen (−20 °C) for later extraction.

4. Remove 100 μl of sap in RNA extraction buffer to a new 1.5 ml microcentrifuge tube and add 1 ml of TriReagent. Sap should be homogenized by gentle shaking before drawing 100 μl, though for tuber sap, after agitation the starch grains should be allowed to settle for ca. 10 min at 4 °C as excessive starch interferes with later steps in the RNA extraction. Shake vigorously or vortex the sap-TriReagent mixture for 15 s then let stand at room temperature for 5–15 min. Samples in TriReagent at this stage can be stored frozen (−20 °C) for several weeks for later extraction.
5. Add 200 μl chloroform and again shake vigorously or vortex for 15 s then let stand at room temperature for 5 min. Centrifuge the tube at 12,000 × *g* for 15 min; the mixture should separate into two distinct phases.
6. Remove 0.5 ml of the upper colorless to yellowish phase and place it into 0.5 ml cold isopropanol in a separate tube. Gently mix by inverting the closed tube several times.
7. Centrifuge the tube at 16,000 × *g* for 10 min, and then pour off isopropanol as waste. RNA should be precipitated as a pellet in the bottom of the tube, though it may not be visible. Add 1 ml 75 % cold ethanol to tube. The extraction can be paused at this ethanol stage or at the preceding isopropanol stage and kept at −20 °C for several days if necessary.
8. Centrifuge the tube at 12,000 × *g* for 10 min, and then pour off ethanol as waste. Dry open but inverted tube on clean paper toweling or KimWipes for 30–45 min in a fume hood or laminar flow hood at room temperature.
9. Dissolve RNA in 100–250 μl (leaf tissue) or 25 μl (tuber tissue) of Ultrapure sterile water containing RNase OUT at 40 U/ml. This RNA solution can be stored frozen (−20 °C) for months to years; multiple freeze–thaw cycles and extended time above 0 °C should be avoided (*see* **Note 6** about handling RNA).

3.2 Reverse Transcription of Viral RNA

1. Heat the tube containing RNA (from the extraction protocol above) to 55 °C to ensure solubilization of RNA.
2. Remove 2.5 μl of RNA solution and place into a new 1.5 ml microcentrifuge tube for each sample; heat this tube to 65 °C for 8 min to ensure the RNA is denatured, then immediately place on ice.
3. Pipette 7.5 μl Reverse Transcription Master Mix (*see* Table 1) into each RNA sample tube on ice (*see* **Note 7**).

Table 1
Two-step reverse transcription master mix

Component	Volume/sample (μl)
5× first strand buffer	2
5 mM dNTPs	2
Ultrapure water	1.375
Dithiothreitol (DTT)	1
Hexanucleotide mix (diluted)	0.5
M-MLV Reverse Transcriptase	0.5
RNasin	0.125
TOTAL per sample	7.5
	+2.5 μl RNA sample

4. Incubate the sample tubes containing RNA and RT-master mix (10 μl total) at 42 °C for 1 to 1.5 h to generate cDNA from viral RNA.
5. Heat the sample tubes to 95 °C for 3 min to inactivate reverse transcriptase.
6. Cool the tubes on ice, then spin gently in centrifuge (c. 100×*g* for a few seconds) to consolidate liquid in bottom of tube.
7. The sample tube containing cDNA can be used immediately for PCR, or stored for months at −20 °C.

3.3 Two-Step PCR Amplification of Viral cDNA

1. Combine 23 μl of Two-step PCR Master Mix (*see* Table 2) with 2 μl sample cDNA (from Subheading 3.2) in a 200 μl PCR tube, strip tube or 96-well plate.
2. Process in a PCR thermocycler following the program (*see* **Note 8**): single stage 92 °C/120 s, then 5 cycles of 92 °C/30 s, 62 °C/45 s, 72 °C/90 s, 5 cycles of 92 °C/30 s, 60 °C/45 s, 72 °C/90 s, 10 cycles of 92 °C/30 s, 58 °C/45 s, 72 °C/90 s, 10 cycles of 92 °C/30 s, 55 °C/45 s, 72 °C/90 s then a single stage 72 °C/600 s and ending with an indefinite hold at 4 °C (*see* **Note 9**).
3. Combine a subsample of the completed PCR reaction with 0.1 volumes of loading dye and run on a 1× TAE 1.5 % agarose electrophoresis gel prestained with 0.83 μl ethidium bromide solution per ml gel. Specific volumes of sample to run, voltage, and time will vary depending on gel size and electrophoresis equipment and should be determined empirically.

Table 2
Two-step PCR master mix (for multiplex PCR to simultaneously detect multiple strains of PVY, *see* Note 1)

Component	Volume/sample (μl)
Ultrapure water	16.875
GeneAmp 10× PCR Buffer	2.5
25 mM $MgCl_2$	1.5
5 mM dNTPs	1
16 μM sense primer	0.5
16 μM antisense primer	0.5
Amplitaq DNA Polymerase	0.125
TOTAL per sample	23
	+2 μl cDNA sample

Table 3
One-step RT-PCR master mix

Component	Volume/sample (μl)
Ultrapure water	18.136
10× PCR Buffer	2.5
5 mM dNTPs	0.83
16 μM sense primer	0.5
16 μM antisense primer	1
HotStarTaq DNA polymerase	0.167
M-MLV reverse transcriptase	0.167
RNaseOUT	0.2
TOTAL per sample	23.5
	+1.5 μl RNA sample

3.4 One-Step RT-PCR

1. Combine 23.5 μl One-Step RT-PCR Master Mix (*see* Table 3) with 1.5 μl RNA extract in a 200 μl PCR tube, strip tube or 96 well plate. The RNA extract should be heated to 65 °C for 8 min to denature RNA, and otherwise kept on ice while processing.
2. Process in a PCR thermocycler following the program: single stage 42 °C/45 min, then 95 °C/15 min to destroy reverse transcriptase and activate hot start Taq DNA polymerase, 35 cycles of 95 °C/30 s, 55 °C/45 s, 72 °C/90 s then a single stage 72 °C/600 s and ending with an indefinite hold at 4 °C.
3. Run samples on an agarose gel as in two-step RT-PCR above (Subheading 3.3).

Table 4
Real-time PCR master mix

Component	Volume/sample (μl)
Ultrapure water	9.1
ABI Master Mix	12.5
50 μM sense primer	0.2
50 μM antisense primer	0.2
Taqman probe	0.5
TOTAL per sample	22.5
	+2.5 μl cDNA sample

3.5 Real-Time RT-PCR

1. Combine 22.5 μl of Real-time PCR-master mix (*see* Table 4) with 2.5 μl sample cDNA in a 200 μl PCR tube. Controls should be included using plant material known to be virus free (negative control) and infected with PVY (positive control).
2. Select appropriate reporting dye and internal control settings (*see* **Note 8**) on the real-time PCR thermocycler/detector and process samples following the program: single stage 50 °C/120 s, then 95 °C/600 s to activate hot start Taq polymerase, then 35 cycles of 95 °C/15 s, 60 °C/60 s, then ending with an indefinite hold at 4 °C.
3. Determine virus status of the individual samples based on whether they cross a critical CT (reporting fluorescence) threshold (PVY positive versus negative status), or by cycle number at which CT threshold is crossed (*see* **Note 8**).

4 Notes

1. A multiplex competitive RT-PCR variant of this protocol [8] has been developed to simultaneously amplify distinct products from PVY belonging to different strains or geographic origins. The RNA extraction and two-step RT-PCR protocol here is compatible with that variant PCR protocol.
2. The protocols here describe the simplest and most reliable methods we have developed for PCR-based PVY detection. For screening large numbers of samples, several modifications can be made to the protocols to reduce time and cost. First, for detection in sample sets where the PVY incidence is expected to be low, combining aliquots of tissue sap from five samples into a single RNA extraction (20 μl each × 5 = 100 μl composite for extraction) can greatly reduce the number of samples screened. In this case, PVY-negative reactions indicate all five

Table 5
Cost analysis of the described RT-PCR and real time RT-PCR protocols

	Reagents and equipment			
96-Sample run	Two-step RT-PCR	One-step RT-PCR	Real-time RT-PCR	Labor hours[a]
Leaf/tuber preparation and RNA extraction	\$129.00	\$129.00	\$129.00	5.5–6.5
cDNA synthesis	\$97.00		\$97.00	2
PCR	\$68.00			1.5
One-step RT-PCR		\$121.00		1.5
Real-Time PCR			\$150.00	1.5
Gel electrophoresis	\$20.00	\$20.00		1.5
TOTAL two-step RT-PCR	\$314.00			10.5–11.5
TOTAL one-step RT-PCR		\$270.00		8.5–9.5
TOTAL real-time RT-PCR			\$376.00	9–10

[a]Does not include incubation times of cDNA synthesis (1.5 h) or PCR (c. 2–3 h)

original samples in the composite are negative; PVY-positive reactions require the five original samples in the composite to subsequently be individually tested to determine their status. Appropriate positive controls for composite samples are made by combining 20 μl sap from a known-positive plant with 80 μl sap from a known negative plant. This compositing technique is only economical if less than ca. 20 % of samples in the set are expected to be PVY-positive. The risks and benefits of this compositing technique have been reviewed by Chiang et al. [11]. For screening large sets of samples, the RNA extraction protocol can be made more economical by cutting all volumes listed by 50 % without affecting the quality of RNA produced. Due to the expense of the critical TriReagent solvent, this can save c. \$500 per thousand samples, but at the expense of producing very little RNA from tubers (12.5 μl final RNA solution).

3. Comparative costs of the protocols described in this paper for routine processing of 96 samples, compatible with the 96 well capacity of standard thermocyclers, is displayed in Table 5. Values are in Canadian dollars (\$1 CDN≈\$1 USD at time of publication).
4. For extraction from foliage, at least two entire leaves are generally picked from different stems of the plant. The developmental status of the plant, source of infection (virus from seed tuber or from primary infection of foliage by aphid vector), and length of time since infection may affect the distribution

of PVY in the plant in the field. Usually the second or third youngest, but fully expanded leaves on each stem are picked. Similarly, usually at least two tubers are used to better ensure detection of the virus within a single field plant as PVY may not be detectable in all tubers of an infected plant. The multiple leaves or tuber slices are pulverized together into a single sample tube at the Pollähne press stage. Prior to extraction, leaf or tuber tissue must be free of soil or other contaminants. Typically, field-collected leaves are clean enough to be directly extracted, but tubers must be thoroughly washed before extraction.

5. Between samples on the Pollähne press, the rollers are typically washed exhaustively with water, then liberally coated with a 20 % solution of commercial bleach (final concentration ca. 1 % sodium hypochlorite [NaHClO]) from a spray bottle, then rinsed again with water. Positive control followed by negative control samples should be processed at each extraction session to ensure that roller washing is sufficient to eliminate carryover contamination, and if not, stronger means such as detergents should be considered. Carryover contamination is typically more problematic with leaf tissue than with tubers, perhaps because of higher titre of virus, or some intrinsic difference in the extraction between the two tissues.
6. Extracted RNA must be handled carefully to avoid contamination and degradation. Starting from the isopropanol stage of the extraction protocol and through all subsequent handling, tubes and pipette tips should be clean and sterile and/or certified DNase/RNase free. It is also advisable to use only filtered pipette tips for transferring RNA or preparing solutions that will come into contact with sample RNA.
7. To maximize RNA transfer to the reverse transcription master mix, the RNA solution should be stirred or pumped with the pipette to mix; also, if a translucent scale-like pellet remains in the RNA sample after elution, it is likely due to starch contamination during tuber extractions and will greatly hamper RNA solubilization. Complementary DNA yield can be increased by replacing the 1.375 μl of water per sample in the reverse transcription master mix with additional RNA solution, combining 3.875 μl RNA sample with 6.125 μl reverse transcription master mix without water. Additionally, cDNA yield can be increased by dissolving the RNA in less water during extraction.
8. Thermocycler used for any protocol should have a heated lid, or samples should be overlaid with a drop of mineral oil to prevent evaporation during cycling. Operations to program thermocycling conditions in traditional or real-time PCR, or to select reporter and control dyes and analyze output for real-time PCR, will be instrument specific will require following of the manufacturer's instructions.

9. The "touchdown" PCR program in the two-step RT-PCR protocol follows the method of Nie and Singh [8] and is used to minimize amplification of nonspecific PCR products while maximizing the likelihood of amplification of all PVY strains.

References

1. Crosslin JM (2013) PVY: An old enemy and a continuing challenge. Am J Potato Res 90:2–6
2. Karasev AV, Gray SM (2013) Continuous and emerging challenges of *Potato virus Y* in potato. Annu Rev Phytopathol 51:571–586
3. Nie X, Singh M, Pelletier Y, McLaren D (2013) Recent advances on *Potato virus Y* research in Canada. Am J Potato Res 90:14–20
4. Singh RP (1998) Reverse-transcription polymerase chain reaction for the detection of viruses from plants and aphids. J Virol Methods 74:125–138
5. Singh M, Singh RP, Fageria MS, Nie X, Coffin R, Hawkins G (2013) Optimization of a real-time RT-PCR assay and its comparison with ELISA, conventional RT-PCR and the grow-out test for large scale diagnosis of *Potato virus Y* in dormant potato tubers. Am J Potato Res 90:43–50
6. Karasev AV, Gray SM (2013) Genetic diversity of *Potato virus Y* complex. Am J Potato Res 90:7–13
7. Nie B, Singh M, Sullivan A, Murphy A, Xie C, Nie X (2012) Response of potato cultivars to five isolates belonging to four strains of *Potato virus Y*. Plant Dis 96:1422–1429
8. Nie X, Singh RP (2002) A new approach for the simultaneous differentiation of biological and geographical strains of *Potato virus Y* by uniplex and multiplex RT-PCR. J Virol Methods 104:41–54
9. Mackay IM, Arden KE, Nitsche A (2002) Real-time PCR in virology. Nucleic Acids Res 30: 1292–1305
10. Nie X, Singh RP (2001) A novel usage of random primers for multiplex RT-PCR detection of virus and viroid in aphids, leaves, and tubers. J Virol Methods 91:37–49
11. Chiang K, Lai S, Chung W, Lin S, Yang T (2010) Cost considerations in choosing group size for group testing in the seed potato certification program. Am J Potato Res 87:472–478
12. Nie X, Singh RP (2003) Specific differentiation of recombinant PVY^{N} and PVY^{NTN} strains by multiplex RT-PCR. J Virol Methods 113: 69–77
13. Nie B, Singh M, Sullivan A, Singh RP, Xie C, Nie X (2011) Recognition and molecular discrimination of severe and mild PVY^{O} variants of *Potato virus Y* in potatoes in New Brunswick. Can Plant Dis 95:113–119

Chapter 3

A New Method to Isolate Total dsRNA

Go Atsumi, Ken-Taro Sekine, and Kappei Kobayashi

Abstract

When a diseased plant is suspected to be infected with unknown viruses, the approach of isolating double-stranded RNA (dsRNA) from diseased tissues and analyzing the sequence has been useful for detecting the viruses. This procedure owes its success to the majority of plant pathogenic viruses being RNA viruses, which accumulate dsRNAs as copies of their genome or as a replicative intermediate in infected cells. Conventional dsRNA isolation methods (e.g., chromatography using CF-11 cellulose) require a significant amount of plant material and are laborious and time consuming. Therefore, it has been impractical to isolate dsRNA from many samples at the same time. To overcome these problems, we developed a novel dsRNA isolation method involving a recombinant dsRNA-binding protein. Using this method, we can readily isolate viral dsRNA from a small amount of plant material, and can process numerous samples simultaneously. Purified dsRNA can be used as a template for cDNA synthesis and sequencing, enabling detection of both known and unknown viruses.

Key words Diagnosis, RNA virus, Replicative intermediate, dsRNA isolation, dsRNA-binding protein

1 Introduction

Many diagnostic methods such as ELISA, (RT-)PCR, and LAMP have been established for viral diseases [1–3]. However, these techniques are useless for unknown viruses because they rely on either antigenic or sequence specificity and therefore are dependent upon prior identification of the target viruses. The vast majority of plant pathogenic viruses are RNA viruses [4]. RNA viruses have single-stranded RNA (ssRNA) or double-stranded RNA (dsRNA) as their genome. In infected cells, ssRNA viruses accumulate a significant amount of dsRNAs as their replicative intermediate. Therefore, detection of high molecular weight dsRNA has been used as a marker for RNA virus infection [5–7]. Many procedures for dsRNA isolation and detection have been established; chromatographic separation (for example on hydroxyapatite, diethylaminoethyl cellulose, or epichlorohydrin triethanolamine cellulose), enzymatic degradation of ssRNA and DNA, and salt fractionation using lithium chloride [5].

Ichiro Uyeda and Chikara Masuta (eds.), *Plant Virology Protocols*, Methods in Molecular Biology, vol. 1236,
DOI 10.1007/978-1-4939-1743-3_3, © Springer Science+Business Media New York 2015

Chromatography using CF-11 cellulose powder may be the most practical method, and has been widely utilized for purification of viral dsRNA. However, it requires a significant amount of plant tissue to obtain sufficient dsRNA for detection by electrophoresis. Furthermore, it is laborious and time-consuming.

Here we describe an efficient and labor-saving method by which dsRNA can be efficiently isolated from total RNA extracted from a small amount of plant tissue (ca. 100 mg) [8]. This method uses a dsRNA-binding protein, which has been widely found in both eukaryotes and prokaryotes, to isolate viral dsRNA [9, 10]. We compared four members of *Arabidopsis thaliana* HYL1/DRB family proteins [11] and selected DRB4*, an artificial mutant of DRB4, for viral dsRNA isolation [8]. Our system has the advantages that samples can be handled throughout the purification step in conventional 1.5 ml microcentrifuge tubes, and that a dsRNA fraction for electrophoresis can be obtained within 1.5 h after starting with total RNA. This small-scale, labor-saving, and rapid method enables isolation and analysis of viral dsRNA in high throughput. Therefore, many plant samples can be readily processed simultaneously using our method, resulting in highly efficient diagnosis of plant viral diseases.

2 Materials

2.1 Plasmid DNAs for Production of Recombinant Proteins

pGST-DRB4*-CaL: A derivative of pGEX6P-2 with DRB4* cDNA introduced (*see* **Notes 1** and **2**). pGST-DRB4*-CaL is available from K. Kobayashi (kappei@ehime-u.ac.jp) or K.-T. Sekine (k-sekine@ibrc.or.jp).

2.2 Preparation of Recombinant Proteins

1. *Escherichia coli* Rosetta competent cells (Merck, Whitehouse Station, NJ, USA; *see* **Note 3**).
2. LB medium: 1 % Bacto tryptone, 0.5 % Bacto yeast extract, 1 % NaCl. Autoclaved in aliquots. For solid (plate) medium, add 1.5 % agar before autoclaving.
3. 100 mg/ml carbenicillin dissolved in water. Store at –25 °C.
4. 0.1 M isopropyl β-D-1-thiogalactopyranoside (IPTG) dissolved in water and filter-sterilized. Store at –25 °C.
5. P buffer: 100 mM Tris–HCl (pH 8.0), 500 mM NaCl, 5 mM EDTA. Store at 4 °C.
6. 10 mg/ml RNase A.
7. Glutathione Sepharose 4B (GE Healthcare, Little Chalfont, UK). Store at 4 °C.
8. Glutathione Sepharose resin equilibrated with P buffer: Gently shake the bottle of Glutathione Sepharose 4B to homogeneity and transfer the slurry (1.5-fold of the resin bed volume is

required) to an appropriate conical tube using a pipette with wide-mouthed tip. Pellet the resin by centrifugation for 2 min at 500 × *g*. Wash the resin by adding at least 10 column volumes of P buffer, invert the tube several times, and briefly centrifuge. Repeat this wash step twice. Add P buffer to the resin to give a 50 % slurry so that it can be handled easily by pipette.

9. Elution buffer: 50 mM reduced glutathione, 50 mM Tris–HCl (pH 8.0). Prepare immediately before use or store at −25 °C in aliquots for single experiments (avoid repetitive freezing and thawing).
10. SDS-PAGE: Denature proteins in NuPAGE LDS sample buffer and NuPAGE sample reducing agent (Life Technologies, Carlsbad, CA, USA) at 70 °C for 10 min as described in the manufacturer's instructions. Separate proteins on 12 % NuPAGE Bis-Tris gels (Life Technologies) with MES SDS running buffer (50 mM MES, 50 mM Tris base, 0.1 % SDS, 1 mM EDTA). Proteins in the gel are stained with Coomassie Brilliant Blue.
11. Amicon Ultra-15 centrifugal filter units (Merck).
12. Dialysis tubing.
13. 0.1 M sodium phosphate buffer (pH 7.0).

2.3 In Vitro Transcription for Preparation of Artificial dsRNA

1. Template DNA: To add a T7 promoter to the end the DNA fragment (partial sequence [ca. 500 bp] of firefly luciferase) by PCR, amplify using the primers described below. Primers for templates of sense and antisense RNA transcription were as follows: For template of sense RNA transcription, T7-Luc_s, 5′-CtaatacgactcactatagggagaATGGAAGACGCCAAAAACATAAAG-3′ and Luc500_as, 5′-AACGTGTACATCGACTGAAATC-3′. For template of antisense RNA transcription, T7-Luc_as, 5′-CtaatacgactcactatagggagaAACGTGTACATCGACTGAAATC-3′ and Luc_s, 5′-ATGGAAGACGCCAAAAACATAAAG-3′. The lowercase letters in the primer sequences indicate T7 promoter sequence.
2. A kit for gel purification of PCR-amplified DNA fragments, such as LaboPass gel (Cosmo Genetech, Seoul, Korea).
3. CUGA 7 in vitro transcription kit (NIPPON GENE, Tokyo, Japan): 5× transcription buffer, 0.1 M DTT, 100 mM CTP, UTP, GTP, and ATP, CUGA7 enzyme solution.
4. Phenol–chloroform–isoamyl alcohol (25:24:1).
5. Ethanol.
6. 3 M sodium acetate (pH 5.2).
7. TURBO DNase I, RNase-free (Life Technologies).
8. RNase-free TE buffer: 10 mM Tris–HCl (pH 8.0), 1 mM EDTA.

2.4 Total RNA Extraction

1. D solution: 4 M guanidine thiocyanate, 0.5 % sodium lauroyl sarcosinate, and 25 mM sodium citrate (pH 7.0). 2-Mercaptoethanol is added at a final concentration of 0.1 M immediately before use. Store at 4 °C.
2. Acid phenol: Phenol saturated with citrate buffer (pH 4.2) (NIPPON GENE).
3. 3 M sodium acetate (pH 5.2).
4. Chloroform.
5. 2-propanol.
6. RNase-free TE buffer: 10 mM Tris–HCl (pH 8.0), 1 mM EDTA.
7. A tissue disruptor such as a Micro Smash system (TOMY, Tokyo, Japan).
8. Screw-cap tubes for tissue disruption. 2 ml microtubes (catalog number 72.693, Sarstedt, Nümbrecht, Germany) are suitable for the Micro Smash.
9. Stainless steel beads: 2 beads/tube, 5 mm diameter.

2.5 dsRNA Isolation

1. dRBB (dsRNA binding buffer): RNase-free 0.1 M MES-KOH (pH 6.5), 0.1 M NaCl, 10 mM $MgCl_2$. Store at room temperature.
2. dRBB containing 0.1 % Tween 20. Store at room temperature.
3. Glutathione Sepharose 4B suspended in dRBB containing 0.1 % BSA. Store at 4 °C.
4. TE buffer containing 0.5 % SDS, 5 mM EDTA, and loading dye. Store at room temperature.
5. Ethachinmate (NIPPON GENE): a co-precipitant for ethanol precipitation of dsRNA.

3 Methods

3.1 Expression of GST-DRB4* Protein in E. coli

1. Transform *E. coli* Rosetta (*see* **Note 3**) with pGST-DRB4*-CaL. Inoculate 2 ml LB liquid medium containing 100 mg/l carbenicillin with a single colony of the transformant and culture overnight at 37 °C with vigorous shaking. Dilute the overnight culture in 200 ml of fresh LB liquid medium containing 100 mg/l carbenicillin. Culture at 30 °C with vigorous shaking until the optical density reaches about 0.5.
2. Add IPTG to a final concentration of 1 mM to induce GST-DRB4* protein expression and culture for an additional 1.5 h at 30 °C with vigorous shaking.

3.2 Purification of GST-DRB4*

1. Collect bacterial cells by centrifugation for 5 min at 5,000 × *g* at 4 °C and suspend in 10 ml (1/20 volume of the culture) of P buffer containing 100 mg/l RNase A to remove RNA (*see* **Note 4**).
2. Sonicate the cells to complete cell lysis and centrifuge for 30 min at 20,000 × *g* at 4 °C to remove cell debris.
3. Transfer the supernatant to 15-ml conical tubes containing Glutathione Sepharose resin (1 ml bed volume) equilibrated with P buffer, and mix gently by rotation for 1 h at 4 °C to allow GST-DRB4* to bind to the resin.
4. Transfer the resin to an appropriate column and wash the resin with P buffer. Check the protein in the flowthrough by measuring the absorbance at 280 nm (A_{280}) or staining with Coomassie Brilliant Blue solution. Continue washing until no protein is detectable in the flowthrough.
5. Elute GST-DRB4* protein with 5 ml elution buffer and collect each 0.5 ml fraction.
6. Analyze the fractions by SDS-PAGE and collect fractions that contain GST-DRB4* (Fig. 1; *see* **Note 5**).
7. Check the A_{280} and roughly estimate the protein yield. Concentration of proteins is calculated as $A_{280} = 1.0$ being equivalent to 1 mg/ml.
8. Concentrate the collected solutions to more than 2 mg/ml using Amicon Ultra-15 centrifugal filter units (*see* **Note 6**).
9. Dialyze against 0.1 M phosphate buffer, pH 7.0, for at least 2 days with three buffer changes.
10. Check the A_{280} and A_{260}. The A_{280}/A_{260} ratio should be greater than 1.5. A protein concentration greater than 1 mg/ml is recommended (*see* **Note 6**).
11. Store the proteins at −25 °C in aliquots.

3.3 Preparation of Artificial dsRNA for Control Experiment

Artificial dsRNA for control experiments is prepared by an in vitro transcription of mixed PCR-amplified DNA fragments containing a single T7 promoter at the end (the mixture of two fragments for synthesizing sense and antisense RNA, respectively; *see* **Note 7**). PCR products should be gel-purified using a kit.

1. Set up in vitro transcription reaction as follows:

   4 μl 5× transcription buffer

   2 μl 0.1 M DTT

   1.5 μl each 100 mM CTP, UTP, GTP, and ATP

   1 μl CUGA 7 enzyme solution

   0.05–0.25 pmol of templates obtained as PCR products (sense: antisense = 1:1)

   RNase-free water up to 20 μl.

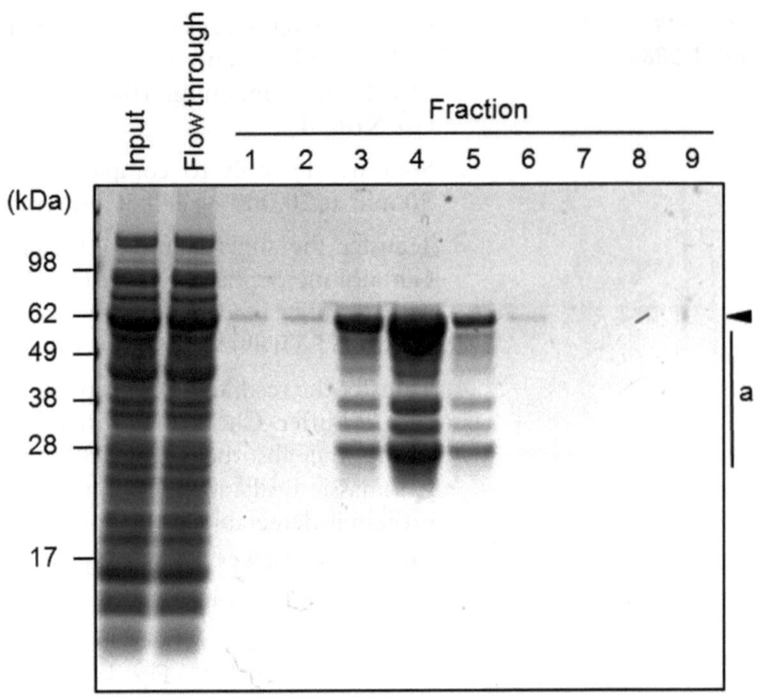

Fig. 1 Expression and purification of recombinant GST-DRB4* protein. pGST-DRB4*-CaL was introduced to *E. coli* Rosetta strain. Expression of GST-DRB4* was induced, and proteins were purified using Glutathione Sepharose resin. Input = protein lysates for input of purification. *Arrowhead* indicates full-length GST-DRB4*, and "a" indicates the proteins that may be degradation or premature translational termination products of GST-DRB4*. In this experiment, fractions 3, 4, and 5 were collected as purified protein

2. Incubate the reaction mixture at 37 °C for 2 h.
3. Denature RNA in the transcription reaction mixture at 75 °C for 15 min. Gradually cool the mixture to 20 °C at a rate of −0.5 °C/min to allow dsRNA formation.
4. Add DNase I and incubate at 37 °C for 30 min to remove template DNA.
5. Purify dsRNA by phenol–chloroform extraction and ethanol precipitation. Dissolve dsRNA in RNase-free TE buffer and store at −80 °C.

3.4 Total RNA Extraction from Virus-Infected Plants (See Note 8)

1. Place the plant tissue sample (100 mg, *see* **Note 9**) into a 2-ml screw-cap microtube together with two stainless steel beads, tightly seal the cap, and place the tube in liquid nitrogen. Grind the sample to fine powder using a cell disrupter (e.g., Micro Smash). Work quickly to ensure that the tissue powder remains frozen.

2. Add 500 µl D solution and mix thoroughly by vigorous shaking.
3. Add 500 µl acid phenol and 50 µl 3 M sodium acetate (pH 5.2), and centrifuge at 15,000 ×*g* for 10 min at 4 °C.
4. Collect supernatant in a new tube, add 0.1 ml chloroform, mix by vigorous shaking, and incubate for 10 min on ice.
5. Centrifuge at 15,000 ×*g* for 5 min and collect the upper aqueous phase in a new tube.
6. Add an equal volume of chloroform and mix by vigorous shaking.
7. Centrifuge at 15,000 ×*g* for 5 min and collect the upper aqueous phase in a new tube.
8. Precipitate RNA by adding an equal volume of 2-propanol, shaking vigorously and centrifuging at 15,000 ×*g* for 10 min at 4 °C. Wash RNA pellet with 70 % ethanol, air-dry, and dissolve in TE buffer. The RNA preparation can be used for dsRNA isolation or stored at −80 °C.

3.5 dsRNA Isolation

1. Add total RNA from the plants, or the artificial dsRNA for a control experiment, (up to one-tenth the volume of dRBB) to 0.5–1 ml dRBB in a 1.5 ml tube and mix thoroughly (*see* **Note 10**).
2. Add 10–20 µg GST-DRB4* to samples (up to 1/50 volume). Incubate at room temperature for 10 min.
3. Add 60 µl of a 35 % slurry of Glutathione Sepharose resin suspended in dRBB containing 0.1 % BSA (added to suppress nonspecific binding of RNA to the resin). Rotate the tube for 30 min to 1 h at 4 °C.
4. Pellet the resin by centrifugation at 2,000 ×*g* for 5 s, remove supernatant, and resuspend the resin in 1 ml dRBB containing 0.1 % Tween 20; repeat this washing step three times.
5. Pellet the resin by centrifugation at 2,000 ×*g* for 5 s, remove supernatant, and resuspend the resin in 1 ml dRBB without 0.1 % Tween 20; repeat this washing step twice. After the second wash with dRBB alone, briefly centrifuge the resin and remove as much of the supernatant as possible using a micropipette.

6a. For gel electrophoresis analysis alone, add 20 µl TE buffer containing 0.5 % SDS, 5 mM EDTA, and loading dye. Incubate for 5 min, pellet the resin by centrifugation and load the supernatant in the agarose gel for electrophoresis (Fig. 2a, b).

6b. For the analysis of dsRNA by PCR and/or sequencing, add 100 µl TE buffer containing 0.5 % SDS, 5 mM EDTA and incubate for 5 min at room temperature.

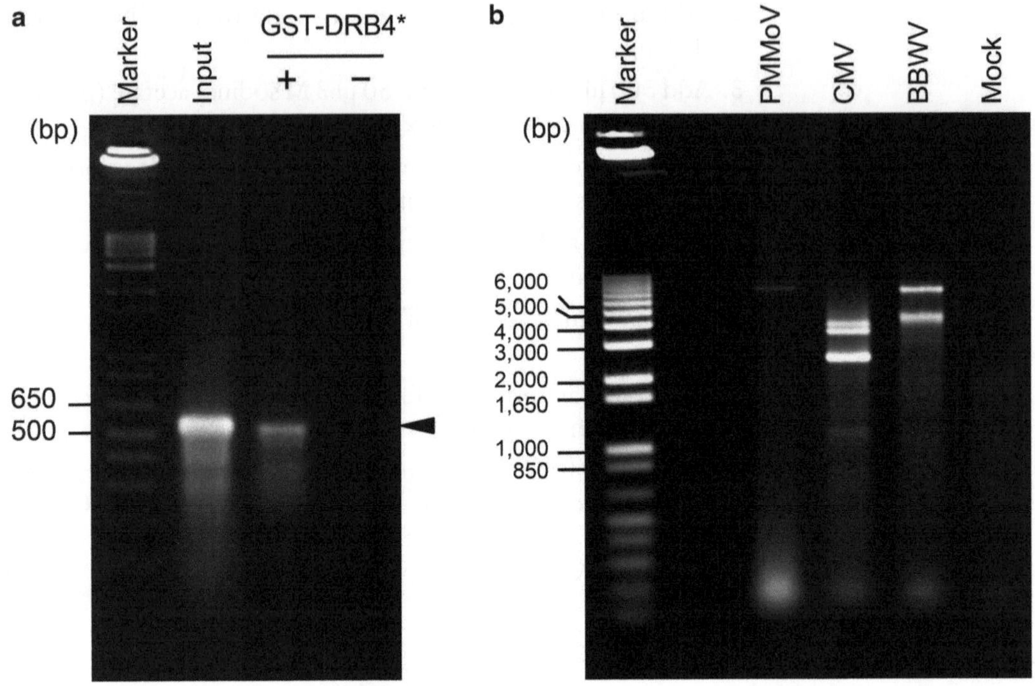

Fig. 2 Electrophoretic analysis of dsRNA isolated from virus-infected tissues. (**a**) Artificial dsRNA of partial sequence (ca. 500 bp) of firefly luciferase (dsLuc500) was synthesized by in vitro transcription. dsLuc500 was incubated with or without 10 μg GST-DRB4* protein. RNA was purified and analyzed by 2 % agarose gel electrophoresis. Input = dsLuc500 used for dsRNA purification. Marker = 1 Kb Plus ladder (Life Technologies). (**b**) Total RNA was isolated from healthy *Nicotiana benthamiana* leaves and from leaves infected with *Broad bean wilt virus-2* (BBWV), *Cucumber mosaic virus* (CMV), and *Pepper mild mottle virus* (PMMoV). Total RNA was extracted from 0.1 g leaf tissue and used for dsRNA isolation. RNA (ca. 150 μg) was incubated with 10 μg GST-DRB4* protein, and dsRNA that bound GST-DRB4* was purified and analyzed by 1 % agarose gel electrophoresis. *Arrowhead* indicates dsLuc500. Marker = 1 Kb Plus ladder

7. Centrifuge at 2,000 × *g* for 10 s and collect supernatant in a new tube.
8. Add an equal volume of phenol–chloroform–isoamyl alcohol (25:24:1), mix vigorously and centrifuge at 15,000 × *g* for 2 min.
9. Collect supernatant in a new tube, add an equal volume of chloroform–isoamyl alcohol (24:1), mix vigorously and centrifuge at 15,000 × *g* for 2 min.
10. Collect supernatant in a new tube and add 1 μl ethachinmate, 1/10 volume 3 M sodium acetate (pH 5.2), and 2–2.5 volumes ethanol.
11. Centrifuge at 15,000 × *g* for 15 min. Wash the pellet with 70 % ethanol and air-dry for 10 min. Add appropriate buffer for the downstream experiment (*see* **Note 11**).

4 Notes

1. DRB4* has an internal 16 amino acid deletion (accession number AB455097). DRB4* was obtained during cloning of the cDNA sequence of *A. thaliana* DRB4 ORF (accession number NM_116145). We used DRB4* instead of wild-type DRB4 because we could obtain more GST-DRB4* protein than GST-DRB4 protein. The dsRNA binding properties did not differ significantly.
2. We could not obtain GST-DRB4* protein using vector pGEX6P-2, probably because the dsRNA binding property of DRB4* was toxic to *E. coli*. Expecting to alleviate the toxicity, highly structured RNA was simultaneously expressed using vector pGEX6P-2CaL. pGEX6P-2CaL contains the 35S leader sequence of *Cauliflower mosaic virus* downstream of the multi-cloning site. Transcripts of 35S leader sequence are known to be highly structured and form a dsRNA structure, which may lower the amount of host RNA that binds GST-DRB4*. Using this vector, we could express and purify a sufficient amount of GST-DRB4* protein.
3. DH5α and BL21 strains can also be used.
4. GST-DRB4* binds host RNA and transcripts of the CaMV 35S leader sequence, which strongly affects the dsRNA binding property of purified GST-DRB4*. Hence, as much RNA as possible should be removed from this protein by adding RNase A to the buffer.
5. Multiple bands are often observed besides GST-DRB4*, which may originate from degradation or premature translational termination (Fig. 1). However, empirically, these products do not seriously affect the dsRNA binding property of GST-DRB4*.
6. In the dsRNA isolation step, 10–20 μg GST-DRB4* per reaction is required. We usually add protein solution up to 1/50 volume of the binding reaction mixture (*see* Subheading 3.5, **step 2**). Thus, the concentration of GST-DRB4* protein should be higher than 1 mg/ml. Because the protein solution is often diluted during dialysis (*see* Subheading 3.2, **step 9**), we recommend concentrating the protein solution to more than 2 mg/ml at this step.
7. Although it is possible to transcribe both the sense and antisense strands in a single tube using a template with T7 promoter sequences on both ends, we do not recommend this method because the yield of transcripts is lower than in the protocol described.
8. Total RNA should be extracted by the acid guanidinium thiocyanate–phenol–chloroform method [12] or commercial

reagents based on it (e.g. TRIzol reagent from Life Technologies). Methods using spin columns (e.g. RNeasy kits from Qiagen, Venlo, The Netherlands or RNAqueous kits from Life Technologies) have not worked as well for dsRNA isolation purposes.

9. Young, newly developing tissues that show symptoms are better for the analysis.
10. We recommend using high-quality tubes with a hydrophilic surface, because the resin sticks to the side wall of cheaper tubes when the resin is washed with a buffer that does not contain any detergent.
11. As an application of the dsRNA isolation method using GST-DRB4*, we developed a system that effectively identifies viral species infecting plants (the DECS [*d*sRNA isolation, *e*xhaustive amplification, *c*loning and *s*equencing] method; [8]). In the DECS method, dsRNA was converted to cDNA, amplified exhaustively and cloned into a plasmid vector, and then the sequences of randomly selected clones were determined. Because the DECS method does not require any sequence information in advance, it can become a powerful tool to identify unknown viruses. In actuality, we successfully identified a new virus using the DECS method [13, 14]. Isolated dsRNA can be used as a template for deep sequencing [15].

Acknowledgements

We thank Reiko Tomita for technical advice.

References

1. Saiki RK, Gelfand DH, Stoffel S, Scharf SJ, Higuchi R, Horn GT, Mullis KB, Erlich HA (1988) Primer-directed enzymatic amplification of DNA with a thermostable DNA polymerase. Science 239:487–491
2. Notomi T, Okayama H, Masubuchi H, Yonekawa T, Watanabe K, Amino N, Hase T (2000) Loop-mediated isothermal amplification of DNA. Nucleic Acids Res 28:E63
3. Clark MF, Adams AN (1977) Characteristics of the microplate method of enzyme-linked immunosorbent assay for the detection of plant viruses. J Gen Virol 34:475–483
4. Fauquet CM, Mayo M, Maniloff J, Desselberger U, Ball LA (2005) Virus taxonomy: VIIIth report of the International Committee on Taxonomy of Viruses. Online via Elsevier, Access
5. Dodds JA, Morris TJ, Jordan RL (1984) Plant viral double-stranded RNA. Annu Rev Phytopathol 22:151–168
6. Morris TJ, Dodds JA, Hillman B, Jordan RL, Lommel SA, Tamaki SJ (1983) Viral specific dsRNA: diagnostic value for plant virus disease identification. Plant Mol Biol Rep 1:27–30
7. Valverde R, Nameth S, Jordan RL (1990) Analysis of double-stranded RNA for plant virus diagnosis. Plant Dis 74:255–258
8. Kobayashi K, Tomita R, Sakamoto M (2009) Recombinant plant dsRNA-binding protein as an effective tool for the isolation of viral replicative form dsRNA and universal detection of RNA viruses. J Gen Plant Pathol 75:87–91
9. Fedoroff NV (2002) RNA-binding proteins in plants: the tip of an iceberg? Curr Opin Plant Biol 5:452–459
10. Eamens AL, Curtin SJ, Waterhouse PM (2011) The *Arabidopsis thaliana* double-stranded RNA binding (DRB) domain protein family. In: Erdmann VA and Barciszewski J

(eds) Non coding RNAs in plants. Springer-Verlag, Berlin, Heidelberg, pp 385–406

11. Hiraguri A, Itoh R, Kondo N, Nomura Y, Aizawa D, Murai Y, Koiwa H, Seki M, Shinozaki K, Fukuhara T (2005) Specific interactions between Dicer-like proteins and HYL1/DRB-family dsRNA-binding proteins in *Arabidopsis thaliana*. Plant Mol Biol 57:173–188
12. Chomczynski P, Sacchi N (1987) Single-step method of RNA isolation by acid guanidinium thiocyanate-phenol-chloroform extraction. Anal Biochem 162:156–159
13. Kobayashi K, Atsumi G, Iwadate Y, Tomita R, Chiba K, Akasaka S, Nishihara M, Takahashi H, Yamaoka N, Nishiguchi M, Sekine KT (2013) Gentian Kobu-sho-associated virus: a tentative, novel double-stranded RNA virus that is relevant to gentian Kobu-sho syndrome. J Gen Plant Pathol 79:56–63
14. Atsumi G, Tomita R, Kobayashi K, Sekine KT (2013) Prevalence and genetic diversity of an unusual virus associated with Kobu-sho disease of gentian in Japan. J Gen Virol 94: 2360–2365
15. Kobayashi K, Atsumi G, Yamaoka N, Sekine KT (2012) Sequencing-based virus hunting and virus detection. Jpn Agric Res Q 46: 123–128

Chapter 4

Multiplex RT-PCR Method for the Simultaneous Detection of Nine Grapevine Viruses

Giorgio Gambino

Abstract

Viral diseases are a serious pathological problem for grapevines, and in recent years the need for increasingly specific and rapid diagnostic methods for the selection of propagation materials has grown. *Arabis mosaic virus*, *Grapevine fanleaf virus*, *Grapevine virus A*, *Grapevine virus B*, *Grapevine rupestris stem pitting-associated virus*, *Grapevine fleck virus*, and *Grapevine leafroll-associated viruses* 1, 2, and 3 are nine of the most widespread viruses that naturally infect grapevines. A multiplex RT-PCR was developed for simultaneous detection of these nine grapevine viruses, in combination with a plant RNA internal control used as an indicator of the effectiveness of the reaction. One to ten fragments specific for the viruses and an internal control were simultaneously amplified from infected samples and identified by their specific molecular sizes in agarose gel. The protocol reported is an update of previously published protocols for RNA extraction and multiplex diagnosis of viruses. After several years of use and hundreds of samples tested, and following validation in several laboratories, this multiplex RT-PCR provides a reliable and rapid method for detecting grapevine viruses from a large number of samples.

Key words Diagnosis, Grapevine, Multiplex detection, RNA extraction, Viruses

1 Introduction

Grapevines (*Vitis* spp.), along with other vegetatively propagated plants, are affected by many viral diseases: the number of known grapevine-infecting viruses has recently risen to 63 [1]. In recent years, the development of the high-throughput sequencing technologies has given a great impetus to the discovery of unknown viruses [2]. The most important viral diseases infecting grapevine, in terms of spreading and economic impact, are the infectious degeneration caused by *Grapevine fanleaf virus* (GFLV) and *Arabis mosaic virus* (ArMV), which belongs to the *Nepovirus* genus [3]. Grapevine leafroll is caused by at least 12 serologically distinct viruses known as *Grapevine leafroll-associated viruses.* Ten of these viruses belong to the *Ampelovirus* genus, and GLRaV-2 belongs to the *Closterovirus*

Ichiro Uyeda and Chikara Masuta (eds.), *Plant Virology Protocols*, Methods in Molecular Biology, vol. 1236,
DOI 10.1007/978-1-4939-1743-3_4,

genus. One species is unassigned (GLRaV-7) [4]. The most widespread leafroll viruses are GLRaV-1 and GLRaV-3, followed by GLRaV-2. Rugose wood complex, which consists of four distinct diseases, is associated with different etiological agents: rupestris stem pitting, caused by *Grapevine rupestris stem pitting-associated virus* (GRSPaV); Kober stem grooving, caused by *Grapevine virus A* (GVA) [5]; corky bark, caused by *Grapevine virus B* (GVB) [6]; and LN33 stem grooving, whose etiological agent is still unknown. GRSPaV is perhaps the most prevalent grapevine virus, and it can occur as distinct variants that showed different symptoms on diverse *Vitis* spp. [7]. For the fleck complex, which consists of several diseases and viruses, the most widespread is the *Grapevine fleck virus* (GFkV), which is latent in *Vitis vinifera* but induces specific foliar symptoms in the indicator *Vitis rupestris* [8].

The most important method for fighting viral diseases is prevention with the production of virus-free propagation material. This strategy needs efficient and reliable systems for detecting the viruses associated with the most common and harmful diseases. In grapevines, viral diagnosis can be performed by visual observation of symptoms directly in the field or by biologically indexing susceptible indicators. However, bioassays are time consuming and require glasshouse facilities. Serological tests (ELISA) are also used routinely, but the low concentration of viruses and their erratic distribution in the host tissues do not always enable their satisfactory and reproducible detection. Molecular techniques, such as RT-PCR have been developed and widespread in recent years with the potential for increased sensitivity and specific or broad-spectrum detection. The simultaneous detection of several viruses in a multiplex RT-PCR, a variant of RT-PCR in which two or more loci are amplified simultaneously, provides quick, reliable, and cost-effective routine diagnosis. Development of a multiplex assay is often complex for the interference and competition that may occur between the individual reactions. In literature there are many examples of simultaneous amplification of RNAs plant viruses by multiplex RT-PCR, but few cases have been reported in which more than five plant viruses or viroids were amplified in a single multiplex reaction [9–11].

We describe the simultaneous amplification of nine grapevine-infecting viruses: ArMV, GFLV, GVA, GVB, GRSPaV, GFkV, GLRaV-1, GLRaV-2, and GLRaV-3. An RNA internal control (18S rRNA) was used as indicator of RNA quality and RT-PCR effectiveness. This protocol is an update of the previously published protocols for rapid and cost-effective RNA extraction [12] and multiplex RT-PCR in grapevine [13]. Several years after the publication [13], this is still the only published and validated protocol [14] to our knowledge that allows simultaneous amplification of nine RNA plant viruses and an RNA internal control.

2 Materials

2.1 RNA Extraction

1. Pestles, mortars, and all glassware are kept overnight at 180 °C.
2. Plastic ware is autoclaved before use.
3. Tissue Lyser (Qiagen).
4. Diethyl pyrocarbonate (DEPC)-treated water: 1 ml of DEPC is added to 1 L of double distilled water, incubated overnight and autoclaved at 120 °C for 20 min. Store at room temperature (*see* **Note 1**).
5. Extraction buffer: 100 mM Tris–HCl (pH 8.0), 25 mM EDTA (pH 8.0), 2 % w/v cetyltrimethylammonium bromide (CTAB), 2 % w/v PVP (MW 40,000), 2.5 M NaCl, autoclaved at 120 °C for 20 min. Store at room temperature.
6. SSTE buffer: 10 mM Tris–HCl (pH 8.0), 1 mM EDTA (pH 8.0), 1 % SDS, 1 M NaCl. Do not autoclave; instead, filter-sterilize. Store at room temperature.
7. β-Mercaptoethanol.
8. Chloroform–isoamyl alcohol (24:1 v/v).
9. LiCl is dissolved in DEPC-treated water at 10 M and autoclaved at 120 °C for 20 min before the storage at 4 °C.
10. Isopropanol. Store at −20 °C.
11. Sodium acetate is dissolved in DEPC-treated water at 3 M; the pH was adjusted to 5.8 with acetic acid before autoclaving at 120 °C for 20 min. Store at room temperature.
12. 70 % Ethanol prepared with DEPC-treated water. Store at −20 °C.
13. 10 mg/ml Glycogen.
14. 2-ml and 1.5-ml micro-centrifuge tubes.

2.2 cDNA Synthesis

1. DNase I (Life Technologies) (*see* **Note 2**).
2. Random primers 50 μM (Sigma).
3. dNTP mix: 10 mM of each dNTP.
4. MMLV (Life Technologies).
5. RNase OUT (Life Technologies).

2.3 Multiplex PCR

1. Primers specific for viruses and endogenous control (*see* Table 1 and **Note 3**).
2. dNTP mix: 10 mM of each dNTP.
3. PlatinumTaq polymerase (Life Technologies) (*see* **Note 4**).
4. Agarose.
5. 0.5× TBE: 90 mM Tris–borate, 2 mM EDTA.
6. 100-bp ladder.

Table 1
Primer sequences, RT-PCR product size, and volume required for setting up the primer mix

Target	Primer	Primer Sequences 5 –3′	Volume in µl for setting up primer mix (concentration of starting solution)	Product size (bp)	References
18S rRNA	Forward	CGCATCATTCAAATTTCTGC	0.075 (10 µM)	844	[13]
	Reverse	TTCAGCCTTGCGACCATACT	0.075 (10 µM)		
GLRaV-2	Forward	GGTGATAACCGACGCCTCTA	0.1 (100 µM)	543	[13]
	Reverse	CCTAGCTGACGCAGATTGCT	0.1 (100 µM)		
GVB	Forward	GTGCTAAGAACGTCTTCACAGC	0.2 (10 µM)	460	[17]
	Reverse	ATCAGCAAACACGCTTGAACCG	0.2 (10 µM)		
ArMV	Forward	ATGCTGATGTTGGAGTTGCTCT	0.375 (10 µM)	416	This chapter
	Reverse	AGTGGCACTAAACCACAAAGC	0.375 (10 µM)		
GLRaV-3	Forward	TACGTTAAGGACGGGACACAGG	0.1 (100 µM)	336	[13]
	Reverse	TGCGGCATTAATCTTCATTG	0.1 (100 µM)		
GVA	Forward	GAGGTAGATATAGTAGGACCTA	0.075 (100 µM)	272	[18]
	Reverse	TCGAACATAACCTGTGGCTC	0.075 (100 µM)		
GLRaV-1	Forward	TCTTTACCAACCCCGAGATGAA	0.05 (100 µM)	232	[13]
	Reverse	GTGTCTGGTGACGTGCTAAACG	0.05 (100 µM)		
GFkV	Forward	TGACCAGCCTGCTGTCTCTA	0.05 (100 µM)	179	[13]
	Reverse	TGGACAGGGAGGTGTAGGAG	0.05 (100 µM)		
GRSPaV	Forward	GGGTGGGATGTAGTAACTTTTGA	0.075 (100 µM)	155	[13]
	Reverse	GCAAGTGAAATGAAAGCATCACT	0.075 (100 µM)		
GFLV	Forward	ATGCTGGATATCGTGACCCTGT	0.075 (100 µM)	118	[13]
	Reverse	GAAGGTATGCCTGCTTCAGTGG	0.075 (100 µM)		
			2.35 Total volume		

3 Methods

The multiplex RT-PCR protocol described below allows the reliable detection of nine grapevine viruses within 1 day (Fig. 1). In the years following the publication of the original work [13], the protocol was further optimized by varying reaction components and cycling conditions. This avoided the extra bands that were sometimes produced in samples with low yields of poor-quality RNA. The original primers for ArMV [13] were changed in favor of new primers (Table 1) that ensure the amplification of a large number of ArMV strains. In an Italian project (ARNADIA) for the validation and standardization of serological and molecular diagnostic protocols of economically important grapevine viruses [14], the multiplex protocol has been proven effective in several laboratories.

The detection limits of multiplex RT-PCR were lower than those of single RT-PCR [13], because in the multiplex assay the cocktail of primers compete for all the templates rather than just for one. In addition, the sensitivity of other recently developed molecular methods, e.g., real-time RT-PCR [15], are greater of this multiplex protocol. However, the level of sensitivity is regarded as adequate for the main purposes of this multiplex method: it provides an effective and rapid diagnostic protocol that is repeatable, with limited cost, and it is easy to use even for laboratories without expertise and without the need for expensive equipment.

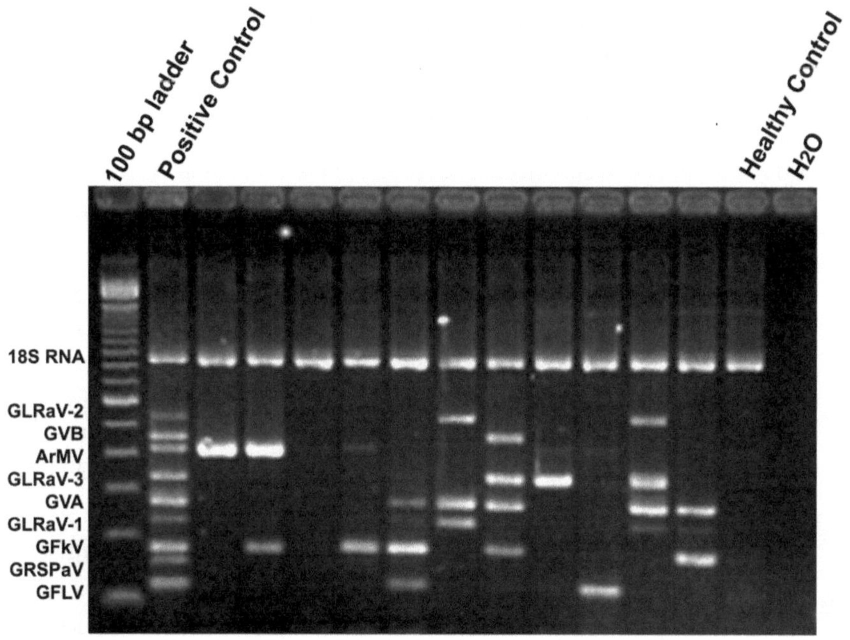

Fig. 1 Agarose gel electrophoretic analysis of DNA fragments amplified by multiplex RT-PCR from naturally infected grapevines

Accordingly, the sensitivity of this protocol allowed detection of the viruses in one positive sample from the bulked extract of several plants (up to 50 plants), which could be useful for large scale indexing of routine samples. The choice of a diagnostic technique always depends on the aim of the work.

In addition to the multiplex protocol, an RNA extraction method has been suggested. This method reduces the time and cost of extraction without reducing the quality and yield of RNA extracted from difficult tissues, in which the commercial kits not always give optimal results [12].

3.1 RNA Extraction

1. Pre-warm extraction buffer at 65 °C. Add β-mercaptoethanol (20 μl for 1 ml of extraction buffer) right before use (*see* **Note 5**).
2. Grind grapevine tissue (0.15–0.2 g) in liquid nitrogen using a Tissue Lyser in 2-ml micro-centrifuge tubes or using a mortar and pestle.
3. Transfer ground tissue to 800 μl of pre-warmed β-mercaptoethanol containing extraction buffer in a 2- micro-centrifuge tube.
4. Incubate the samples at 65 °C for 10 min, vortexing every 2–3 min.
5. Add 800 μl of chloroform–isoamyl alcohol (24:1) and immediately invert and vortex vigorously the micro-centrifuge tube. Centrifuge at 11,000 × *g* for 10 min at 4 °C.
6. Recover the aqueous phase and re-extract adding one volume of chloroform–isoamyl alcohol (24:1), vortex vigorously and centrifuge at 11,000 × *g* for 10 min at 4 °C.
7. Transfer the aqueous phase to a new 1.5-ml micro-centrifuge tube. Add 0.5 volume of ClLI 10 M. Mix by inverting the tube and precipitate in ice for 30 min (*see* **Note 6**).
8. Centrifuge at 21,000 × *g* for 30 min at 4 °C.
9. Remove the supernatant, and dissolve the pellet in 500 μl of SSTE pre-warmed at 65 °C by vortexing.
10. Add 500 μl of chloroform–isoamyl alcohol (24:1) and immediately invert and vortex vigorously the micro-centrifuge tube. Centrifuge at 11,000 × *g* for 10 min at 4 °C.
11. Recover the aqueous phase and add 0.7 volume of Isopropanol, 0.1 volume of 3 M pH 5.2 sodium acetate, and 2 μl of glycogen. Mix by inverting the tube (*see* **Note 7**).
12. Centrifuge at 21,000 × *g* for 20 min at 4 °C.
13. Wash the pellets with 70 % ethanol and centrifuge at 21,000 × *g* for 5 min at 4 °C.
14. Remove the ethanol using a pipet after repeated pulse centrifugations and briefly air-dry the pellet.

15. Dissolve the pellet in 20 μl of DEPC-treated water and store at −80 °C.
16. Assess the RNA purity and concentration using NanoDrop or spectrophotometer.

3.2 cDNA Synthesis

1. RNA is treated with DNase in order to remove the contamination of DNA. 500 ng of total RNA is dissolved in 4 μl of H_2O-DEPC. The amount of starting RNA can be increased if there is the need for greater amounts of RNA free of DNA contamination.
2. Add 0.5 μl of 10× reaction buffer and 0.5 μl of DNase I.
3. Incubate at room temperature for 15 min.
4. Add 0.5 μl of EDTA solutions.
5. Incubate at 65 °C for 10 min and transfer on ice or at −80 °C for long-term storage.
6. Add to 5 μl of total RNA treated with DNase a mix containing: 0.5 μl of dNTP mix, 0.5 μl of Random Primers, 2 μl of 5× First Strand buffer, 1 μl of 0.1 M DTT, 0.5 μl of RNaseOUT, and 0.5 μl of MMLV reverse transcriptase. The reaction volume can be increased if there is the need for greater amounts of cDNA.
7. Incubate at 25 °C for 10 min, 37 °C for 60 min, and 70 °C for 15 min.
8. Transfer on ice or at −20 °C for long-term storage.

3.3 Multiplex PCR

1. Multiplex reaction: 2.75 μl of 10× Reaction Buffer, 1.6 μl of MgCl (50 mM), 0.75 μl of dNTP mix, 2.35 μl of primer mix (*see* Table 1), 0.23 μl of PlatinumTaq polymerase, 15.32 μl of sterile H_2O, and 2 μl of cDNA (*see* **Notes 8** and **9**).
2. Cycling conditions: Initial denaturation at 94 °C for 2 min, followed by 35 cycles at 94 °C for 30 s, 55 °C for 1 min, and 72 ° C for 1 min and 30 s. Final extension at 72 °C for 10 min.
3. Analyze reaction products by electrophoresis on 2.7 % agarose gels buffered in 0.5× TBE and visualized by UV light after staining with ethidium bromide (*see* **Note 10**).

4 Notes

1. Alternatively, RNase-free water purchased from a number of different companies can be used.
2. The DNase treatment of total RNA is not necessary for virus detection but can be important for the amplification of grapevine 18S rRNA used as internal control. However, if it has been used a RNA extraction method with limited contamination of DNA, and using the small concentration of primers indicated for 18S rRNA amplification, the DNase treatment can be omitted.

3. In order to obtain the best results from this multiplex RT-PCR diagnosis, it is important to use only the primers for viruses and endogenous control indicated in Table 1 at the specific concentrations reported for each oligonucleotide.
4. The Taq polymerase indicated is the enzyme that gives the best performance with the protocol reported. It is theoretically possible to use other Taq polymerases from different companies, but it will probably be necessary to re-calibrate the primer concentrations and the PCR reaction parameters.
5. The rapid CTAB protocol indicated gives high-quality RNA in only 4 h (for the extraction of 15–20 samples) at low cost with equal or higher efficiency than that obtained with other time-consuming and expensive protocols. However, any protocol can be used for RNA extraction, including commercial extraction kits. Nevertheless, it is very important to start the reaction with the amount indicated of RNA, and to use RNA that is relatively free of protein and polysaccharide and phenolic compounds with good A_{260}/A_{280} and A_{260}/A_{230} ratios and without RNA degradation. It is strongly recommended not to use crude plant extracts without purification.
6. The protocol can be interrupted in this phase, and RNA precipitation in LiCl can be extended to overnight at 4 °C.
7. The addition of glycogen is optional. However, it can help the precipitation and the individuation of the pellet for the extraction of some difficult plant materials, such as woody canes or berries. It is possible to improve the RNA yield with an overnight incubation at −20 °C before centrifugation.
8. For the best performance with this Taq polymerase, use exactly the concentration indicated for each component of PCR mix and these cycling conditions.
9. For each multiplex PCR experiment, it is very important to use several controls. A positive control contained all 9 potentially detectable viruses. Since it is very difficult to find a single grapevine infected with all of these viruses, a positive control can be created by unifying the plant material, the RNA or the cDNA extracted from plants infected with different viruses. In addition to water used as a negative control, it is appropriate to include a healthy control without these viruses. The best healthy controls are grapevine plants regenerated from somatic embryos. Somatic embryogenesis, usually practiced to regenerate plantlets in biotechnological breeding programs, has been shown to efficiently eliminate several phloem-limited viruses, nepoviruses, and viroids from grapevines [16].
10. For a good separation of the 10 amplification products obtained from the multiplex reaction, use agarose with a concentration not less than 2.5 % and gels of at least 8–10 cm in length. Alternatively, polyacrylamide gels can be used. It is advisable to

put the positive control both at the beginning and end of the gel to make it easier to understand the results. Ethidium bromide can be replaced with safer staining molecules.

References

1. Martelli GP (2012) Grapevine Virology Highlights: 2010–2012. Proceedings of the 17th Congress of the International Council for the Study of Virus and Virus-like Diseases of the Grapevine (ICVG), Davis, California, USA, October 7–14, 2012, 13–31
2. Giampetruzzi A, Roumi V, Roberto R, Malossini M, Yoshikawa N, La Notte P, Terlizzi F, Saldarelli P (2011) A new grapevine virus discovered by deep sequencing of virus- and viroid-derived small RNAs in cv Pinot gris. Virus Res 163:262–268
3. Martelli GP (1993) Grapevine de generation: in Graft-transmissible diseases of grapevines. In: Martelli GP (ed) Handbook for detection and diagnosis. FAO, Rome, pp 9–27
4. Martelli GP, Abou Ghanem-Sabanadzovic N, Agranovsky AA, Al Rwahnih M, Dolja VV, Dovas C, Fuchs M, Gugerli P, Hu JS, Jelkmann W, Katis NI, Maliogka VI, Melzer MJ, Menzel W, Minafra A, Rott ME, Rowhani A, Sabanadzovic S, Saldarelli P (2012) Taxonomic revision of the family Closteroviridae with special reference to the grapevine leafroll-associated members of the genus Ampelovirus and the putative species unassigned to the family. J Plant Pathol 94:7–19
5. Chevalier S, Greif C, Clauzel J-M, Walter B, Fritsch C (1995) Use of an immunocapture polymerase chain reaction procedure for the detection of Grapevine Virus A in Kober stem grooving-infected grapevines. J Phytopathol 143:369–373
6. Boscia D, Savino V, Minafra A, Namba S, Elicio V, Castellano MA, Gonsalves D, Martelli GP (1993) Properties of a filamentous virus isolated from grapevines affected by corky bark. Arch Virol 130:109–120
7. Meng B, Rebelo AR, Fisher E (2006) Genetic diversity analyses of grapevine Rupestris stem pitting-associated virus reveal distinct population structures in scion versus rootstock varieties. J Gen Virol 87:1725–1733
8. Sabanadzovic S, Abou-Ghanem N, Castellano MA, Digiaro M, Martelli GP (2000) Grapevine fleck virus-like viruses in Vitis. Arch Virol 145:553–565
9. Ito T, Ieki H, Ozahi K (2002) Simultaneous detection of six citrus viroids and Apple stem grooving virus from citrus plants by multiplex reverse transcription polymerase chain reaction. J Virol Methods 106:235–239
10. Sánchez-Navarro JA, Aparicio MC, Herranz MC, Minafra A, Myrta A, Pallás V (2005) Simultaneous detection and identification of eight stone fruit viruses by one-step RT-PCR. Eur J Plant Pathol 111:77–84
11. Hajizadeha M, Navarro B, Bashira NS, Torchetti EM, Di Serio F (2012) Development and validation of a multiplex RT-PCR method for the simultaneous detection of five grapevine viroids. J Virol Methods 179:62–69
12. Gambino G, Perrone I, Gribaudo I (2008) A rapid and effective method for RNA extraction from different tissues of grapevine and other woody plants. Phytochem Anal 19:520–525
13. Gambino G, Gribaudo I (2006) Simultaneous detection of nine grapevine viruses by multiplex RT-PCR with coamplification of a plant RNA as internal control. Phytopathology 96: 1223–1229
14. Faggioli F, Anaclerio F, Angelini E, Antonelli MG, Bertazzon N, Bianchi G, Bianchedi P, Bianco PA, Botti S, Bragagna P, Cardoni M, Casati P, Credi R, De Luca E, Durante G, Gianinazzi C, Gambino G, Gualandri V, Luison D, Luvisi A, Malossini U, Mannini F, Saldarelli P, Terlizzi F, Triolo E, Trisciuzzi N, Barba M (2012) Validation of diagnostic protocols for the detection of grapevine viruses covered by phytosanitary rules. Proceedings of the 17th Congress of the International Council for the Study of Virus and Virus-like Diseases of the Grapevine (ICVG), Davis, California, USA, October 7-14, 2012: 260–261
15. López-Fabuel I, Wetzel T, Bertolini E, Bassler A, Vidal E, Torres LB, Yuste A, Olmos A (2013) Real-time multiplex RT-PCR for the simultaneous detection of the five main grapevine viruses. J Virol Methods 188:21–24
16. Gambino G, Navarro B, Vallania R, Gribaudo I, Di Serio F (2011) Somatic embryogenesis efficiently eliminates viroid infections from grapevines. Eur J Plant Pathol 130:511–519
17. Minafra A, Hadidi A (1994) Sensitive detection of grapevine virus A, B, or leafroll-associated III from viruliferous mealybugs and infected tissue by cDNA amplification. J Virol Methods 47:175–188
18. Goszczynski DE, Jooste AEC (2003) Identification of divergent variants of Grapevine virus A. Eur J Plant Pathol 109:397–403

Chapter 5

Detection Methods for Rice Viruses by a Reverse-Transcription Loop-Mediated Isothermal Amplification (RT-LAMP)

Takahide Sasaya

Abstract

Developing a quick and accurate method to diagnose rice viruses in host plants and in vector insects is very important to control virus diseases of rice. A reverse-transcription loop-mediated isothermal amplification (RT-LAMP) assay, one of the most promising molecular diagnostic methods, was established to detect nine viruses, including eight RNA viruses and one DNA virus, in infected rice plants and the viruliferous vector insects. The sensitivities of the assays were either higher than or similar to those of one-step RT-PCR. With a combination of rapid RNA extraction and a RT-LAMP assay, these nine viruses were detected within 2 h from infected rice plants and the viruliferous insects without expensive or unusual equipment. This RT-LAMP method for rice viruses can therefore be adopted not only for diagnosis but also to study the epidemiology and molecular pathology of rice viruses.

Key words RT-LAMP, Rice virus, Detection

1 Introduction

Rice (*Oryza sativa* L.), one of the most important crops in the worlds, is cultivated in more than 100 countries in tropical and semitropical regions of nearly all of Asia, South and Central America, and central and eastern Africa. The world's annual rice production totals 650 million tons and more than 90 % of rice production comes from Asian countries [1].

As the global population continues to increase, it is important to ensure stable rice production in Asian countries. However, viral diseases have become a menace to rice production. For example, more than 485,000 hectares of paddy fields were severely affected by infection with rice grassy stunt virus (RGSV) or co-infection with RGSV and rice ragged stunt virus (RRSV), resulting in the loss of 828,000 tons of rice valued at US $120 million, and directly affecting millions of rice farmers in southern Vietnam during 2006–2007

Ichiro Uyeda and Chikara Masuta (eds.), *Plant Virology Protocols*, Methods in Molecular Biology, vol. 1236,
DOI 10.1007/978-1-4939-1743-3_5,

[2]. Furthermore, a new virus, southern rice black-streaked dwarf virus, that was first discovered in 2001 in Guangdong, China, has rapidly spread throughout southern and central China and has now spread all over Asian countries [3].

Of the 15 viruses known to affect rice production, 10 cause severe diseases in Asian countries. All these rice viruses are transmitted by leafhoppers on plant hoppers in persistent or semi-persistent manners, and some are transmitted transovarially [4] (*see* Table 1). Their vector insects are distributed widely in Asian countries and migrate long distances, even across the ocean and, invade other countries [5].

We need diagnostic systems that are rapid and accurate for detecting the causal viruses in diseased rice plants and viruliferous vector insects, not only for epidemiological studies but also to monitor and control outbreaks of the viral diseases. Such an assay system has to be specific, sensitive, and adaptable to detect newly emerged virus strains. Actually, to aid in forecasting the viral diseases and surveying disease spread, the diagnostic methods prevail widely for virus detection in rice plants and their insects at many agricultural research institutes, plant protection stations, and plant quarantine stations [6].

From the early 2000s, when the sequences of rice many viruses were determined, several molecular methods such as dot blot hybridization, reverse transcription-polymerase chain reaction (RT-PCR), real-time RT-PCR, and reverse transcription-loop-mediated isothermal amplification (RT-LAMP) have been developed that are sensitive, rapid, and able to differentiate closely related viruses. RT-PCR assay is considered the most popular molecular diagnostic method. However, RT-PCR requires an expensive thermo-cycler and is relatively time-consuming because of its longer amplification time and an additional visualization step on agarose gels involving the use of the mutagen ethidium bromide [7]. In contrast, the LAMP assay, first reported in 2000 by Notomi et al. [8], employs two pairs of primers that recognize six regions of a target sequence, and amplify specifically and isothermally the target sequence, eliminating the need for thermo-cycling and agarose gel electrophoresis. The RT-LAMP reaction is run in a water bath or on a heat block at 60–65 °C for 60 min, and the RT-LAMP products are evaluated by direct inspection of the reaction tube after the addition of SYBR Green I fluorescent detection reagent. To develop quick and accurate methods to diagnose rice viruses in the host plants and their insects, we have established molecular diagnostic assays based on RT-LAMP principles to detect the major nine rice viruses that occur in Asia [9].

Table 1
Overview of rice viruses, vector insects, and primer sets for RT-LAMP and one-step RT-PCR

Virus (abbreviation)/genus/family	Vector insect (Transmission mode)	Targeted segment or gene (Accession numbers)	Primer sequences (5′ to 3′)
Rice black-streaked dwarf virus (RBSDV)/*Fijivirus*/*Reoviridae*	*Laodelphax striatellus* (Persistent)	Segment 10 (GU322365)	F3: CCCCAGAGACTTTCCGATAC B3: GGTCTTTAAGTTGCGTGATGT FIP: CGTGGGGTGCTTTGACAATGTTTCAACCGACCAACAATCACTC BIP: TCGCAAACATTTGTTGACCCGACGGTAAAGTGCTAGTTTCTACG
Rice dwarf virus (RDV)/*Phytoreovirus*/*Reoviridae*	*Nephotettix cincticeps* (Persistent)	Segment 8 (D13773.1)	F3: ATTCCAGCCGGGGCATAT B3: CCCACCACCAAGTGAGAAC FIP: AACGCCAGCTATTGTCGTTCCAGGGCATCAGTGCTAAGTGT BIP: CTACTGCAACTGCCGCAGACGTCCGTTTGGACAGGGAGG
Rice gall dwarf virus (RGDV)/*Phytoreovirus*/*Reoviridae*	*Nephotettix cincticeps* (Persistent)	Segment 8 (D13410.1)	F3: AATCAGATTGCGCGCTTC B3: TTTTCGGGATGCAAATGG FIP: CCTGATTAGCTGGCATATATTGCCTAATTTTTAGTCAGTCGATGAACAC BIP: TTGCTTTTGTATCACCCTGGTACGCAGGGGTTGGTTAAACG
Rice ragged stunt virus (RRSV)/*Oryzavirus*/*Reoviridae*	*Nilaparvata lugens* (Persistent)	Segment 8 (AF486811.1)	F3: GACTAGGGATGTGCGTTC B3: TGTAATCGACGTTCGCTC FIP: TGTTATTCTGCCTTGTTCTTTCAACTTCTGATTTGATTGTTTTGAGCA BIP: TCGACTTGGTTTAGCCAAGATG-TTGTTCAGTGATGATTCGC
Rice grassy stunt virus (RGSV)/*Tenuivirus*/-	*Nilaparvata lugens* (Persistent)	RNA5 (AB000403.1)	F3: AAAGACCAACTCAGAGGCA B3: TCTAGAGCAGTTTCCTGTAGTC FIP: CTGACTTAGTGTGGACACTGTGCTTTTGTGTTACCAAGTCTGTGTG BIP: CACTGCATGGGTTTTGTCAACCTGGAGATCATCCTTCTACCAGCT
Rice stripe virus (RSV)/*Tenuivirus*/-	*Laodelphax striatellus* (Persistent)	RNA3 (DQ333944.1)	F3: GTGACCTTTGCTGGTCAGAT B3: ACCGAGGACACTATCCCAT FIP: GGCCAGTGTGTCACCACCTTGGCTATGATGCTGCAACTCT BIP: GAGAGGCACTGGCTTTGTGAGACCAAGGTTGAAGCCTCTGTG

(continued)

Table 1
(continued)

Virus (abbreviation)/genus/family	Vector insect (Transmission mode)	Targeted segment or gene (Accession numbers)	Primer sequences (5′ to 3′)
Rice tungro bacilliform virus (RTBV)/*Tungrovirus*/*Caulimoviridae*	*Nephotettix virescens* (Semi-persistent)	ORF3 (D10774.1)	F3: ACTCTTTGATAGACTACCAGAAG B3: GGATTTTTCGTTTCTTATAATCTCC FIP: GCTATTCCTATTCCTGCTTCATAGGGGAAAGGTAGTAAAAGCGGA BIP: CATGGATGAGAGCAAAATGCATTAAGATCTACAGAATGCTAAGGATG
Rice tungro spherical virus (RTSV)/*Waikavirus*/*Sequiviridae*	*Nephotettix virescens* (Semi-persistent)	ORF1 (GU723290)	F3: CCGTACTGTGCAAGAACAGA B3: GCTCTTGATGTCATCCGCG FIB: GGCACCGCTACGCAAATCAAGTCCCAAAGGCTTATGCGTCTA BIP: TTGTCTCGATCGCTGGGGGAGTCACTCACTGAGCCACATT
Rice transitory yellowing virus (RTYV)/*Nucleorhabdovirus*/*Rhabdoviridae*	*Nephotettix cincticeps* (Persistent)	Nucleoprotein gene (AB011257.1)	F3: GGACGACCATCAAGACAGC B3: GCAACAGGTGTACCACTGTA FIP: GCCCCTGAGGTTGCATGCTATCACAACACTTTCAGCGAGACA BIP: TGGCAGCACCCCTTTGTTGGCATCAGTTGACGGAGCGG

2 Materials

2.1 Viruses and Insects

Nine viruses: rice black-streaked dwarf virus (RBSDV), rice dwarf virus (RDV), rice gall dwarf virus (RGDV), rice ragged stunt virus (RRSV), rice grassy stunt virus (RGSV), rice stripe virus (RSV), rice tungro bacilliform virus (RTBV), rice tungro spherical virus (RTSV), and rice transitory yellowing virus (RTYV) (*see* Table 1). Four vector insects: green rice leafhoppers (*Nephotettix cincticeps* and *N. virescens*), brown plant hopper (*Nilaparvata lugens*), and small brown plant hopper (*Laodelphax striatellus*) (*see* Table 1).

2.2 Rice Cultivar

1. Rice cultivars Nipponbare and Taichung Native 1 (*see* **Note 1**).
2. Bonsol potting soil.
3. Kadan D insecticide (*see* **Note 2**).
4. Mass breeding box for leafhopper.

2.3 Design of the RT-LAMP Primer Sets

It is most important to select the most conserved regions of the nucleotide sequence for each virus. Thus, RT-LAMP primer sets for detection of the viruses (*see* Table 1) have been designed based on the highly conserved regions in the genes coding for structural proteins of each virus with the LAMP Primer Explorer Web interface (available at http://primerexplorer.jp/e/).

2.4 RNA Extraction and RT-LAMP Reaction

1. 0.5 N NaOH solution.
2. 100 mM Tris–HCl (pH 8.0).
3. Multi-bead shaker (Yasui Kikai Co., Japan).
4. 2× Reaction mixture: 40 mM Tris–HCl (pH 8.8), 20 mM KCl, 16 mM $MgSO_4$, 20 mM $(NH_4)_2SO_4$, 0.2 % Tween-20, 1.6 M betaine, and 1.4 mM dNTP mix.
5. Primer mixture: 10 μM each of two outer primers (F3 and B3), and 80 μM each of two inner primers (FIP and BIP) (*see* Table 1).
6. Enzyme mixture: 8 U/μl of *Bst* DNA polymerase and 1.25 U/μl of AMV reverse transcriptase.
7. SYBR Green 1 fluorescent detection reagent (Eiken Chemical Co., Japan).
8. The pENTR/D-TOPO cloning plasmids (Invitrogen Life Technologies Co., Japan) harboring the complete nucleotide region for the viral targeted segments or genes (*see* **Note 3**).

2.5 One-Step RT-PCR Reaction

1. PrimeScript One Step RT-PCR Kit ver.2 (Takara Bio. Inc., Japan).
2. One-step PCR primer mixture: 10 μM each of F3 and B3 primers (*see* Table 1).

3 Methods

The majority of rice viruses are only transmitted by their vector insects, so mechanical or graft inoculation cannot be used to inoculate host rice plants. It is very important despite the large amount of time and effort required to maintain the rice viruses and their insects (*see* **Note 4**).

RT-LAMP assays with the nine primers sets detect the nine individual rice viruses in the virus-infected rice plants and the viruliferous insects were established (*see* **Note 5**). All primer sets were consistent for detecting the presence of the corresponding virus RNAs from the infected rice plants and their viruliferous insects in a fashion similar to that for plasmids harboring the target sequences (Fig. 1, tubes 4 and 5) (*see* **Note 6**). The assays appeared to be specific as because no amplification occurred in samples with RNA extracts from either

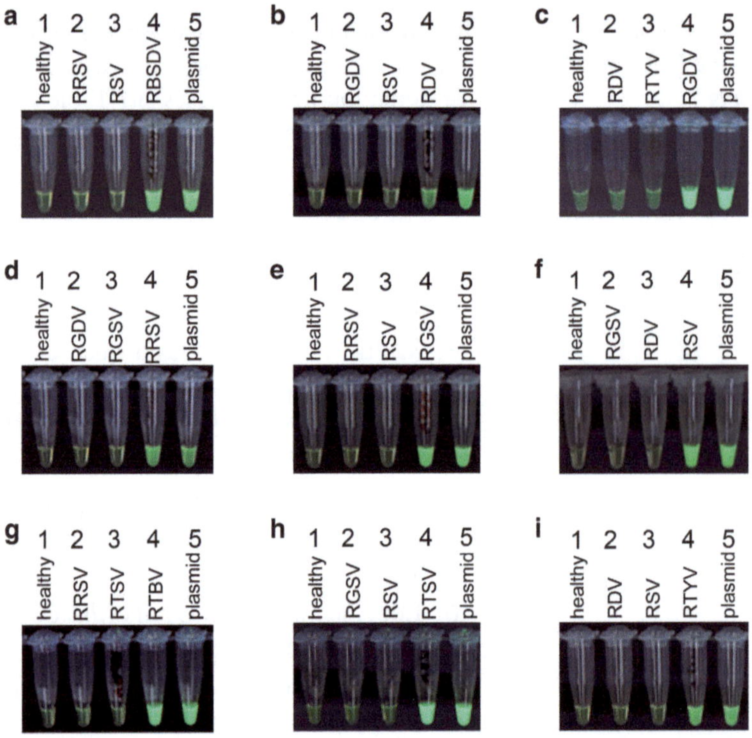

Fig. 1 Specificity of RT-LAMP assays to detect rice viruses from infected rice plants. (**a–i**) RT-LAMP assay for *rice black-streaked dwarf virus* (RBSDV), *rice dwarf virus* (RDV), *rice gall dwarf virus* (RGDV), *rice ragged stunt virus* (RRSV), *rice grassy stunt virus* (RGSV), *rice stripe virus* (RSV), *rice tungro bacilliform virus* (RTBV), *rice tungro spherical virus* (RTSV), and *rice transitory yellowing virus* (RTYV) as targeted viruses. The RT-LAMP reaction was conducted using RNA preparations from healthy rice plants (tube 1), rice plants infected with nontargeted viruses (tubes 2 and 3), and rice plants infected with targeted viruses (tube 4). Plasmid DNAs harboring the targeted viral sequence were used as positive controls (tube 5)

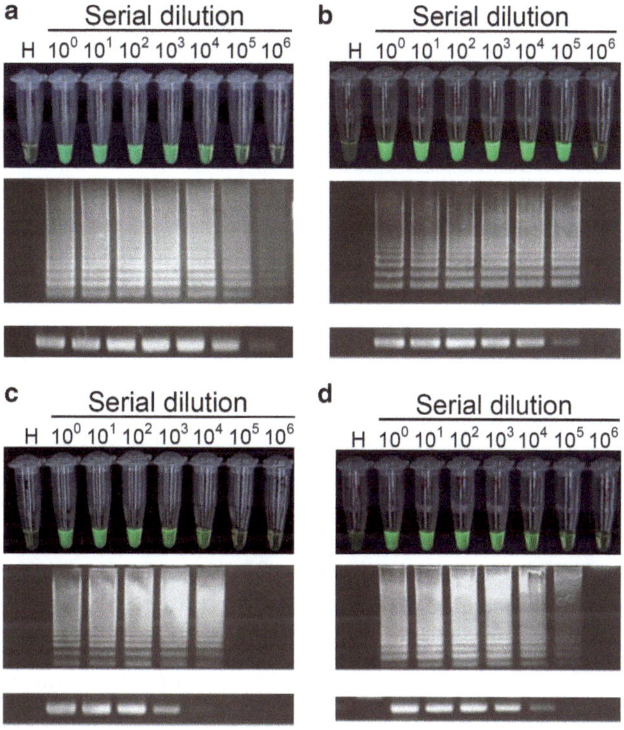

Fig. 2 Sensitivity of RT-LAMP (*top* and *middle*) and RT-PCR (*bottom*) using serial dilutions of RNA preparations from RDV (**a**)-, RSV (**b**)-, RTBV (**c**)-, and RTYV (**d**)-infected rice plants as templates. The reaction mixture (3 μl) was electrophoresed on a 2 % agarose gel. *Lane 1*: RNA preparations from healthy rice plants, *lanes 2–8*: 10^0–10^6 dilutions of RNA preparations from infected rice plants. In the RT-LAMP assays, these four rice viruses were detected in the RNA preparations from virus-infected rice plants diluted up to 10^4–10^5-fold. The sensitivity of RT-LAMP was similar to that with the RT-PCR assay for RDV, and higher than those with the RT-PCR assays for RSV, RTBV, and RTYV

healthy rice plants (Fig. 1, tubes 1) or rice plants infected with the nontargeted viruses (Fig. 1, tubes 2 and 3). By combining the rapid RNA extraction method [10] and the RT-LAMP assay, we detected these nine viruses within 2 h of sampling the infected rice plants and the viruliferous insects. When we further compared the sensitivity of RT-LAMP for detection of the respective viruses with that of conventional RT-PCR using the Takara One-step RT-PCR Kit, the sensitivity of RT-LAMP assays was higher than or at least similar to that of the RT-PCR assays (Fig. 2).

3.1 Acquisition Feeding of Vector Insects

1. Dig the soil away from the roots of inoculum rice plants with typical symptoms 2–3 months after inoculation (*see* **Note 7**).
2. Wrap the roots of the inoculum in wet paper and place them in a mass breeding box for leafhoppers.

3. Place 2–3-day-old young instar nymphs of vector insects on the rice plants for a 1–2-day acquisition access feeding (*see* **Note 8**).
4. After the acquisition access feeding, transfer and raise the insects in healthy rice seedlings until the imago stage (inoculation period; ~2 weeks) in an air-conditioned room at 25 ± 3 °C (*see* **Note 9**).

3.2 Inoculation and Maintenance of the Viruses

1. Expose seedlings of rice cultivars Nipponbare and Taichung Native 1 at the three-leaf stage to ~10–15 viruliferous insects per plant in a mass breeding box for leafhoppers.
2. Incubate seedlings for 1–2 days in the mass breeding box; gently brush insects off the plants five times a day to obtain higher rates of infection.
3. After the inoculation period, kill the insects with insecticide and transfer plants to an insect-free greenhouse at 25 ± 3 °C under natural sunlight for symptom development (*see* **Note 10**).

3.3 Rapid RNA Extraction Method from Rice Plants and Insects

1. Take ~100 mg of leaves or one insect in the 2-ml tubes containing ball beaters, and then add 0.4 ml of 0.5 N NaOH (*see* **Note 11**).
2. Shake the tubes using a multi-bead shaker at 2,000 rpm for 20 s ON, 10 s OFF for a total of 10 cycles.
3. Take 10 μl of the resulting solution and add 490 μl of 100 mM Tris–Cl buffer (pH 8.0).
4. Use 1.5 μl of the diluted sample directly in the RT-LAMP or one-step RT-PCR reaction (*see* **Note 5**).

3.4 RT-LAMP Assay

1. Set up the RT-LAMP reaction mixture (total volume 50 μl) as follows: 25 μl of 2× reaction mixture, 1 μl of LAMP primer mixture, 1 μl of enzyme mixture, and dH_2O up to 48.5 μl.
2. Add 1 μl of SYBR Green 1 fluorescent detection reagent for visualization under UV (254–366 nm) (*see* **Note 12**).
3. Add 1.5 μl of the template RNA preparation or the positive control plasmid sample (*see* **Note 13**).
4. Incubate the solution at 63 °C for 60 min, except for RBSDV at 61 °C for 60 min, and then heat at 80 °C for 5 min to terminate the LAMP reaction (*see* **Note 14**).

3.5 One-Step RT-PCR Assay

1. Set up the one-step RT-LAMP reaction mixture (total volume 50 μl) as follows: 25 μl of 2× one-step buffer, 1 μl of one-step PCR primer mixture, 2 μl of PrimeScript one-step enzyme mixture, and dH_2O up to 48.5 μl.
2. Add 1.5 μl of the template RNA preparation.
3. Use the following RT-PCR: reverse transcription time of 30 min at 50 °C and heat-denaturation time of 2 min at 94 °C, followed by 30 cycles of 94 °C/15 s, 55 °C/15 s, 68 °C/1 min, and a final extension of 10 min at 68 °C (*see* **Note 5**).

4 Notes

1. Rice cultivar Taichung Native 1 is preferred for the inoculation experiment for RTBV and RTSV because the viruses are more sensitive to Taichung Native 1 than to Nipponbare.
2. The insecticide chosen should have a short residual effectiveness. The Kadan D insecticide whose active ingredients are 0.19 % allethrin and 0.60 % 2,4,5,6-tetrachloroisophthalonitri le usually loses its efficacy in 1 week.
3. Entire targeted segments or genes of the viruses were amplified by RT-PCR using the primers designed on the base of nucleotide sequence deposited in the database (*see* Table 1). The PCR products were cloned directly into the pENTR/D-TOPO cloning plasmids using Gateway technology according to the manufacturer's instruction. The pENTR/D-TOPO cloning plasmids harboring the target sequences are purified using a Qiagen Plasmid Midi Kit according to the manufacturer's instruction. The diluted plasmid samples (~1 ng DNA/μl water) are used as positive controls in the RT-LAMP reaction.
4. Except for RSV, the eight rice viruses can be maintained in the rice plants as inoculum. RSV infection leads to death of rice plants, and RSV cannot be maintained in the rice plants as inoculum. RSV should be maintained in its vector insect, the small brown plant hopper (*L. striatellus*).
5. The RT-LAMP assay and one-step RT-PCR assay described here are applicable for the detection of RTBV, a double-stranded DNA virus, in the virus-infected rice plants and the viruliferous insects.
6. The positive controls are important and essential for the diagnosis of rice viruses. Virus-infected rice plants are usually used as positive controls. Unfortunately, not all virus-infected rice plants are available at many agricultural research institutes, plant protection stations, and plant quarantine stations, the sites where the diagnosis is done. The RT-LAMP assay demonstrated that the plasmids harboring the virus target sequences can be used as positive controls and seems to eliminate the need for verified infected rice plants as positive controls.
7. Do not apply insecticide to inoculum rice plants for more than 1 week when they are used as inoculum.
8. In the case of RTBV and RTSV, which are transmitted in a semi-persistent manner, adult insects (*N. virescens*) are allowed 1–2 days of acquisition access feeding on the virus-infected rice plants, and the insects can transmit the viruses without inoculation periods. In addition, RTBV does not have an ability of the insect-transmission and RTBV can be transmitted by the insects with assistance of RTSV. The rice plants that are infected

with RTBV and RTSV should be used for the acquisition feeding of RTBV.

9. Insects cannot acquire RSV effectively using this acquisition feeding method. The viruliferous insect itself should be maintained as inoculum.
10. The symptoms caused by virus infection usually appear 2–4 weeks after virus inoculation.
11. Young, developing leaves with symptoms should be used for the rapid RNA extraction. Sometimes the targeted viruses cannot be detected in the old leaves.
12. The RT-LAMP products can also be visualized with agarose gel electrophoresis after the RT-LAMP reactions (*see* Fig. 2). However, extreme care should be taken when open the tubes for electrophoresis.
13. The great sensitivity of LAMP results in cross-contamination being a common occurrence. Extreme care should be taken with template preparation and setting up RT-LAMP, e.g., opening of tubes and ejecting tips used for pipetting. In our laboratory, the RT-LAMP reaction mixture is set up and SYBR Green 1 fluorescent detection reagent is added in clean benches, and the template RNA preparation or the positive control plasmid sample is added in the separated work areas. The used tips, tubes, and gloves are promptly discarded. Never open the tubes after the RT-LAMP reactions.
14. The RT-LAMP reactions are monitored in real time by the turbidity of the RT-LAMP reaction mixture using a Loopamp Real Time Turbimeter, LA-200 or Realoop-30. The addition of the fluorescent detection regent does not affect this real-time monitoring.

Acknowledgment

The author is grateful to Dr. Dung Tien Le for the experimental research and Drs. Toshihiro Omura and Hiroyuki Hibino for technical advice on rice virus research. This work was supported by a grant from Postdoctoral Fellowships for Foreign Researchers of the Japan Society for the Promotion of Science (JSPS).

References

1. Zeigler R, Barclay A (2008) The relevance of rice. Rice 1:3–10
2. Cabauatan PQ, Cabunagan RC, Choi I-R (2009) Rice viruses transmitted by the brown planthopper *Nilaparvata lugens* Stål. In: Heong KL, Hardy B (eds) Planthoppers: new threats to the sustainability of intensive rice production systems in Asia. International Rice Research Institute, Los Baños, pp 357–358
3. Hoang AT, Zhang HM, Yang J, Chen JP, Hébrard E, Zhou GH, Vinh VN, Cheng JA (2011) Identification, characterization, and

distribution of southern rice black-streaked dwarf virus in Vietnam. Plant Dis 95:1063–1069
4. Hibino H (1996) Biology and epidemiology of rice viruses. Annu Rev Phytopathol 34:249–274
5. Kishimoto R (1971) Long distance migration of plant hopper, *Sogatella furcifera* and *Nilaparvata lugens*. Trop Agric Res Ser 5:201–216
6. Uehara-Ichiki T, Shiba T, Matsukura K, Ueno T, Hirae M, Sasaya T (2013) Detection and diagnosis of rice-infecting viruses. Front Micro 4: doi: 10.3389/fmicb.2013.00289
7. Periasamy M, Niazi FR, Malathi VG (2006) Multiplex RT-PCR, a novel technique for the simultaneous detection of the DNA and RNA viruses causing rice tungro disease. J Virol Methods 134:230–236
8. Notomi T, Okayama H, Masubuchi H, Yonekawa T, Watanabe K, Amino N, Hase T (2000) Loop-mediated isothermal amplification of DNA. Nucl Acids Res 28:e63
9. Le DT, Netsu O, Uehara-Ichiki T, Shimizu T, Choi IR, Omura T, Sasaya T (2010) Molecular detection of nine rice viruses by a reverse-transcription loop-mediated isothermal amplification assay. J Virol Methods 170:90–93
10. Wang H, Qi M, Cutler AJ (1993) A simple method of preparing plant samples for PCR. Nucl Acids Res 31:4153–4154

Chapter 6

Real-Time PCR Protocols for the Quantification of the Begomovirus *Tomato Yellow Leaf Curl Sardinia Virus* in Tomato Plants and in Its Insect Vector

Emanuela Noris and Laura Miozzi

Abstract

Tomato yellow leaf curl Sardinia virus (TYLCSV) (*Geminiviridae*) is an important pathogen, transmitted by the whitefly *Bemisia tabaci*, that severely affects the tomato production in the Mediterranean basin. Here, we describe real-time PCR protocols suitable for relative and absolute quantification of TYLCSV in tomato plants and in whitefly extracts. Using primers and probe specifically designed for TYLCSV, the protocols for relative quantification allow to compare the amount of TYLCSV present in different plant or whitefly samples, normalized to the amount of DNA present in each sample using endogenous tomato or *Bemisia* genes as internal references. The absolute quantification protocol allows to calculate the number of genomic units of TYLCSV over the genomic units of the plant host (tomato), with a sensitivity of as few as ten viral genome copies per sample. The described protocols are potentially suitable for several applications, such as plant breeding for resistance, analysis of virus replication, and virus-vector interaction studies.

Key words Real-time PCR, Begomovirus, TYLCSV, *Bemisia tabaci*, Relative and absolute quantification, Internal reference genes

1 Introduction

Phytopathogenic geminiviruses have a circular single-stranded (ss) DNA genome enclosed in geminated particles. In the last 20 years, they have caused particularly severe damages all over the world, attacking fiber and crop plants. They can be transmitted by leafhoppers, plant hoppers, or whiteflies, in a circulative persistent manner.

The genus *Begomovirus* is the largest within the *Geminiviridae* family and includes viral species infecting economically important dicotyledonous plants, such as cassava, cotton, and tomato. They are transmitted by the whitefly *Bemisia tabaci*, an invasive species that thrives worldwide in tropical, subtropical, and less predominantly in temperate habitats.

Ichiro Uyeda and Chikara Masuta (eds.), *Plant Virology Protocols*, Methods in Molecular Biology, vol. 1236,
DOI 10.1007/978-1-4939-1743-3_6,

Begomoviruses include viral species causing the Tomato Yellow Leaf Curl Disease, one of the top ten most relevant phytopathogenic pests of tomato [1]. In the Mediterranean basin, four species are currently present that cause severe epidemics on cultivated tomato, frequently limiting its production; these are *Tomato yellow leaf curl Sardinia virus* (TYLCSV), *Tomato yellow leaf curl virus* (TYLCV), *Tomato yellow leaf curl Malaga virus*, and *Tomato yellow leaf curl Axarquia virus* [2–4]. They all have a monopartite genome of 2.8 kb encoding six proteins. These viruses spread rapidly, for the wide diffusion of their vector and its high transmission efficiency. Monitoring campaigns are required to control the sanitary state of the commercialized plantlets and of the insect population.

Current detection methods for these pathogens are based on molecular hybridization, but PCR-based assays have the advantage to allow a more precise quantification, as recently reviewed [5]. Real-time PCR offers a greater sensitivity than end-point conventional PCR [6–8] and for this it has become the method of choice to detect RNA or DNA viruses in insect vectors [9–12].

Real-Time PCR assays can be based on TaqMan or SYBR Green chemistry. The TaqMan system employs a specific fluorogenic oligonucleotide probe that binds to the target DNA sequence, between the two PCR primers, generating a specific fluorescent signal [13]. The TaqMan probe has a reporter dye at the 5′ end and a quencher at the 3′ end; when the probe is intact, no fluorescence is detected, but when it is cleaved by the 5′–3′ exonuclease activity of the Taq polymerase during PCR, reporter and quencher are spatially separated and the fluorescent signal becomes detectable, allowing quantitative measurements of the target sequence. The use of a TaqMan probe significantly increases the specificity of the detection. With SYBR Green PCR, no specific probes are needed and costs are consequently reduced; the SYBR Green dye binds to the double-stranded amplified DNA allowing to quantify the target sequence [14]. A drawback of this system is that the dye is nonspecific, possibly generating false positive signals and reducing the sensitivity of the assay.

In both cases, accurate quantification of the target sequence is done by directly correlating the fluorescent signal to the amount of input DNA, using reactions performed in parallel with known amounts of reference DNA. Standard curves generated by plotting the threshold cycle (Ct) versus the logarithm (log) of the amount of the starting quantities (SQ) are used to calculate the amount of DNA that is present in the experimental samples and to evaluate the reaction efficiency.

Here we describe protocols for the absolute and relative quantification of viruses causing TYLCD using Taqman probes and for the relative quantification with SYBR Green. These methods are useful to determine the abundance of these viruses in tomato leaves or in their natural insect vector. The relative quantification methods allow to compare the amount of virus in biological samples, i.e. in different infected plants, plant tissue samples, or populations of viruliferous insects. The absolute quantification protocol allows to determine, as a final result, the number of viral genomic copies (genomic units, GU) present in an experimental sample and ultimately to refer this value to the number of plant GUs.

According to experimental needs, these protocols can be adapted to evaluate the quantity of any begomovirus and more generally of any DNA virus in a plant or insect.

2 Materials

2.1 Biological Materials

1. Tomato (*Solanum lycopersicum*, cultivar Moneymaker) seeds (*see* **Note 1**).
2. Adult *B. tabaci* (Gennadius) (*Homoptera*, *Aleyrodidae*), biotype B [15] (*see* **Note 2**).
3. Methacrylate cages for insect growth.
4. Plant growth rooms.
5. pBin19/TYLCV-S1.8 clone [16] (GenBank Acc. No. X61153) (*see* **Note 3**).
6. *Agrobacterium tumefaciens* LBA4404.
7. YEB medium: 1 g beef extract Difco/1.1 g yeast extract/l, 5 g peptone (casein- hydrolysate)/1.5 g sucrose/1.2 mM magnesium sulfate); pH brought to 7.2 with 1 M NaOH. Autoclaved for 15 min.
8. Kanamycin sodium salt solution prepared in water at 100 mg/ml and stored at −20 °C.
9. 12-ml round-bottomed culture tubes.
10. Bench centrifuge.
11. 25 μl-glass syringe sterilized with chlorine and rinsed with sterile water.

2.2 DNA Preparation

1. Extraction buffer (EB): 0.5 M NaCl, 0.1 M Tris–HCl (pH 8), 50 mM EDTA, 10 mM beta-mercaptoethanol, 1 % sodium dodecyl sulfate (SDS).
2. Potassium acetate (K-Ac) dissolved in water at 5 M and sterilized by autoclaving for 20 min.
3. Isopropanol, kept at −20 °C.
4. Sterile mortars and pestles.
5. Liquid N_2.
6. Ice.
7. RNase A solution (20 mg/ml).

8. TE buffer: 10 mM Tris–HCl, brought to pH 8.0 with concentrated HCl, 1 mM EDTA.
9. 70 % ethanol, kept at −20 °C.
10. PCI (water-saturated phenol–chloroform–isoamyl alcohol = 25:24:1).
11. Water-saturated phenol (liquid phenol–water = 4:1).
12. CIA (chloroform–isoamyl alcohol = 24:1).
13. Sodium acetate (NaOAc) dissolved in water at 3 M concentration, brought to pH 5.2 and sterilized by autoclaving.
14. Thermal incubator set at 65 °C.
15. Minicentrifuge for Eppendorf tubes.
16. Sterile razor blades.
17. 1.5-ml microcentrifuge tubes.

2.3 Primers

Oligonucleotide primers and probe for real-time PCR were designed using the Primer 3 software (http://frodo.wi.mit.edu/cgi-bin/primer3/primer3www.cgi).

1. Virus template: Oligonucleotides to amplify the viral DNA fragment (~100 bp) were as follows: TYLCSV-2256 (+) 5′-CGTCCGTCGATCTGGAAAGT-3′, TYLCSV-2355 (−) 5′-ATCCGAACATTCAGGGAGCTA-3′. The TYLCSV probe was 5′-CCCATTCAAGAACATCTCCGTCCTTGTC-3′ (*see* **Note 4**).
2. Plant endogenous template: Oligonucleotides to amplify a fragment of the nuclear-encoded large subunit ribosomal RNA gene (Tomato 25S ribosomal RNA gene, GenBank Acc. no. X13557), used as reference gene, were as follows: 25S-rRNA PLA(+) 5′-ATAACCGCATCAGGTCTCCA-3′, 25S-rRNA PLA(−) 5′- CCGAAGTTACGGATCCATTT-3′. The 25S PLA probe was 5′-CAATGTAGGCAAGGGAAGTCGGCA-3′ (*see* **Note 5**).
3. Insect endogenous template: Oligonucleotides to amplify a fragment of a highly conserved region of the 18S rRNA gene common to leafhoppers and *B. tabaci* (GenBank Acc. No. U20401) were as follows: 18S-rRNA INS(+) 5′-AACGG CTACCACATCCAAGG-3′, 18S-rRNA INS(−) 5′-GCCTCGG ATGAGTCCCG-3′. The 18S INS probe was 5′-AGGCA GCAGGCACGCAAATTACCC-3′ (*see* **Note 6**).

Oligonucleotide primers were purchased from Invitrogen, while probes were obtained from Eurogentec; the 5′- and 3′-ends of the probes were labeled with the fluorescent dyes FAM (6-carboxyfluorescein, excitation wavelength = 494 nm, emission wavelength = 521 nm) and TAMRA (6-carboxy-tetramethyl-rhodamine), respectively.

2.4 Real-Time PCR

1. Platinum Quantitative PCR SuperMix UDG (Invitrogen).
2. iCycler iQTM Real-Time PCR Detection System (Bio-Rad).

3 Methods

3.1 Preparation of Infected Plants

1. Grow tomato plants in soil in a plant growth room under a 16-h light (2,500 lx) and 8-h dark regime at 23 °C.
2. Multiply the pBIN19/1.8 TYLCV plasmid in *Agrobacterium tumefaciens* LBA4404 in YEB medium (50 ml) supplemented with 50 μg/ml kanamycin, at 28 °C for 2 days. Pellet bacteria by centrifugation at 3,800 × *g* for 30 min and resuspend them in 1.5 ml sterile water. Inoculate plants by injecting 40 μl of bacterial suspension in the leaf axils (*see* **Note 7**).
3. Maintain plants in growth rooms for 3–6 weeks (*see* **Note 8**).

3.2 Preparation of Viruliferous Whiteflies

1. Maintain adult *B. tabaci* on cucumber plants in insect cages under a 16-h light and 8-h dark regime at 25 °C.
2. Transfer *B. tabaci* onto TYLCSV-infected tomato plants for 24 h for virus acquisition (*see* **Note 9**).
3. Transfer *B. tabaci* to healthy cucumber plants; collect batches of viruliferous whiteflies for DNA extraction and virus quantification at the appropriate times after acquisition (*see* **Notes 10** and **11**).

3.3 Preparation of Total DNA

3.3.1 Plants

1. Excise 150 mg of leaf tissue from plants using sterile razor blades. Grind the material in a mortar with liquid N_2 and resuspend it in 0.5 ml EB. Transfer the suspension in 1.5-ml Eppendorf tubes and incubate at 65 °C for 5 min. Add 150 μl 5 M K-Ac, incubate on ice for 10 min and centrifuge at 15,000 × *g* for 10 min. Transfer 0.5 ml of supernatant to new tubes and add 350 μl cold isopropanol. Centrifuge at 15,000 × *g* for 10 min, wash pellet with 70 % cold EtOH and finally dilute it in 0.5 ml TE.
2. To remove RNA, add 1 μl RNase A (20 mg/ml) and incubate at 37 °C for 30 min (*see* **Note 12**).
3. Purify DNA by adding an equal volume of PCI, vortex and centrifuge at 15,000 × *g* for 2 min at room temperature. Transfer the aqueous phase to a new tube and add an equal volume of CIA. Vortex and centrifuge at 15,000 × *g* for 2 min at room temperature. Transfer the aqueous phase to a new tube and precipitate DNA by adding a 1/15 volume of 3 M NaOAc (pH 5.2) and 2.5 volumes of EtOH. Keep at −80 °C for 30 min, centrifuge at 15,000 × *g* for 15 min, wash the pellet with 70 % EtOH, and centrifuge at 15,000 × *g* for 2 min.
4. Pour off the supernatant, air-dry the DNA, dissolve it in TE buffer, and either use it immediately or store it at −20 °C.

3.3.2 Insects

1. Pool approximately 50 insects (females and males in equal numbers) and extract their DNA following the protocol described in Subheading 3.3.1 (*see* **Note 13**).

3.4 Preparation of Standard Curves

1. To prepare the standard curves for the relative quantification of TYLCSV in plants or insects, quantify the extracted RNAse-free DNA by UV-spectrophotometry; prepare threefold serial dilutions ranging from 30 to 0.37 ng/μl.
2. To prepare the standard curves for the absolute quantification of TYLCSV in plants, prepare tenfold serial dilutions of a plasmid containing the cloned full-length TYLCSV genome dissolved in 2.3 ng/μl of DNA extracted from healthy tomato plants; dilutions are made in order to range from 10^6 to 10 copies (genomic units) of viral DNA.

3.5 PCR Reactions

1. For real-time PCR of virus template, design primers to amplify a ~100-bp fragment within a region conserved on the basis of the alignment of three different TYLCSV isolates, e.g., GenBank Acc. No. TYLCSV X61153, TYLCSV-[Sicily] Z28390, and TYLCSV-[Spain1] Z25751.
2. For real-time PCR of the endogenous plant gene template selected as reference gene, design primers to amplify a ~150-bp fragment from the 25S rRNA tomato gene. This gene was selected for its relatively high copy number, allowing to obtain comparable Ct values for viral and endogenous plant templates.
3. Design a probe on the 25S rRNA amplicon.

3.5.1 Real-Time PCR Using Taqman Probes

1. Set up a real-time PCR reaction (total volume 25 μl) as follows: 12.5 μl of Platinum Quantitative PCR SuperMix UDG, 2.5 μl of each primer (0.3 μM each), 2.5 μl of Taqman probe (0.2 μM), 1 μl of UV-quantified DNA (within the range of the standard curves), and 4 μl water.
2. Use the following PCR cycling parameters: 1 cycle at 50 °C for 3 min (activation of UNG), 1 cycle at 95 °C for 5 min (DNA polymerase activation), and 45 cycles, each consisting of 15 s at 95 °C (denaturation) and 1 min at 60 °C (annealing and extension). Perform all reactions with three technical replicates and three biological replicates [17].

3.5.2 Real-Time PCR Using SYBR Green

1. Set up a real-time PCR reaction (total volume 10 μl) as follows: 5 μl of EvaGreen Mix, 1 μl of each primer (0.3 μM each), 0.3 μl of ROX, 1 μl of UV-quantified DNA (~40 ng), and 1.7 μl water.
2. Use the following PCR cycling parameters: 1 cycle at 95 °C for 5 min (DNA polymerase activation) followed by 40 cycles, each consisting of 15 s at 95 °C (denaturation) and 1 min at 60 °C (annealing and extension). Perform all reactions with three technical replicates and three biological replicates.

3. At the end of run, in order to assess the specificity of the amplification product, perform a melting curve from 60 to 95 °C, with an increment of 0.5 °C every 5 s.

3.6 Standard Curves Calculation

1. For each PCR system, obtain a standard curve by linear regression analysis of the value of the threshold cycle (Ct) plotted over the amount of DNA of each of the three standard-dilutions replicates. Data acquisition and analysis are performed automatically by the PCR equipment software.
2. Calculate the PCR efficiency (E) by the formula:

$$E = e^{\ln 10/-s} - 1$$

where a slope (s) = −3.3 (E = 2) represents a 100 % efficiency.

3. A PCR efficiency of 100 ± 5 % using the standard curve constructed with serial dilutions of genomic DNA is sufficient for further quantification.

3.7 Virus Quantification

3.7.1 Relative Quantification Using Taqman Probes or SYBR Green

This protocol is intended to provide the way to measure the relative amount of viral DNA present in a sample compared to another sample (herein defined as samples 1 and 2).

1. For each sample, calculate the average Ct (Ct_{av}) of the three technical replicates for both the virus and the endogenous reference gene, as follows:

$$Ct_{av} = \left(Ct_1 + Ct_2 + Ct_3\right)/3$$

where Ct_1, Ct_2, Ct_3 are the values of the three technical replicates.

2. Calculate the relative amount of virus for each sample using the ΔCt method [18] as follows:

$$\Delta Ct_{av} = \left[Ct_{av}(end) - Ct_{av}(virus)\right]$$

where *end* is the endogenous gene used as reference.

3. To compare the viral quantities present in different samples, calculate the $\Delta\Delta Ct_{av}$ as follows:

$$\Delta\Delta Ct_{av} = \Delta Ct_{av}\text{sample } 1 - \Delta Ct_{av}\text{sample } 2.$$

4. Calculate the fold change (FC) of the viral DNA amount between different samples as follows:

$$FC = 2^{-\Delta\Delta Ctav}$$

Representative results of relative quantifications with either the Taqman probe system (Fig. 1) or the SYBR Green protocol (Fig. 2) are reported.

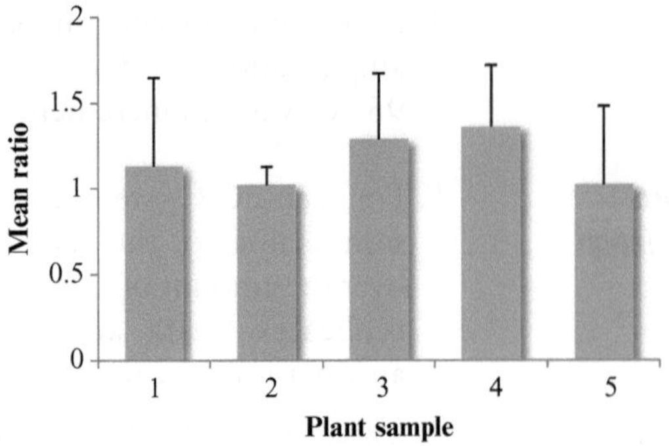

Fig. 1 Relative quantification of TYLCSV in different tomato plants, using a real-time PCR protocol with Taqman probes. *Bars* represent the ratio between the mean SQ of the virus and the mean SQ of the internal reference gene 25S rRNA, i.e., the relative amount of virus present in each plant sample, normalized to the amount of plant DNA represented by the endogenous 25S rRNA tomato gene. *Vertical lines* on each bar represent standard deviations. The figure represents the mean of four uppermost leaves collected from five different plants. No statistical significance was found by analyzing the data with a two-way ANOVA

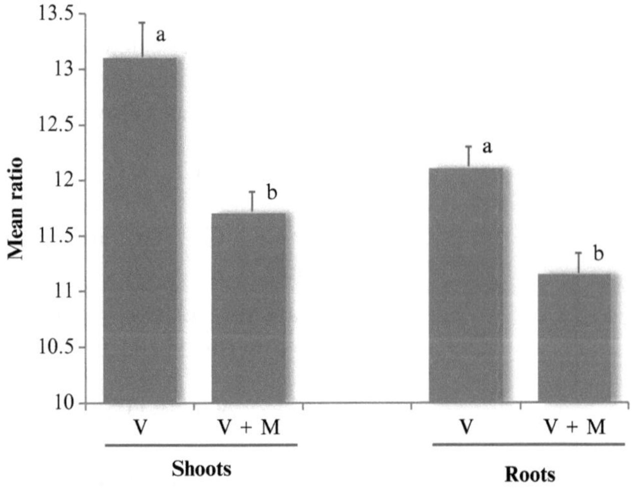

Fig. 2 Relative quantification of TYLCSV in shoots and roots of tomato plants (second uppermost leaf of a plant) in the presence or absence of mycorrhizal colonization, using the real-time PCR protocol with SYBR Green. Values on *Y*-axis represent the difference between the Ct values of TYLCSV and that of the reference gene (in this case the Tomato APX gene, SGN Acc. No. SGN-U579973). In this case we measured the viral concentration in different batches of samples, TYLCSV-infected plants (V) and TYLCSV-infected plants with mycorrhizal colonization (M + V). *Vertical lines* on each bar represent standard error. *Different letters* indicate statistically significant differences within the same organ ($p < 0.05$, ANOVA)

3.7.2 Absolute Quantification Using Taqman Probes

The absolute quantification protocol here described allows to calculate the number of viral copies (genomic units, GUs) present in a tomato sample and to refer this value to the number of plant host GUs (herein tomato plants).

1. To calculate the number of TYLCSV GUs, consider that the size of the cloned TYLCSV genome is 2,773 bp [16]. Since the mean weight of one bp is 660 Da, calculate the molecular weight of TYLCSV expressed in Dalton as follows:

$$2,773\text{bp} \times 660\text{Da} = 1.83 \times 10^{6}\,\text{Da}$$

Since each mole contains 6.02×10^{23} molecules (Avogadro's number), calculate the weight of the TYLCSV genome expressed in grams as follows:

$$1.83 \times 10^{6}\,\text{Da} / 6.02 \times 10^{23} = 3.04 \times 10^{-18}\,\text{g}$$

corresponding to 3.04×10^{-6} pg.

2. To calculate the number of tomato plant GU, consider that the tomato diploid genome size is approximately 1.8×10^{9} bp [19]. Since the mean weight of one bp is 660 Da, calculate the molecular weight of the tomato genome in Dalton as follows:

$$1.8 \times 10^{9}\,\text{bp} \times 660\text{Da} = 1.188 \times 10^{12}\,\text{Da}.$$

Since each mole contains 6.02×10^{23} molecules (Avogadro's number), calculate the weight of the tomato genome expressed in grams as follows:

$$1.188 \times 10^{12}\,\text{Da} / 6.02 \times 10^{23} = 1.973 \times 10^{-12}\,\text{g}$$

corresponding to 1.973 pg $\approx$ 2 pg.

3. For each sample, calculate the average Ct (Ct_{av}) of the three technical replicates for both the virus and the endogenous reference gene, as follows:

$$Ct_{av} = \left(Ct_1 + Ct_2 + Ct_3\right) / 3$$

where Ct_1, Ct_2, Ct_3 are the values of the three technical replicates.

4. Using the Ct_{av} value obtained for each sample in the standard curve formula, calculate the weight of the viral and tomato DNA present in each sample, expressed in grams.
5. Convert in number of GUs the weight (in grams) of the viral and tomato plant DNA obtained in each sample.
6. Calculate the number of viral GUs present in each sample referred to the number of tomato plants GUs.

Figure 3 is an example of the standard curves used for the absolute quantification.

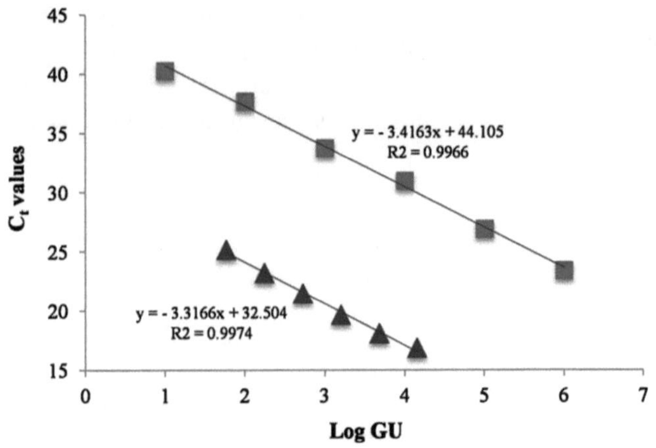

Fig. 3 Standard curves generated for the absolute quantification of TYLCSV in tomato plants samples. The graph shows the equations obtained by plotting the Ct values against the log of the TYLCSV GU (*squares*) or of tomato GU (*triangles*), assuming that its genome weights 1.973 pg [18]. To obtain the TYLCSV curve (*squares*), a full-length genomic clone of TYLCSV was tenfold diluted in 2.3 ng/μl DNA extracted from healthy plants, in order to have from 10^6 to 10 copies of viral DNA

4 Notes

1. As plant host, any tomato cultivar can be used. The plant material analyzed with these protocols can derive from greenhouse or cultivated fields.
2. Other biotypes of *B. tabaci* can be used. The PCR system designed was originally described for leafhoppers, in particular *Macrosteles quadripunctulatus* [20]. Provided that the sequence of the 18S rRNA gene is available and that the gene is 100 % homologous to U20401 here described, this protocol can be directly applied to other insect vectors.
3. Source plants can be inoculated using either agroinfection or viruliferous whiteflies.
4. The quantification protocols here described are suitable for the three begomoviruses affecting tomato crops in Italy, i.e., GenBank Acc. No. TYLCSV X61153, TYLCSV-[Sicily] Z28390, and TYLCSV-[Spain1] Z25751. The primers and the probe for TYLCSV have been designed based on a sequence stretch conserved among these three isolates. However, other primers and probes, specific for a single isolate can be designed, depending on the experimental purpose.
5. The choice of an endogenous reference gene can be cumbersome. The 25S rRNA reference gene here described for virus quantification in plants has been selected among others for its high copy number, since a comparable amount of viral DNA and

endogenous reference DNA guarantees a more accurate quantification. However, other genes can be used (*see* [21]).

6. As for plants, the choice of the 18S rRNA as endogenous reference gene for insects was done based on its high copy number.
7. The plant inoculation protocol here described allows to inoculate approximately 35 plants. As appropriate, the volume of agrobacterial culture can be scaled up.
8. Due to the complexity of the virus quantification protocol using real time, it is advisable to evaluate the state of the infection of each plant to be tested using a tissue-print hybridization method before proceeding with DNA extraction and virus quantification. Evaluation of symptoms alone may not be reliable, depending on the aggressiveness of a virus isolate, the growing conditions, or the host plant genotype.
9. Acquisition access period of TYLCSV by *B. tabaci* can vary between 18 and 48 h. Shorter acquisition periods do not guarantee a sufficiently high level of virus acquired by the insect population, while longer acquisition periods do not imply an increase in virus acquisition.
10. Once insects have acquired the virus, they maintain it until their death (about 20 days). Depending on the purpose of the experiment, it is advisable to analyze insects no more than 10 days after acquisition, to avoid high mortality levels.
11. The protocol described has been applied to virus retention studies, but it can be used for other studies, such as different whitefly retention kinetics of mutant viruses, or different whitefly retention kinetics of viruses in the presence of a chemical or biological treatment.
12. The RNase treatment is absolutely necessary since only DNA, quantified by a spectrophotometer, must be present in samples where a virus with a DNA genome has to be quantified, using host/vector DNA as internal reference.
13. The sensitivity of the real-time procedure would be sufficient to detect the virus in a single insect. However, we do not advise to use this protocol for single whiteflies, due to the high variability occurring in the DNA extraction efficiency among insects.

Acknowledgment

The authors thank Piero Caciagli and Gian Paolo Accotto for their support in preparing these protocols.

References

1. Scholthof KB, Adkins S, Czosnek H, Palukaitis P, Jacquot E, Hohn T, Hohn B, Saunders K, Candresse T, Ahlquist P, Hemenway C, Foster GD (2011) Top 10 plant viruses in molecular plant pathology. Mol Plant Pathol 12: 938–954
2. Accotto GP, Navas-Castillo J, Noris E, Moriones E, Louro D (2000) Typing of tomato yellow leaf curl viruses in Europe. Eur J Plant Pathol 106:179–186
3. Monci F, Sànchez-Campos S, Navas-Castillo J, Moriones E (2002) A natural recombinant between the geminiviruses Tomato yellow leaf curl Sardinia virus and Tomato yellow leaf curl virus exhibits a novel pathogenic phenotype and is becoming prevalent in Spanish populations. Virology 303:317–326
4. Garcìa-Andrès S, Monci F, Navas-Castillo J, Moriones E (2006) Begomovirus genetic diversity in the native plant reservoir Solanum nigrum: evidence for the presence of a new virus species of recombinant nature. Virology 350:433–442
5. Accotto GP, Noris E (2007) Detection methods for TYLCV and TYLCSV. In: Czosnek H (ed) Tomato yellow leaf curl virus disease, Management, molecular biology, breeding for resistance. Springer, Dordrecht, pp 241–249
6. Mumford RA, Walsh K, Barker I, Boonham N (2000) Detection of *Potato mop top virus* and Tobacco rattle virus using a multiplex real-time fluorescent reverse-transcription polymerase chain reaction assay. Phytopathology 90: 448–453
7. Korimbocus J, Coates D, Barker I, Boonham N (2002) Improved detection of *Sugarcane yellow leaf virus* using a real-time fluorescent (TaqMan) RTPCR assay. J Virol Methods 103: 109–120
8. Boonham N, Perez LG, Mendez MS, Peralta EL, Blockley A, Walsh K, Barker I, Mumford RA (2004) Development of a RT-PCR assay for the detection of potato spindle tuber viroid. J Virol Methods 116:139–146
9. Lett JM, Granier M, Hippolyte I, Grondin M, Royer M, Blanc S, Reynaud B, Peterschmitt M (2001) Spatial and temporal distribution of Geminiviruses in leafhoppers of the Genus *Cicadulina* monitored by conventional and quantitative polymerase chain reaction. Phytopathology 92:65–74
10. Boonham N, Smith P, Walsh K, Tame J, Morris J, Spence N, Bennison J, Barker I (2002) The detection of Tomato spotted wilt virus (TSWV) in individual thrips using real-time fluorescent RT-PCR (TaqMan). J Virol Methods 101: 37–48
11. Fabre F, Kervarrec C, Mieuzet L, Riault G, Vialatte A, Jacquot E (2003) Improvement of Barley yellow dwarf virus-PAV detection in single aphids using a fluorescent real-time RT-PCR. J Virol Methods 110:51–60
12. Olmos A, Bertolini E, Gil M, Cambra M (2005) Real-time assay for quantitative detection of non-persistently transmitted Plum pox virus RNA targets in single aphids. J Virol Methods 128:151–155
13. Holland PM, Abramson RD, Watson R, Gelfand DH (1991) Detection of specific polymerase chain reaction product by utilizing the $5' \rightarrow 3'$ exonuclease activity of *Thermus aquaticus* DNA polymerase. Proc Natl Acad Sci U S A 88:7276–7280
14. Schneeberger C, Speiser P, Kury F, Zeillinger R (1995) Quantitative detection of reverse transcriptase-PCR products by means of a novel and sensitive DNA stain. PCR Methods Appl 4:234–238
15. Simòn B, Cenis JL, Demichelis S, Rapisarda C, Caciagli P, Bosco D (2003) Survey of *Bemisia tabaci* (*Hemiptera*: *Aleyrodidae*) biotypes in Italy with the description of a new biotype (T) from *Euphorbia characias*. Bull Entomol Res 93:259–264
16. Kheyr-Pour A, Bendahmane M, Matzeit V, Accotto GP, Crespi S, Gronenborn B (1991) *Tomato yellow leaf curl virus* from Sardinia is a whitefly-transmitted monopartite geminivirus. Nucleic Acids Res 19:6763–6769
17. Mason G, Caciagli P, Accotto GP, Noris E (2008) Real-time PCR for the quantitation of Tomato yellow leaf curl Sardinia virus in tomato plants and in Bemisia tabaci. J Virol Methods 147:282–289
18. Rasmussen R (2001) Quantification on the LightCycler. In: Meuer S, Wittwer C, Nakagawara K-I (eds) Rapid cycle real-time PCR. Springer, Berlin, pp 21–34
19. Tomato Genome Consortium (2012) The tomato genome sequence provides insights into fleshy fruit evolution. Nature 485: 635–641
20. Marzachì C, Bosco D (2005) Relative quantification of Chrysanthemum Yellows (16Sr I) phytoplasma in its plant and insect host using real-time polymerase chain reaction. Mol Biotechnol 30:117–127
21. Maffei G, Miozzi L, Fiorilli V, Novero M, Lanfranco L, Accotto GP (2014) The arbuscular mycorrhizal symbiosis attenuates symptom severity and reduces virus concentration in tomato infected by *Tomato yellow leaf curl Sardinia virus* (TYLCSV). Mycorrhiza 24:179–186

Chapter 7

Detection and Analysis of Non-retroviral RNA Virus-Like Elements in Plant, Fungal, and Insect Genomes

Hideki Kondo, Sotaro Chiba, and Nobuhiro Suzuki

Abstract

Endogenous non-retroviral RNA like sequences (NRVSs) have been discovered in the genome of a wide range of eukaryotes. These are considered as fossil RNA viral elements integrated into host genomes by as-yet-known mechanisms, and in many cases, those fossils are estimated to be millions-of-years-old. It is likely that the number of NRVS records will increase rapidly due to the growing availability of whole-genome sequences for many kinds of eukaryotes. Discovery of the novel NRVSs and understanding of their phylogenetic relationship with modern viral relatives provide important information on deep evolutionary history of RNA virus–host interactions. In this chapter, therefore, the common strategies for the identification and characterization of endogenous NRVSs from plants, insects, and fungi are described.

Key words Paleovirology, Molecular fossil record, Non-retrovirus-like sequence, Database search, Whole-genome shotgun, Genomic PCR, Southern blotting, Phylogenetic analysis, Maximum-likelihood

1 Introduction

Paleovirology, the study of endogenous viral elements, provides us with important information about the deep evolutionary history of virus–host interactions [1, 2]. In eukaryotes, their genomes contain numerous sequences that have originated from retroviruses (reverse-transcribing viruses) whose replication requires integration as proviral DNA into the genome of host cells [2]. Endogenous retroviral sequences are the evidence of ancient retroviral infections, thus considered as a kind of molecular fossil record. In contrast, the sequences of non-retroviral RNA viruses, which do not use a DNA intermediate, were until recently considered not to leave such molecular fossils in eukaryotic nuclear genomes. However for the past 5 years, the rapid progress on whole-genome sequencing for large numbers of eukaryotes has led to the discovery of non-retroviral RNA virus-like sequences (NRVSs, syn. endogenous virus elements: EVEs) integrated into the diverse eukaryotic

Ichiro Uyeda and Chikara Masuta (eds.), *Plant Virology Protocols*, Methods in Molecular Biology, vol. 1236,
DOI 10.1007/978-1-4939-1743-3_7,

genomes [1, 2], which are probably the result of heritable horizontal gene transfer (HGT) from viruses to hosts.

The first of these discoveries in vertebrate genomes is a set of NRVSs called EBLN (endogenous bornavirus-like nucleoprotein) which derived from the nucleoprotein gene of an ancient bornavirus [3–5]. In addition, NRVSs originated from ancient filoviruses (Ebola and Marburg viruses) have also been found in vertebrate [4–6]. Both bornaviruses and filoviruses are negative strand (-)ssRNA viruses belonging to the order *Mononegavirales.* Subsequently, the presence of several NRVSs related to (−)ssRNA viruses in plant, fungal, and insect genomes have been reported. NRVSs related to the nucleocapsid protein genes of cytorhabdoviruses and varicosaviruses were found in species of over nine plant families, including *Brassicaceae* and *Solanaceae* [7]. L polymerase-like elements of rhabdoviruses and nyamiviruses (order *Mononegavirales*) were identified in the genomes of black-legged tick, mosquitoes, and several kinds of other insects [4, 8]. In fungi, the first (−)RNA virus infection was evidenced based on a discovery of mononegavirus L-like elements in the genome of a phytopathogenic obligate ascomycete, *Erysiphe pisi* [8]. For dsRNA viruses, the most widespread NRVSs are related to the capsid protein (CP) and RNA-dependent RNA polymerase genes from partitiviruses and totiviruses in the genomes of plants, arthropods, fungi, nematodes, and protozoa [7, 9, 12]. In comparison with dsRNA and (-)ssRNA viruses, there are fewer examples of NRVSs (plant and invertebrate) related to positive-sense (+)ssRNA viruses. These include some members of plant virus genera such as *Benyvirus, Cilevirus, Citrivirus,* and *Tobamovirus,* and one insect virus genus *Flavivirus* [4, 7, 10, 11].

Here we describe the details of the methods for the identification and characterization of novel NRVSs from plant, insect, and fungal genomes (*see* Fig. 1). These methods are also applicable to other organisms for NRVSs searches.

2 Materials

2.1 Searching Tools for Detection of NRVSs in Public Databases

1. BLAST sequence database search at the National Center for Biotechnology Information (NCBI) [13]. A Web server is running at http://blast.ncbi.nlm.nih.gov/Blast.cgi.
2. Phytozome (41 green plant and algal species in version 9.1) (at http://www.phytozome.net). Other local databases for plant genomes: for example Brassica database (BRAD) (at http://brassicadb.org/brad/), Sol genomics network (SGN, Solanaceae Project) (at http://solgenomics.net), Miyakogusa.jp (*Lotus japonicas*) (at http://www.kazusa.or.jp/lotus/blast.html), SoyBase (soybean genetics and genomics database) (at http://soybase.org), MaizeGDB (*Zea mays*) (at http://www.maizegdb.org), Dendrome (a forest tree genome database) (at http://dendrome.ucdavis.edu).

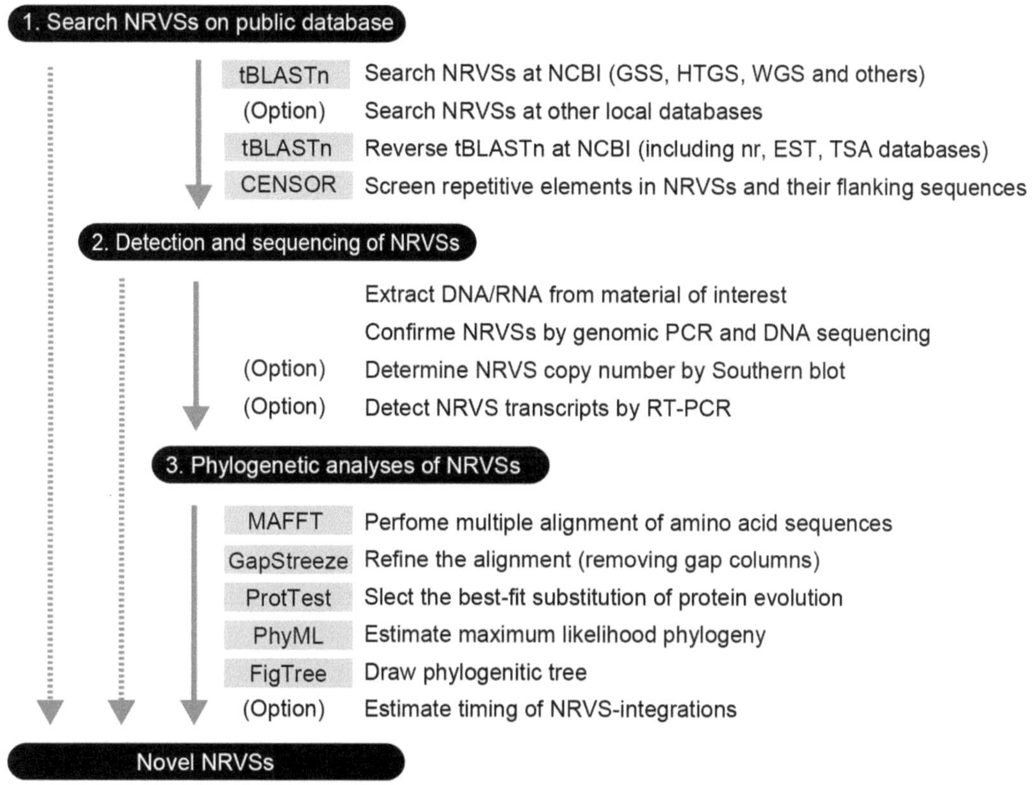

Fig. 1 Scheme for the detection and analysis of endogenous non-retroviral RNA virus-like sequences (NRVSs)

3. FungiDB: an integrated genome database for fungi (46 Fungi and 6 Oomycetes in version 2.3) (at http://fungidb.org/fungidb/) (*see* **Note 1**).
4. Arthropod genome databases: for example AphidBase (the aphid genome database) (at http://www.aphidbase.com), BeetleBase (*Tribolium castaneum*) (at http://beetlebase.org), FlyBase (a database of *Drosophila* genes and genomes) (at http://flybase.org), Hymenoptera Genome Database (HGD) (at http://hymenopteragenome.org), SilkDB (at http://silkworm.genomics.org.cn), VectorBase (Genomic resources for invertebrate vectors of human pathogens) (at https://www.vectorbase.org).

2.2 Detection and Sequencing of NRVSs from Plant, Fungal, and Insect Materials

1. Materials of interest, i.e., plant seeds or fresh leaves, fungal strains, and/or insect individuals.
2. DNA/RNA extraction.
   (a) DNA extraction buffer (200 mM Tris–HCl, 250 mM NaCl, 25 mM ethylene diamine tetraacetic acid [EDTA], and 5 % [w/v] SDS, pH 7.5).
   Isopropanol.
   1× TE buffer: 10 mM Tris–HCl, 1 mM EDTA, pH 8.0.
   (b) DNeasy® Blood and Tissue Kit (Qiagen).

(c) 2× CTAB extraction buffer: 2 % (w/v) hexadecyltrimethylammonium bromide (CTAB), 100 mM Tris–HCl pH 8.0, EDTA 20 mM pH 8.0, 1.4 M NaCl, 1 % (w/v) PVP (polyvinylpyrrolidone, MW 40,000), 0.3 % (v/v) β-mercaptoethanol.

(d) Phenol (water-saturated for RNA).
Chloroform–isoamyl alcohol (24:1).
Absolute ethanol.
0.1 M sodium acetate, pH 5.
75 % Ethanol.

3. Reagents for polymerase chain reaction (PCR), reverse transcription (RT)-PCR, nucleotide sequencing, and Southern blotting.
4. Wizard® SV Gel and PCR Clean-Up System (Promega).
5. Auto Assembler™ DNA Sequence Assembly Software (Applied Biosystems Inc.).
6. GENETYX-MAC/ATSQ (GENETYX Co.) or Enzyme X version 3 (Mek & Tosj) from http://nucleobytes.com/index.php/enzymex
7. CENSOR, a tool for annotation, submission and screening of repetitive elements in Repbase (at http://www.girinst.org/censor/index.php) [14]. CENSOR can be downloaded from the Genetic Information Research Institute (GIRI) for local installation (http://www.girinst.org/censor/download.php) (*see* **Note 2**).

2.3 Phylogenetic Analyses

1. MAFFT (Multiple alignment program for amino acid or nucleotide sequences) version 7 (at http://mafft.cbrc.jp/alignment/server) [15]. MAFFT (version 7.130) is also downloadable from http://mafft.cbrc.jp/alignment/software/.
2. Alignment curing programs.
   (a) Gap Strip/Squeeze (version 2.1.0), a tool for removing gaps in the alignment, in the Los Alamos HIV Sequence Database (http://www.hiv.lanl.gov/content/sequence/GAPSTREEZE/gap.html).
   (b) MEGA (Molecular Evolutionary Genetics Analysis) version 4.02 software (from http://www.megasoftware.net/mega4/mega.html) [16]. A current version for MEGA (version 6 for windows or version 5 for Mac OS) is downloadable from http://www.megasoftware.net
3. ProtTest server (current version is 2.4), a bioinformatics tool for the selection of best-fit models of protein evolution (at http://darwin.uvigo.es/software/prottest_server.html) [17]. ProtTest (version 3.2.1) is downloadable from http://code.google.com/p/prottest3/.

4. PhyML 3.0, a tool for estimating maximum-likelihood (ML) phylogenies (at http://www.atgc-montpellier.fr/phyml/) [18]. A new release of PhyML (version 3.1) is also available at http://www.atgc-montpellier.fr/phyml/versions.php.
5. FigTree version 1.3.1 software, a tool for drawing the tree (from http://tree.bio.ed.ac.uk/software/).

3 Methods

Here we describe a protocol for NRVS searches in Subheading 3.1, a brief procedure for confirmation experiments in Subheading 3.2, and a method of maximum-likelihood phylogenetic analyses of NRVSs in Subheading 3.3, which were basically used in our previous studies (*see* Fig. 1). Examples of results obtained in the above processes (based on discovery of NRVSs) are also provided.

3.1 Genome Sequence Database Search

1. To screen host genomic sequences for NRVSs, prepare query viral sequences. We usually select type species of the genera (non-retroviral RNA viruses) and obtain viral sequences from the NCBI taxonomy browser (http://www.ncbi.nlm.nih.gov/Taxonomy/Browser/wwwtax.cgi?name=Viruses). Queries should be amino acid sequences of the entire or conserved domain region of viral proteins (*see* **Note 3**). Those of viruses of your interest would be also preferable.
2. Conduct BLAST (tBLASTn) searches with prepared query sequences against genome sequence databases available from the NCBI (nucleotide collection, nr/nt; genome survey sequences, GSS; high-throughput genomic sequences, HTGS; whole-genome shotgun contigs, WGS, and others) with the default parameter setting.
3. As an option, other local databases would be used for NRVS searches. Specific sequence data warehouses such as Phytozome, FungiDB, and other arthropod genome databases usually include both updated and newly released genomes (*see* **Note 1**).
4. Genome sequences that matched viral peptides with *E*-values smaller than $1e^{-5}$ (the most common conventional value) are extracted as candidate NRVSs together with their flanking sequences if available (*see* **Note 4**).
5. Analyze candidate sequences and determine the region covering possible viral elements. GSS-, HTGS- and WGS-derived sequences often provide only partial fragments, but in many cases partially or entirely overlapping fragments can be found in a series of search. Thus it is possible to extend given NRVSs by assembling of those.

6. If NRVSs have interrupted ORFs, restore them by adding single or double "N" at frame-shifting sites or by introducing triplet "N" instead of internal stop-codons where they can be inferred by tBLASTn alignment (*see* Fig. 2a and **Note 5**). This step is required to obtain continuous, deduced amino acid sequences of NRVSs for further analyses. Edited residues are shown as "Xs" and these would be taken into account for mutations occurred during the course of evolution.
7. Each extracted candidate is then used as a reverse tBLASTn query against the non-redundant (nr) database (*see* Fig. 2a). This step is helpful to detect related NRVSs as well as more evolutionally close viruses than query viruses. These NRVSs are also use to screen WGSs of other organisms, ESTs (non-human, non-mouse expressed sequence tags) and TSAs (transcriptome shotgun assembly) available from the NCBI (*see* **Note 6**).
8. To examine for the presence of potential transposable elements and repetitive sequences in NRVSs and their flanking sequences, candidates (non-restored sequences) are subjected to the CENSOR program provided by the GIRI (*see* Fig. 2a and **Note** 7).
9. Nomenclature of NRVSs. Currently no fixed rule for nomenclature of fossil viral elements is present, and you may find many terms depending on reports. A system we use is, for example, as follows: "host organism name" + "viral protein" + "-like sequence" + "numbers defining virus species," i.e., *Arabidopsis thaliana* partitivirus CP-like sequence 1 (AtPCLS1), *Erysiphe pisi* mononegavirus L protein-like sequence 1 (EpMLLS1), etc. (*see* legend for Figs. 2, 4 and 5).
10. Deposition of NRVSs with appropriate annotations in GenBank/EMBL/DDBJ would be helpful for further studies. Nevertheless, this should be restricted to ones determined here by genomic PCR and sequencing (see below).

3.2 Detection and Sequencing of NRVSs from Plant, Fungal, and Insect Materials

1. To experimentally confirm the presence of NRVSs in the plant, fungus, and insect chromosomes, obtain each material of interest (i.e., plant seeds, fungal strains, or insect individuals) from commercially available materials, public stock centers, laboratories, and/or fields.
2. DNA/RNA preparation (in our system).
   (a) For plant genomic DNA, seeds or fresh leaf materials are homogenized in a micocentrifuge tube containing DNA extraction buffer [19]. Homogenates are subsequently centrifuged at 15,000×*g* for 10 min. Nucleic acids in the supernatant are then precipitated by centrifugation as above after addition of 0.6 volume of isopropanol. Resuspend the pellet in TE buffer and use as a template for genomic PCR (*see* **Note 8**).

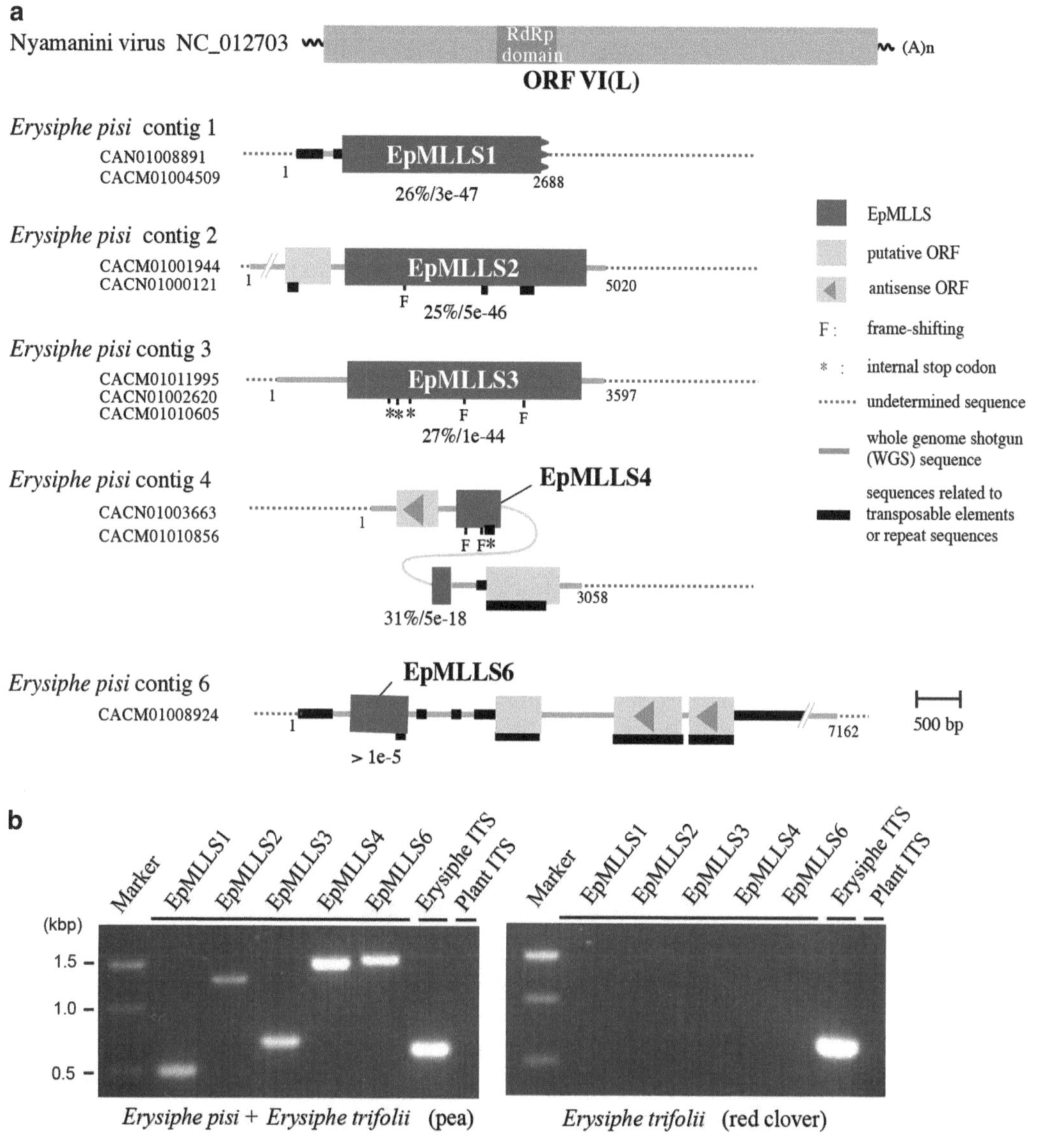

Fig. 2 Negative-strand RNA virus-related sequences from fungal nuclear genome. (**a**) Schematic representation of selected *Erysiphe pisi* (mononegavirus L protein-like sequences) MLLSs and their flanking regions. The corresponding positions of EpMLLSs on the Nyamanini virus (a prototypic member of *Nyavirus*) L-polymerase mRNA. The potential coding regions of EpMLLSs and flanking small ORFs are shown as *boxes*. EpMLLSs are distantly related to Nyamanini viruses (tBLAST % identities and *e*-value are shown). Retrotransposon-like sequences are shown by *thick black lines*. (**b**) Genomic PCR analysis of EpMLLSs. EpMLLSs of the *Erysiphe* spp. field samples isolated from pea (*left panel*) and red clover (*right panel*) were amplified using a primer set specific for each EpMLLS. Primer pairs, EryF (TACAGAGTGCGAGGCTCAGTCG) and EryR (GGTCAACCTGTGATC CATGTGACTGG) and At-IRS-FW and At-IRS-RV, were used for amplification of the *Erysiphe* spp. and plant ribosomal internal transcribed spacer (ITS) regions, respectively. This analysis was in part reported by Kondo et al. [8]

(b) For fungal and insect materials, total genomic DNA are purified using the DNeasy® Blood and Tissue Kit (Qiagen) according to the manufacturer's instructions (*see* **Note 8**).

(c) The genomic DNA isolation for Southern blot analysis is performed following a standard CTAB protocol [20].

(d) Total RNA fractions are obtained from fresh leaf materials of interest by two rounds of phenol–chloroform extraction and one round of chloroform extraction, followed by ethanol precipitation.

3. To amplify the NRVS fragments from DNA samples by PCR, specific primer pairs are designed based on the virus-related sequences and their flanking sequences (*see* Fig. 2b and **Note 9**). As a control, primer sets to amplify a fragment of the ribosomal internal transcribed spacer (ITS) region are also used (*see* **Note 10**).
4. PCR is generally carried out in a final volume of 50 μl containing 25 μl of Quick Taq HS DyeMix (TOYOBO) with the following conditions: an initial 94 °C for 2 min, followed by 35 cycles of 94 °C for 10 s, 58 °C for 30 s, and 68 °C for 1 min, then 68 °C for 10 min (*see* **Note 11**).
5. Subject the PCR product to 1 % agarose gel electrophoresis. Stain the gel with ethidium bromide and visualize DNA bands under UV illumination.
6. Sequence the purified PCR fragment. We use an ABI3100 DNA sequencer (Applied Biosystems) with BigDye® Terminator chemistry and a specific primer (*see* **Note 12**).
7. Assemble resultant sequences into a single fragment. We use the Auto Assembler software for the sequence assembly and analyze the sequences using the DNA processing software packages (GENETYX or Enzyme X). Finally, compare actual NRVSs and counterparts from databases, and confirm the presence of NRVSs on the genome of tested organisms.
8. (Option 1) To know the copy number of the NRVSs in chromosomes, we recommend performing Southern blot analysis as described by Faruk et al. [21]. Digoxigenin (DIG)-11-dUTP-labeled DNA fragments are amplified as probes from genomic DNA according to methods recommended by the manufacturer (Roche Diagnostics).
9. (Option 2) To test whether NRVSs could be expressed in cells, the BLASTx and BLASTp searches against databases for ESTs and TSAs available from the NCBI are a relatively easy way. In addition, RT-PCR detection of NRVS transcripts using an RNA template is more preferable (*see* **Note 13**). A separate reverse-transcription (M-MLV, Invitrogen) and PCR (*see* **step 4**) reactions are conducted with specific primer pairs for each

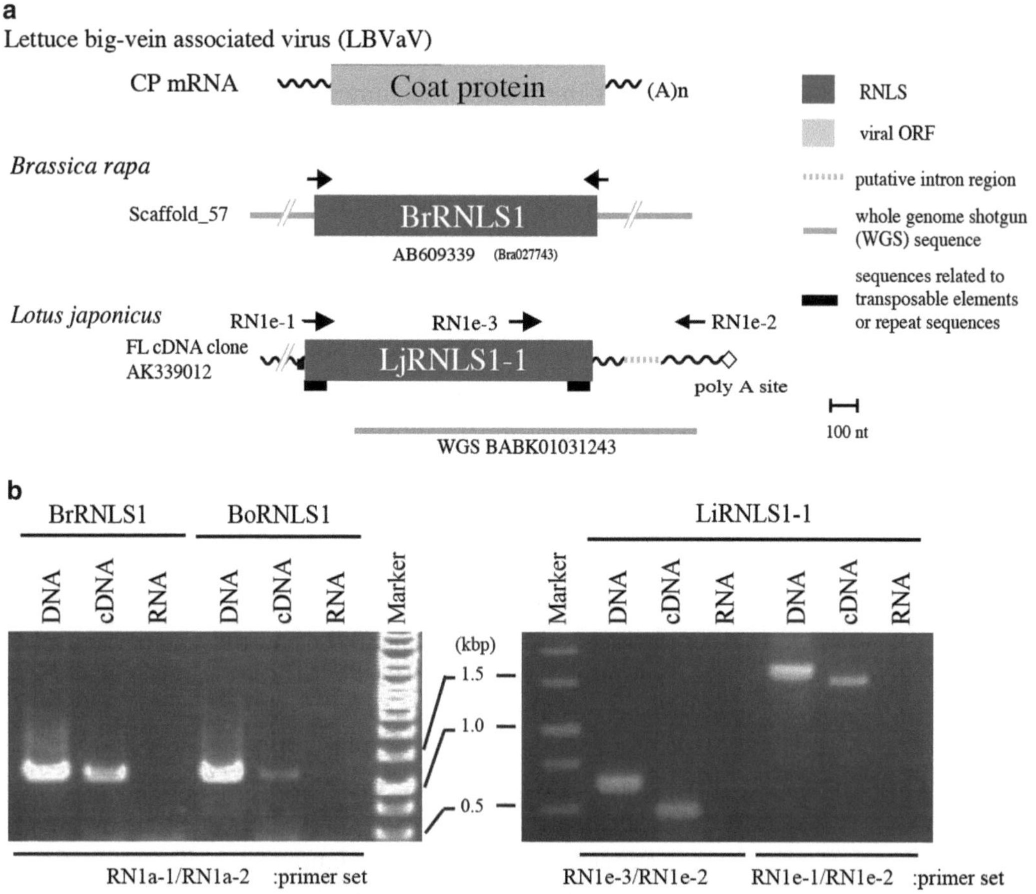

Fig. 3 Expression of negative-strand RNA virus-related sequences from plant nuclear genomes. (**a**) Schematic representation of selected rhabdovirus N-like sequences (RNLSs). RNLSs found in the genome sequence database of *Brassica rapa* (BrRNLS1) and *Lotus japonicus* (LjRNLS1-1) have significant sequence similarity to CP from lettuce big-vein associated virus (LBVaV, Varicosavirus) [7]. Note that varicosaviruses are evolutionally related to the family *Rhabdoviridae*. The positions of the primers used for genomic PCR are shown by *arrows*. (**b**) Molecular detection of RNLSs from *Brassica* and *Lotus* plants. Representative RNLSs from *B. rapa* (BrRNLS1), *B. oleracea* (BoRNLS1) and *L. japonicus* (LjRNLS1-1) were detected by genomic PCR (DNA) as well as by RT-PCR (cDNA) but not by RNA-temperate PCR (RNA). The size differences between genomic PCR (lanes DNA) and RT-PCR (lanes cDNA) products in LjRNLS1-1 verified the presence of an intron region (*dot-line* in **a**)

NRVS (*see* Fig. 3). One-step RT-PCR kits (e.g., OneStep RT-PCR Kit, Qiagen) are also recommended to provide a fast and successful alternative.

3.3 Phylogenetic Analyses

1. The deduced amino acid sequences of NRVSs (in a restored form) are used for the phylogenetic analysis with related protein sequences from extant viruses. Prepare a sequence list of those in FASTA format.
2. Multiple alignments of amino acid sequences are constructed using MAFFT version 7 (*see* **Note 14**). Copy and paste the

prepared sequence list to the window of the program, then run under the default parameters.

3. Reformat the resultant alignment with "ReadSeq" (copyright 1990 by D. G. Gilbert) in the same site to convert between the different sequence file formats used by following programs.
4. Automatic sequence alignments generally contain numerous gaps, and therefore we recommend to refine the alignment by removing gap regions (columns) (*see* **Note 15**).
   (a) Gap columns in the multiple alignments can be removed by using the Gap Strip/Squeeze program online.
   (b) The alignment can be manually edited in MEGA software.
5. To obtain appropriate substitution models for the maximum likelihood (ML) analyses, each of cured data sets (PHYLIP formatted alignment) should be subjected to the "AIC" (Akaike information criterion) calculation using ProtTest server (an example of output, AIC: LG+I+G+F). You may receive an e-mail notification when ProtTest analysis is done.
6. A phylogenetic tree is generated using the appropriate substitution model (see below) in PhyML 3.0 at the ATGC bioinformatics platform.
   (a) Input above cured data (PHYLIP formatted alignment).
   (b) Select the appropriate substitution model and other specific improvements determined above; "F" (amino acid *f*requencies; optimized or empirical), "I" (proportion of *i*nvariable sites; fixed or estimated) and "G" (*g*amma distribution parameter; fixed or estimated). PhyML 3.0 uses the "LG" substitution model as default.
   (c) The tree searching algorithms (the type of tree improvement) provided by PhyML are "NNI" (nearest neighbor interchange, a fast algorithm) and "SPR" (subtree pruning and regraft, a slower but efficient algorithm) [22]. The best option here is probably to use a SPR search (see PhyML-Manual version 3, available http://www.atgc-montpellier.fr/phyml/usersguide.php).
   (d) Selection of the method used to measure branch support. PhyML 3.0 provides "aLRT" (approximate likelihood ratio test, a fast algorithm) for convincing the branching accuracy, which is a good alternative to the bootstrap analysis (time-consuming). The default is to use "SH-like" (Shimodaira–Hasegawa-like) procedure [23] (*see* **Note 16**).
   (e) PhyML generates a tree file and a model parameter file. The estimated maximum likelihood tree is in the standard "Newick" format. You may receive an e-mail with a compressed Zip file containing these materials.

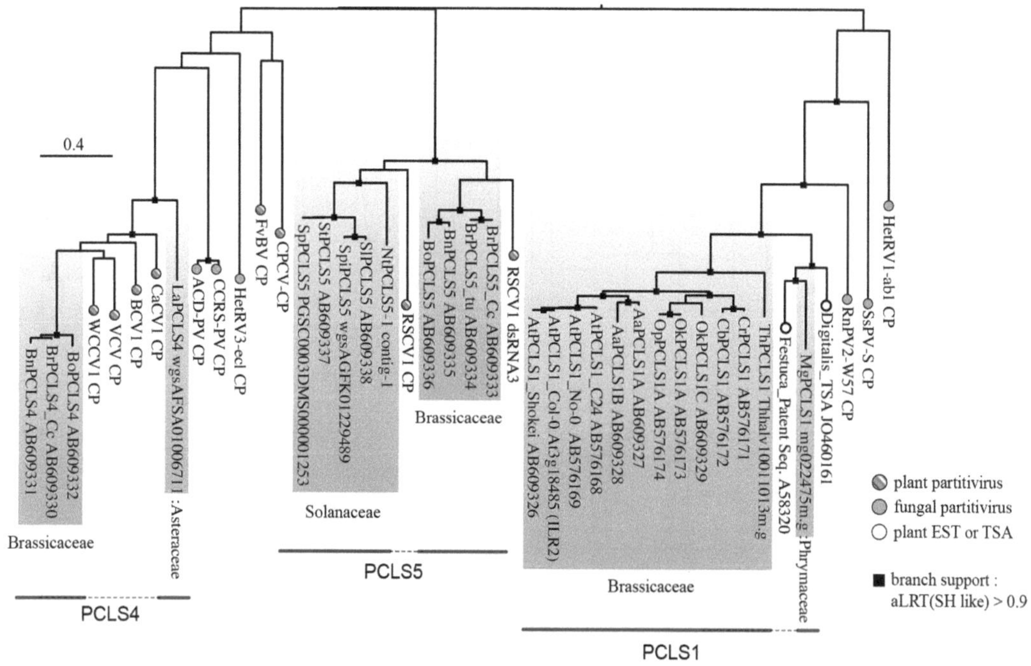

Fig. 4 Phylogenetic tree of selected partitivirus coat proteins (CPs) and partitivirus CP-like sequences (PCLSs) present on plant genomes. An alignment of CP sequences of representative partitiviruses and PCLSs was analyzed by the ML method. The accession numbers are shown next to the sequence names in the figure. *Gray-shaded* sequences represent NRVSs (PCLSs). A part of this analysis was reported by Chiba et al. [7] and Kondo et al. [25]. Plant PCLS1: AtPCLS1 (*Arabidopsis thaliana*), AlPCLS1 (*A. lyrata*), AaPCLS1 (*Arabidopsis arenosa*), OkPCLS1 (*Olimarabidopsis korshinskyi*), OpPCLS1 (*O. pumila*), CrPCLS1 (*Capsella rubella*), CbPCLS1 (*C. bursa-pastoris*), ThPCLS1 (*Thellungiella halophila*), MgPCLS1 (*Mimulus guttatus*); PCLS4: BrPCLS4 (*Brassica rapa*), BoPCLS4 (*B. oleracea*), BnPCLS4 (*B. napus*), LaPCLS4 (*Lactuca sativa*); PCLS5: BrPCLS5 (*B. rapa*), BoPCLS5 (*B. oleracea*), BnPCLS5 (*B. napus*); StPCLS5 (*Solanum tuberosum*), SpPCLS5 (*S. phureja*), SpiPCLS5 (*S. pimpinellifolium*), NtPCLS5 (*Nicotiana tabacum*). Best model according to AIC: WAG + I + G + F. *Closed boxes* on the nodes represent aLRT values derived using an SH-like calculation (only values greater than 0.9 are shown)

7. Open the tree file (Newick format) in the FigTree program. Select "Midpoint Rooting" from the view menu or "Reroot" button to do outgroup rooting if you include appropriate outgroup(s). The node/branches of the tree (aLRT-SH like values) can be labeled from the "Display" in the Node Labels menu. The tree can be exported as a graphic file by selecting "Export Graphic" from the file menu. We are using the Mac OSX version here, but there is also a Windows version. Obtained trees are modified in the illustrator software (or in any drawing tools) by keeping node length and scale bar. *See* Figs. 4 and 5 for the phylogenetic analysis of NRVSs.

8. (Option 3) Timing of NRVS-integrations into host chromosomes is of particular interest. This requires both information on divergence timing of host species within given taxa and on substantial variety of the NRVS locating on the same locus of

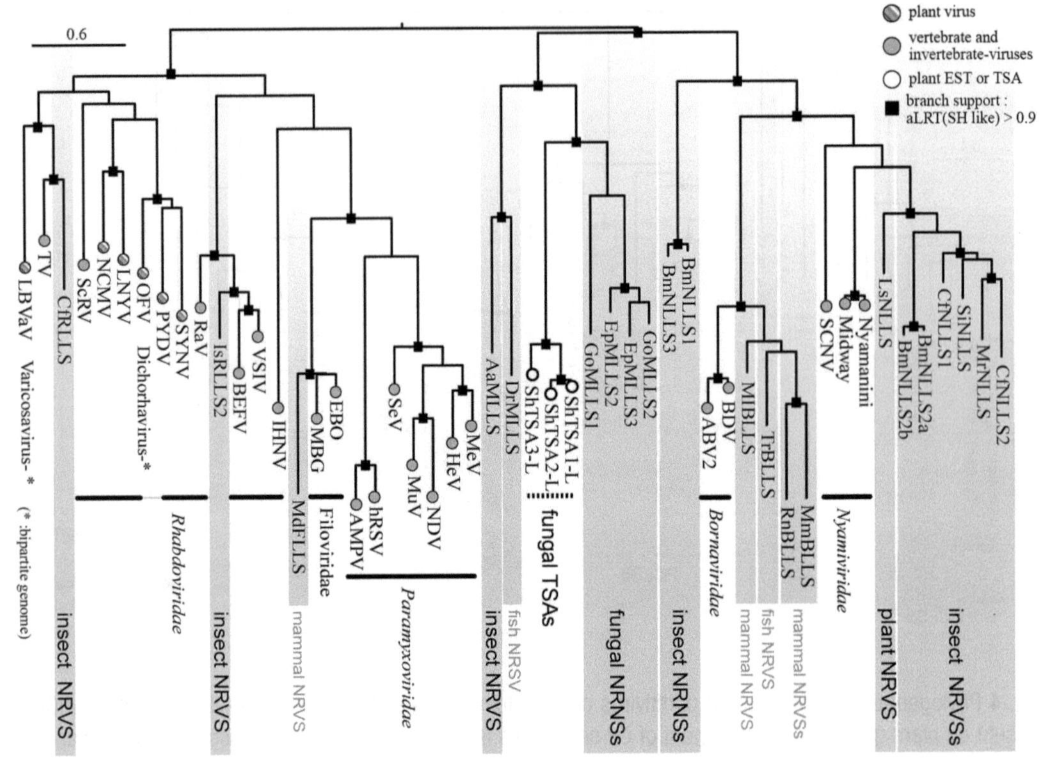

Fig 5 Phylogenetic relationship of L protein sequences of mononegaviruses (filoviruses, paramyxoviruses, rhabdoviruses, bornaviruses and nyamiviruses) [26] and endogenous mononegavirus L protein-like sequences (MLLSs) from fungi, plants, insects, fish, and mammals. This ML-tree was constructed using PhyML 3.0 based on a multiple amino acid sequence alignment of the RdRp polymerase core module. *Gray-shaded* sequences represent NRVSs (MLLSs) [8]. Fungal MLLS: EpMLLS2, 3 (a pea powdery mildew fungus, *Erysiphe pisi*), GoMLLS1, 2, (another powdery mildew fungus, *Golovinomyces orontii*); Insect MLLS: AaMLLS (the yellow fever mosquito, *Aedes aegypti*); Plant NLLS (nyavirus L protein-like sequence, one form of MLLSs): LsNLLS (lettuce, *Lactuca sativa*); insect NLLSs: BmNLLS1, (silkworm, *Bombyx mori*), MrNLLS, (leafcutter bee, *Megachile rotundata*), CfNLLS (carpenter ant, *Camponotus floridanus*), SiNLLS (fire ant, *Solenopsis invicta*); insect RLLSs (rhabdovirus L protein-like sequence): CfRLLS (*C. floridanus*), IsRLLS (black-legged tick, *Ixodes scapularis*); host names for other MLLSs from mammal and fish genomes are not shown. Fungal TSA-derived L-like sequences: ShTSAs (*Sclerotinia homoeocarpa* transcriptome shotgun assembly). Best model according to AIC: LG + G + F. *Closed boxes* on the nodes represent aLRT values derived using an SH-like calculation (only values greater than 0.9 are shown)

chromosomes in related organisms. Based on detection profile of the NRVS in related organism, the integration timing can be estimated from host-divergence timing; the integration should have occurred before branching of NRVS-containing groups but after branching of those with outer groups which lack the NRVS in the same locus. See an example of AtPCLS1 reported by Chiba et al. [7] (*see* also **Note 17**).

4 Notes

1. Searching local databases for the fossil records may also be important for the discovery of novel NRVSs. In our previous study, we identified (–)ssRNA virus-like sequences (NRVSs) in the WGS assemblies of the chromosomes of a plant pathogenic obligate ascomycete (powdery mildew fungus), *Golovinomyces orontii* (order Erysiphales) [8] (*see* Fig. 5), which are available from the Max Planck Institute for Plant Breeding Research Powdery Mildew Genome Project site (http://www.mpipz.mpg.de/24322/Project_Description), but not from the NCBI.
2. RepeatMasker may also useful to identify, characterize, and analyze repetitive elements in genomic sequences. The program can be run on a Web server at http://www.repeatmasker.org/cgi-bin/WEBRepeatMasker.
3. Longer and/or multiple queries for WGS searching (e.g., entire amino-acid sequences for viral polyprotein or replicase) tend to be aborted with the message "Error: CPU usage limit was exceeded." Therefore, we recommend the shorter queries (e.g., conserved domain regions for viral proteins) and/or selecting against the specific plants (taxid:3193), fungi (taxid:4751), or insects (taxid:6960) by using "Choose Search Set" to search for novel NRVSs through tBLASTn.
4. In our previous papers, we use <0.01 as the cut-off value for a "match," because some hits with values smaller than 0.01 show higher values in the reverse BLAST analyses against identified NRVSs [7, 11]. In addition, although some viral genes had sequence similarity with plant helicase or heat shock proteins, we removed those from further considerations, since these are likely originated from horizontal gene transfer events form host organisms to viruses [7].
5. Although several NRVSs in plant chromosomes retain "in-frame" ORFs [7], mononegavirus-like sequences in fungal chromosome and benyvirus-like fragments in plant and insect chromosomes are mostly fragmented [8, 11].
6. Several studies have demonstrated that the transcriptome analysis based on next-generation sequencing technologies is a useful new research tool for the discovery of RNA viruses. Thus, this reverse BLAST searching has another important aspect for discovering potential novel viruses. In fact, during BLAST searching for (–)ssRNA virus-like sequences, we found evidence strongly suggesting the presence of extant (–)ssRNA viruses in the kingdom Fungi for the first time [8].

7. Many NRVSs have distinguishable flanking sequences of the host origin in which these contain trace putative transposable elements [7, 8, 11] (*see* Fig. 2). From these data, it is speculated that progenitor viral segments might have been integrated after reverse-transcription with the aid of retrotransposons.
8. Purified DNA solutions (TE buffer) were stored at 4 °C until use.
9. This step importantly confirms whether candidate NRVSs are of chromosomal sequences or contaminating sequences from exogenous viruses.
10. For plant genome, a primer set, At-IRS-FW (ITS-F: CCGTAGGTGAACCTCGGAGGG) and At-IRS-RV (ITS-R: GGTGATCCCGCCTGACCTGG) [7], are used for amplification of the ITS regions 1 and 2 including the 5.8S rDNA.
11. PCR conditions should be optimized for each DNA template.
12. PCR fragments may have a few varieties in their sequences if NRVSs are multiplied after integration. To analyze multiple copies of NRVSs, PCR fragments from these loci should be cloned in pGEM T-easy (Promega) or other suitable vectors and then sequenced.
13. Some NRVSs having a long "in-frame" ORF appear to be transcribed (*see* Fig. 3b) and they might be translated as cellular proteins. However, further studies are required to determine their molecular function in the cells. It should be noted that the CP of a fungal partitivirus (Rosellinia necatrix partitivirus 2) had the greatest sequence similarity to a plant gene product, *Arabidopsis thaliana* auxin indole-3-acetic acid (IAA) Leucine resistant 2 (ILR2), which is thought to regulate a part of plant hormone homeostasis [7, 9].
14. It is also useful for refining multiple alignments obtained by other methods, e.g., T-Coffee (from http://www.tcoffee.org) and ClustalW (from http://www.clustal.org/clustal2/) programs.
15. We also recommend the removal (masking) of the align parts with low confidence (potentially misaligned) using the Gblocks, a well known program to remove poorly aligned regions (from http://molevol.cmima.csic.es/castresana/Gblocks.html) [24]. However, in our previous studies, some alignments appeared to be too short for the elimination of problematic regions by Gblocks. Thus we used only Gap Strip/Squeeze or MEGA program.
16. The sets of branching support values with bootstrap proportion >0.75 and aLRT >0.9 (SH-like option) tend to be similar (see a user guide at http://www.atgc-montpellier.fr/phyml/usersguide.php?type=online).
17. Using endogenous viral elements, bornavirus fossils were determined to be more than 40 million years old in primates,

and even much older in other mammalians [3]. Endogenous filovirus (Ebola and Marburg viruses) fossils were also identified and estimated to be at least 20 million years old in rodents and 40 million years old in other mammals [5]. PCLS1 (partitivirus-CP) in Arabidopsis-related plants is also estimated to be integrated at 10~24 million years ago [7].

Acknowledgment

This work was supported in part by a Grant-in-Aid for Scientific Research from the Japanese Ministry of Education, Culture, Sports, Science, and Technology (KAKENHI 40263628, 25252011) (H.K. and N.S.).

References

1. Feschotte C, Gilbert C (2012) Endogenous viruses: insights into viral evolution and impact on host biology. Nat Rev Genet 13:283–288
2. Patel MR, Emerman M, Malik HS (2011) Paleovirology - ghosts and gifts of viruses past. Curr Opin Virol 1:304–309
3. Horie M, Honda T, Suzuki Y et al (2010) Endogenous non-retroviral RNA virus elements in mammalian genomes. Nature 463(7277):84–90
4. Katzourakis A, Gifford RJ (2010) Endogenous viral elements in animal genomes. PLoS Genet 6:e1001191
5. Belyi VA, Levine AJ, Skalka AM (2010) Unexpected inheritance: multiple integrations of ancient bornavirus and ebolavirus/marburgvirus sequences in vertebrate genomes. PLoS Pathog 6:e1001030
6. Taylor DJ, Leach RW, Bruenn J (2010) Filoviruses are ancient and integrated into mammalian genomes. BMC Evol Biol 10:193
7. Chiba S, Kondo H, Tani A et al (2011) Widespread endogenization of genome sequences of non-retroviral RNA viruses into plant genomes. PLoS Pathog 7:e1002146
8. Kondo H, Chiba S, Toyoda K, Suzuki N (2013) Evidence for negative-strand RNA virus infection in fungi. Virology 435:201–209
9. Liu HQ, Fu Y, Jiang D et al (2010) Widespread horizontal gene transfer from double-stranded RNA viruses to eukaryotic nuclear genomes. J Virol 84:11876–11887
10. Cui J, Holmes EC (2012) Endogenous RNA viruses of plants in insect genomes. Virology 427:77–79
11. Kondo H, Hirano S, Chiba S et al (2013) Characterization of burdock mottle virus, a novel member of the genus *Benyvirus*, and the identification of benyvirus-related sequences in the plant and insect genomes. Virus Res 177:75–86
12. Taylor DJ, Bruenn J (2009) The evolution of novel fungal genes from non-retroviral RNA viruses. BMC Biol 7:88
13. Altschul SF, Madden TL, Schäffer AA et al (1997) Gapped BLAST and PSI-BLAST: a new generation of protein database search programs. Nucleic Acids Res 25:3389–3402
14. Kohany O, Gentles AJ, Hankus L, Jurka J (2006) Annotation, submission and screening of repetitive elements in Repbase: RepbaseSubmitter and Censor. BMC Bioinform 7:474
15. Katoh K, Toh H (2008) Recent developments in the MAFFT multiple sequence alignment program. Brief Bioinf 9:286–298
16. Tamura K, Dudley J, Nei M, Kumar S (2007) MEGA4: molecular evolutionary genetics analysis (MEGA) software version 4.0. Mol Biol Evol 24:1596–1599
17. Abascal F, Zardoya R, Posada D (2005) ProtTest: selection of best-fit models of protein evolution. Bioinformatics 21:2104–2105
18. Guindon S, Dufayard JF, Lefort V et al (2010) New algorithms and methods to estimate maximum-likelihood phylogenies: assessing the performance of PhyML 3.0. Syst Biol 59:307–321
19. Miura E, Kato Y, Matsushima R et al (2007) The balance between protein synthesis and degradation in chloroplasts determines leaf variegation in *Arabidopsis* yellow variegated mutants. Plant Cell 19:1313–1328
20. Porebski S, Bailey LG, Baum BR (1997) Modification of a CTAB DNA extraction protocol for plants containing high polysaccharide and polyphenol components. Plant Mol Biol Rep 15:8–15

21. Faruk MI, Eusebio-Cope A, Suzuki N (2008) A host factor involved in hypovirus symptom expression in the chestnut blight fungus, *Cryphonectria parasitica*. J Virol 82:740–754
22. Hordijk W, Gascuel O (2005) Improving the efficiency of SPR moves in phylogenetic tree search methods based on maximum likelihood. Bioinformatics 21:4338–4347
23. Anisimova M, Gascuel O (2006) Approximate likelihood-ratio test for branches: a fast, accurate, and powerful alternative. Syst Biol 55:539–552
24. Talavera G, Castresana J (2007) Improvement of phylogenies after removing divergent and ambiguously aligned blocks from protein sequence alignments. Syst Biol 56:564–577
25. Kondo H, Kanematsu S, Suzuki N (2013) Viruses of the white root rot fungus, *Rosellinia necatrix*. Adv Virus Res 86: 177–214
26. Kuhn JH, Bekal S, Caì Y et al (2013). Proposal for a new family in the order. *Arch Virol* 158:2209–2226

Chapter 8

Detection of Plant Viruses in Natural Environments by Using RNA-Seq

Atsushi J. Nagano, Mie N. Honjo, Motohiro Mihara, Masanao Sato, and Hiroshi Kudoh

Abstract

Sequencing of RNA by next generation sequencers, RNA-Seq, is revolutionizing virus detection. In addition to the unbiased detection of various viruses from wild plants in natural environments, RNA-Seq also allows for the parallel collection of host plant transcriptome data. Host transcriptome data are highly valuable for studying the responses of hosts to viral infections, as well as viral host manipulation. When detecting viruses using RNA-Seq, it is critical to choose appropriate methods for the removal of rRNA from total RNA. Although viruses with polyadenylated genomes can be detected by RNA-Seq following mRNA purification using oligo-dT beads, viruses with non-polyadenylated genomes are not effectively detected. However, such viruses can be detected by RNA-Seq using the rRNA selective depression method. The high-throughput and cost-effective method of RNA-Seq library preparation which is described here allows us to detect a broad range of viruses in wild plants.

Key words Virus detection, RNA-Seq, Wild plant, Selective depression of rRNA, Natural environment

1 Introduction

Local and/or systemic symptoms are signs of infection with plant viruses. The existence of a virus(es) in plants displaying such symptoms has been confirmed by the purification of viral particles and electron microscopy. Genome sequences of viruses were determined with purified viruses. In this way, phenotypic diagnoses have been confirmed and characterized with molecular evidences on viral infection [1]. Conventionally, sequences specific to a viral species or strain have been used as a marker to detect the virus by PCR or other methods using available viral genome sequences. Thus, researchers had to decide their target virus species to detect them.

Sequencing of RNA by next generation sequencers, RNA-Seq, is now revolutionizing virus detection. RNA-seq is a comprehensive

Ichiro Uyeda and Chikara Masuta (eds.), *Plant Virology Protocols*, Methods in Molecular Biology, vol. 1236, DOI 10.1007/978-1-4939-1743-3_8, © Springer Science+Business Media New York 2015

technology that allows for the sequence characterization and quantification of RNA molecules in a given sample. The majority of known plant viruses have an RNA genome, which can be detected using RNA-Seq. DNA viruses are also detectable by RNA-Seq as their infection cycles involve production of RNAs via transcription of their DNA genomes. Importantly, RNA-Seq allows for the inclusive screening of a broad range of virus species in a non-discriminatory manner, offering improvements over more traditionally used technologies (e.g., sequence-specific PCR). In addition, it is possible to use RNA-Seq to screen samples without prior information on viral infection. Thus, RNA-Seq is a very powerful tool for the detection and quantification of viruses in plants with and without symptoms.

To date, majority of researchers has focused on viruses in crops, and few studies have investigated viral infection in wild plants, particularly those in natural environments. Plant-virus interactions in plants not exhibiting apparent symptoms of infections have not received much attention from plant pathologists. As a result, the geographical distribution of viruses, as well as the prevalence of latent infections and mixed infections involving multiple viruses in wild plants is poorly understood [2]. The recent development and application of highly sensitive methods for the detection of viruses has shown that viruses responsible for disease in humans and livestock have frequently been found in soil, natural water sources, and wild animals [3–6], demonstrating their importance as reservoirs and carriers of pathogenic viruses. Likewise, wild plants are regarded as carriers of plant viruses responsible for disease in commercial crops [7]. Wild plants are also valuable for studying the ecology of viruses in natural environments.

In previous metagenomic studies of plant viruses, information about host plants could not be obtained simultaneously, because the amplification of viruses of a specific group or a selective concentration of virus particles was used to detect the viruses [8]. In addition to the unbiased detection of viruses in various groups, RNA-Seq allows for collection of host plant transcriptome data in parallel with viral RNA quantification. Host transcriptome data is valuable for studying both host responses to viral infections and the viral manipulation of hosts.

Unlike laboratory experiments, it is difficult to control genetic and environmental conditions in a study using wild plants under natural environments. Therefore, it is important to measure many samples in a factorial experimental design and to analyze the data with appropriate statistical methods [9]. Unfortunately, no commercially available kit for RNA-Seq library preparation has sufficient throughput to satisfy this requirement. We herein describe a high-throughput and cost-effective method for RNA-Seq library preparation that can be used for the detection of viruses in wild plants.

2 Materials

2.1 Sampling in Fields

1. Liquid nitrogen or RNA later solution (Ambion by Life Technologies).

2.2 RNA Extraction

1. RNeasy Plant Mini Kit (QIAGEN).
2. RNA-specific fluorescent dye (e.g., Qubit RNA assay kit (Life Technologies) and Quant-iT RiboGreen RNA assay kit (Life Technologies)).

2.3 RNA Quantity and Quality Assessment

1. Agilent 2100 Bioanalyzer (Agilent Technologies).

2.4 Selective Depression of rRNA (SDRNA) with Thermostable RNaseH

All buffers should be prepared using RNase free water.

1. SDRNA oligo pool: 252 tubes of 60-mer DNA covering rRNA (100 μM) are mixed in equal amount. The final concentration of the SDRNA oligo pool is approximately 0.4 μM each, totaling 100 μM.
2. ERCC RNA Spike-In control mixes (Ambion by Life Technologies).
3. 10× hybridization buffer: 1 M Tris–HCl (pH 7.4) and 2 M NaCl.
4. 10× RNaseH digestion buffer: 500 mM Tris–HCl (pH 7.4), 1 M NaCl, and 200 mM $MgCl_2$.
5. Hybridase Thermostable RNaseH (Epicentre).
6. AMpure XP (Beckman Coulter).
7. DNase I (RNase-free) (Takara).
8. 100 mM DTT (dithiothreitol) (Invitrogen).
9. AMpureXP buffer: 20 % polyethylene glycol-8000 and 2.5 M NaCl.

2.5 RNA-Seq Library Preparation and Sequencing

1. Random hexamer (Invitrogen).
2. RNasin Plus (Promega).
3. dNTP (25 mM each) (Promega).
4. M-MuLV Reverse Transcriptase (MMLV RTase) (Enzymatics).
5. 99.5 % ethanol.
6. dUTP/NTP mix (2 mM dA, dC, dG, and 4 mM dU) (Fermentas).
7. RNase H (Enzymatics).
8. DNA polymerase I (Enzymatics).
9. dNTP (2.5 mM each).
10. End-Repair Mix LC (Enzymatics).
11. 10 mM dATP mix (Promega).

12. Klenow 3′→5′ exo- (Enzymatics).
13. T4 DNA Ligase (Enzymatics).
14. Y-shape adapter: anneal 5′-A*A*TGATACGGCGACCACCG AGATCTACACTCTTTCCCTACACGACGCTCTTCCGAT *C*T-3′ and 5′-/5Phos/-G*A*TCGGAAGAGCACACGTC TGAACTCCAGTC*A*C-3′ (*see* **Note 15**). * signifies a phosphorothioate bond./5Phos/signifies a phosphorylation.
15. 12P beads (*see* **Note 1**).
16. Uracil DNA glycosylase (Enzymatics).
17. Index primer: 5′-CAAGCAGAAGACGGCATACGAGATXX XXXXXXGTGACTGGAGTTCAGACGTGT-3′. XXXXXXXX represents the 8-mer index sequences (*see* **Note 2**).
18. Universal primer: 5′-AATGATACGGCGACCACCGAGAT CTACACTCTTTCCCTACACGACGCTCTTCCGATCT-3′.
19. 5× Phusion HF buffer (Finnzymes).
20. Phusion DNA polymerase (Finnzymes).
21. E-Gel electrophoresis systems (Life Technologies).
22. E-Gel SizeSelect Agarose Gels, 2 % (Life Technologies).

3 Methods

Overview

The RNA-Seq library preparation described in this protocol is based on Wang et al. and Li et al. with some modifications [10, 11]. Libraries constructed using this protocol are compatible with Illumina TruSeq v3 chemistry, and can be used for strand-specific sequencing using HiSeq and MiSeq.

For the detection of plant viruses using RNA-Seq, it is critical to choose an appropriate method for the removal of rRNA from total RNA. Purification methods using oligo-dT beads are widely used for the removal of rRNA. The RNA genomes of some plant viruses, such as Potexviruses and Potyviruses, are polyadenylated, whereas other viruses have non-polyadenylated genomes. Although viruses with polyadenylated genomes can be detected by RNA-Seq following purification using oligo-dT beads, viruses with non-polyadenylated genomes are not effectively detected. However, viruses with non-polyadenylated genomes can be detected using RNA-Seq paired with the rRNA selective depression method [12]. For example, Turnip mosaic virus (TuMV), which has a polyadenylated genome, was detected at comparable levels in RNA solutions purified using either the oligo-dT beads method or the rRNA selective depression method (Fig. 1a). However, Cucumber mosaic virus (CMV), which has a non-polyadenylated genome, reduced using the oligo-dT beads method compared to using the rRNA selective depression method (Fig. 1b).

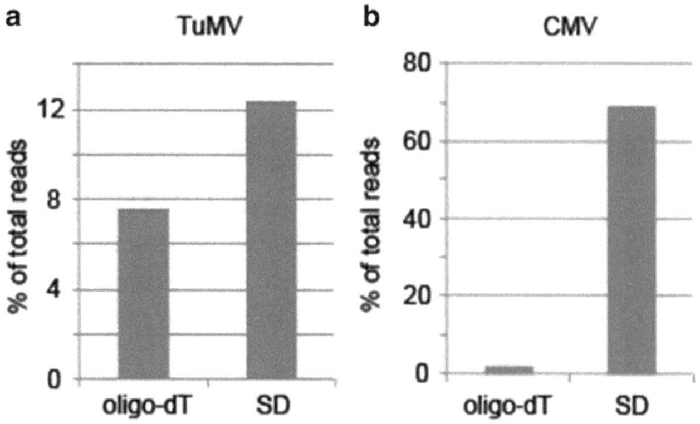

Fig. 1 Fraction of virus reads in RNA-Seq data with two different methods for removing rRNA. Oligo-dT is the mRNA purification by oligo-dT beads. SD is the selective depression of rRNA with thermostable RNaseH. (**a**) Fractions of reads mapped to a *Turnip mosaic virus* (TuMV) genome in RNA solution extracted from a leaf of wild *Arabidopsis halleri.* (**b**) Fractions of reads mapped to a *Cucumber mosaic virus* (CMV) genome in another RNA sample

3.1 Sampling in Fields

Plant tissues collected in fields (*see* **Note 3**) should be frozen immediately after harvest in liquid nitrogen and stored at −80 °C until RNA isolation. When the use of liquid nitrogen is not possible at a field site (*see* **Note 4**), the use of RNA later solution is also a feasible option, although less optimal. Samples should be fully submerged in the RNA later solution, and waxy leaves and thick roots should be cut into small pieces to ensure that the RNA later solution is able to fully penetrate into the tissues.

3.2 RNA Extraction

Total RNA is isolated from each sample using the RNeasy Plant Mini Kit (QIAGEN) (*see* **Note 5**).

3.3 RNA Quantity and Quality Assessment

The extracted RNA is first quantified using an RNA specific fluorescent dye (the Qubit RNA assay kit or the Quant-iT RiboGreen RNA assay kit etc.), and the quality of the sample is determined using the Agilent 2100 Bioanalyzer (*see* **Note 6**).

3.4 Selective Depression of rRNA (SDRNA) with Thermostable RNaseH

Mix total RNA (typically 500 ng), the ERCC RNA Spike-In control mixes, 1 μl of the SDRNA oligo pool, and RNase-free water in a PCR tube, with a total volume of 10 μl. The mixture is then annealed using a thermal cycler with the following program: 95 °C for 2 min, slow-cooled to 45 °C (0.1 °C/s), followed by 45 °C for 5 min. Add the following premix (pre-warmed) and mix by pipetting: 1 μl of the Hybridase Thermostable RNaseH, 2 μl of 10× RNaseH digestion buffer, and 7 μl of RNase-free water. Incubate at 45 °C for 30 min to selectively digest RNA of the DNA/RNA hybrid double strand. Add 24 μl of the AMpure XP beads and perform purification using the magnetic stand (*see* **Note 7**).

Resuspend the washed beads with 10 μl of RNase free water. Prepare the DNaseI premix as follows: 1 μl of DNaseI (5 unit/μl), 1 μl of 10× DNaseI buffer (*see* **Note 8**), 1 μl of 100 mM DTT, and 7 μl of RNase free water. Add the DNaseI premix and incubate at 37 °C for 30 min to degrade the SDRNA oligo pool (*see* **Note 9**). Then add 24 μl of the AMpure XP buffer and perform the purification using the magnetic stand (*see* **Note** 7). The AMpure XP magnetic beads with the purified RNA should be used for the next step immediately without over drying.

3.5 RNA-Seq Library Preparation and Sequencing

1. Prepare fragmentation master mix as follows: 2 μl of 10× MMLV buffer (*see* **Note 10**), 1 μl of 100 mM DTT, and 7 μl of RNase free water. Mix the beads with the purified RNA and 10 μl of the fragmentation master mix by pipetting. Heat the mixture at 94 °C for 4.5 min and immediately cool at 4 °C (*see* **Notes 11** and **12**). Place the tube on the magnetic stand until the supernatant becomes clear. Transfer the supernatant to a new tube.
2. Add 2 μl of the random hexamer and 0.5 μl of the RNasin Plus to 10 μl of the fragmented RNA solution. Incubate it at 50 °C for 5 min and immediately place it on ice to relax the secondary structures of the RNA/DNA. Prepare reverse transcription master mix as follows: 1 μl of 100 mM DTT, 0.1 μl of dNTP (25 mM each), 0.5 μl of MMLV RTase, and 5.9 μl of RNase free water. Mix the fragmented RNA with the random hexamer and the reverse transcription master mix. For the reverse transcription step, incubate the solution at 25 °C for 10 min, followed by 50 min at 42 °C. Add 36 μl of AMpure XP beads and 18 μl of 100 % ethanol and perform the purification using the magnetic stand (*see* **Note 7**). Resuspend the washed beads with 10 μl of nuclease free water and mix well. Move the supernatant (purified DNA/RNA hybrid solution) to a new tube using the magnetic stand.
3. Prepare the second strand synthesis master mix as follows: 2 μl of 10× Blue Buffer (*see* **Note 13**), 1 μl of dUTP/NTP mix, 0.5 μl of 100 mM DTT, 0.5 μl of RNase H, 1 μl of DNA polymerase I, and 5 μl of nuclease free water. Add 10 μl of the second strand synthesis master mix to 10 μl of the purified DNA/RNA hybrid solution without beads. Incubate it at 16 °C overnight. Add 25 μl of AMpure XP beads and perform the purification using the magnetic stand (*see* **Note** 7). Resuspend the washed beads with 10 μl of nuclease free water.
4. Prepare end-repair master mix as follows: 2 μl of 10× End-Repair Buffer (*see* **Note 14**), 1 μl of dNTP (2.5 mM each), 1 μl of End-Repair Mix LC, and 6 μl of nuclease free water. Add 10 μl of the end-repair master mix to 10 μl of the purified double-stranded DNA solution with beads. Incubate it at 20 °C

for 30 min (*see* **Note 9**). Add 25 μl of AMpure XP buffer and perform purification using the magnetic stand (*see* **Note** 7). Resuspend the washed beads with 10 μl of nuclease free water.

5. Prepare dA-tailing master mix as follows: 2 μl of 10× Blue Buffer (*see* **Note 13**), 1 μl of 10 mM dATP, 0.5 μl of Klenow 3′→5′ exo-, and 6.5 μl of nuclease free water. Add 10 μl of the dA-tailing master mix to 10 μl of the purified end-repaired DNA solution with beads. Incubate it at 37 °C for 30 min (*see* **Note 9**). Add 25 μl of AMpure XP buffer and perform purification using the magnetic stand (*see* **Note** 7). Resuspend the washed beads with 10 μl of nuclease free water and mix well. Move the supernatant (A-tailed DNA solution) to a new tube using the magnetic stand.
6. Add 1 μl of 0.1 μM Y-shape adapter (*see* **Note 15**) to 10 μl of the A-tailed DNA solution. Prepare ligation master mix as follows: 12 μl of 2× Rapid Ligation Buffer (*see* **Note 16**), and 1 μl of T4 DNA Ligase. Mix the ligation master mix, the universal adapter, and the A-tailed DNA solution by pipetting. Incubate it at 20 °C for 20 min.
7. Add 15 μl of 12P beads (*see* **Note 1**) to 12 μl of the ligation product and perform the purification using the magnetic stand (*see* **Note** 7). Resuspend the washed beads with 10 μl of nuclease free water. Mix 10 μl the purified DNA with beads and 10 μl of AMpure XP buffer. Perform the purification using the magnetic stand again (*see* **Note** 7). Resuspend the washed beads with 15 μl of 10 mM Tris–HCl (pH 8.0) and mix well. Move the supernatant (purified ligation product) to a new tube using the magnetic stand.
8. Add 1 μl of uracil DNA glycosylase to 15 μl of the purified ligation product. Incubate it at 37 °C for 30 min.
9. Mix 5.4 μl of the UDG digested DNA, 1 μl of 10 μM index primer (*see* **Note 2**), 1 μl of 10 μM universal primer, 2 μl of 5× Phusion HF buffer, 0.3 μl of 10 mM dNTP, and 0.3 μl of Phusion DNA polymerase. DNA fragments with adapters and an index sequence are amplified using a thermal cycler with the following program: denature at 94 °C for 2 min, 10–20 cycles at 98 °C for 10 s, 65 °C for 30 s, 72 °C for 30 s as an amplification step, and 72 °C for 5 min for the final extension (*see* **Note 17**). Add 10 μl of AMpure XP beads to PCR products. If you want to sequence multiple (*N*) samples, mix each of the PCR products and then add 10× *N*μl of AMpure XP beads. Perform purification using the magnetic stand (*see* **Note** 7). Resuspend the washed beads with 25 μl of nuclease free water. Transfer the supernatant to a new tube using the magnetic stand.
10. To remove small (<200 bp) and large (>600 bp) fragments, perform an additional purification using E-Gel SizeSelect

Agarose Gels (2 %), following manufacturer instructions. Typically, five fractions are obtained from 250 bp at 30 s intervals. Combine them and store at −20 °C until sequencing. 1 μl of the combined library is used for electrophoresis using the Bioanalyzer to check the quality (*see* **Note 18**). The library can then be sequenced using HiSeq or MiSeq technology.

3.6 Mapping and Quantification

Summary of sequencing results are created by FastQC and RSeQC. Low quality reads and adapter sequences are removed using the programs tagdust, samtools, and FASTX-Toolkit. Reads passing the quality filter are then mapped to reference sequences using bowtie2 (*see* **Note 19**). Counts of reads are calculated using RSEM.

We used the genomes of 40 viruses reported to infect Brassicaceae plants, as well as the transcriptome of the host plant (*Arabidopsis halleri*) as reference sequences. The virus genome sequences were downloaded from NCBI/GenBank (*see* **Note 20**).

4 Notes

1. The 12P beads are prepared as follows. Remove buffer of AMpure XP on the magnetic stand. Add 12P buffer (12 % PEG-8000, 2.5 M NaCl) equal to the volume of the removed AMpure XP buffer and resuspend the magnetic beads well.
2. The index primers contain variable 8-mer index sequences that allow multiple samples to be read in a single HiSeq lane. To distinguish samples, different index primers should be used for different samples. We used 96 sequences that contained at least three nucleotides differences.
3. Contamination from other organisms should be minimized as much as possible. For example, insects, epiphytic fungi, and the pollens of other plants can be found on/in wild plants, and should be avoided.
4. Do not transport liquid nitrogen by car without appropriate equipment, to avoid risk of suffocation caused by spilled liquid nitrogen.
5. The TRIzol reagent can also be used. Modifications of extraction protocols and additional purification may be needed for some samples, depending on plant species and tissue type.
6. RIN values may not reach 9–10 even if intact RNA is measured, and is dependent on the species and tissue used.
7. AMpure XP purification method. After adding AMpure XP beads or AMpure XP buffer, hold the mixture for 6 min at room temperature and remove the supernatant using the magnetic stand. Add 75 % ethanol and remove the ethanol after the supernatant becomes clear. Repeat the wash procedure.

8. 10× DNaseI buffer supplied with DNaseI enzyme (Takara). The composition of 10× DNaseI buffer is 400 mM Tris–HCl, pH 7.5, 80 mM $MgCl_2$, and 50 mM DTT.
9. This reaction is performed using the AMpure XP beads in the reaction solution.
10. 10× M-MuLV RT buffer is supplied with the MMLV RTase (Enzymatics). The composition of the 10× M-MuLV RT buffer is 500 mM Tris–HCl (pH 8.3), 750 mM KCl, 30 mM $MgCl_2$, and 100 mM DTT.
11. To strictly control temperature, it is recommended to perform the reaction using a thermal cycler.
12. The length of the incubation at 94 °C should be optimized in your laboratory, depending on the plant species investigated, the heating equipment available, and insert length of library.
13. The 10× blue buffer is supplied with enzymes (e.g., DNA polymerase I; Enzymatics). The composition of the 10× blue buffer is 500 mM NaCl, 100 mM Tris–HCl (pH 7.9), 100 mM $MgCl_2$, and 10 mM DTT.
14. The 10× end-repair buffer is supplied with the end-repair mix LC (Enzymatics). The composition of the 10× end-repair buffer is 1 mM Tris–HCl (pH 7.5), 500 mM NaCl, 100 mM $MgCl_2$, 50 mM DTT, and 0.25 % Triton-X 100.
15. The Y-shape adapter is prepared by annealing two partially complementary oligo-DNAs. A mixture of 100 μM adapter F and R is annealed using a thermal cycler with the following program: 95 °C for 2 min, slow-cooled to 25 °C (0.1 °C/s), followed by 30 min at 25 °C. The annealed adapter (50 μM) is stored at −20 °C. It should be diluted to the working concentration (0.1 μM) just before use.
16. The 2× rapid ligation buffer is supplied with the T4 DNA Ligase rapid (Enzymatics). The composition of the 2× rapid ligation buffer is 132 mM Tris–HCl (pH 7.6), 20 mM $MgCl_2$, 2 mM DTT, 2 mM ATP, and 15 % PEG 6000.
17. The PCR cycles should be optimized depending on the samples being prepared.
18. The contamination of small (<200 bp) and large (>600 bp) fragments, especially adapter dimers (121 bp), reduces sequencing efficiency.
19. Other short-read aligners (e.g., BWA) are also acceptable.
20. The genome sequences of many viruses are deposited in GenBank. If you have no information about virus species possibly detected in your sample, all virus genomes in GenBank can be used as references.

Acknowledgement

This work was supported by JST PRESTO (to A.J.N.) and the Next Program (GS013) (to H.K.).

References

1. Hull R (2002) Matthews' plant virology, 4th edn. Elsevier, Amsterdam
2. Malmstrom CM, Melcher U, Bosque-Pérez NA (2011) The expanding field of plant virus ecology: historical foundations, knowledge gaps, and research directions. Virus Res 159: 84–94
3. Le Guyader F, Dubois E, Menard D, Pommepuy M (1994) Detection of Hepatitis-A virus, rotavirus, and enterovirus in naturally contaminated shellfish and sediment by reverse transcription-seminested PCR. Appl Environ Microbiol 60(10):3665–3671
4. Munster VJ, Baas C, Lexmond P, Waldenstrom J, Wallensten A, Fransson T et al (2007) Spatial, temporal, and species variation in prevalence of influenza A viruses in wild migratory birds. PLoS Pathog 3:e61
5. Dovas CI, Papanastassopoulou M, Georgiadis MP, Chatzinasiou E, Maliogka VI, Georgiades GK (2010) Detection and quantification of infectious avian influenza A (H5N1) Virus in environmental water by using real-time reverse transcription-PCR. Appl Environ Microbiol 76:2165–2174
6. Honjo MN, Minamoto T, Kawabata Z (2012) Reservoirs of Cyprinid herpesvirus 3 (CyHV-3) DNA in sediments of natural lakes and ponds. Vet Microbiol 155:183–190
7. Fargette D, Fauquet C, Grenier E, Thresh J (1990) The spread of African cassava mosaic-virus into and within cassava fields. J Phytopathol 130:289–302
8. Roossinck MJ (2012) Plant virus metagenomics: biodiversity and ecology. Annu Rev Genet 46:359–369
9. Nagano AJ, Sato Y, Mihara M, Antonio BA, Motoyama R, Itoh H et al (2012) Deciphering and prediction of transcriptome dynamics under fluctuating field conditions. Cell 151: 1358–1369
10. Wang L, Si Y, Dedow LK, Shao Y, Liu P, Brutnell TP (2011) A low-cost library construction protocol and data analysis pipeline for Illumina-based strand-specific multiplex RNA-seq. PLoS One 6(10):e26426
11. Zhong S, Joung JG, Zheng Y, Chen YR, Liu B, Shao Y et al (2011) High-throughput illumina strand-specific RNA sequencing library preparation. Cold Spring Harb Protoc 2011:940–949
12. Morlan JD, Qu K, Sinicropi DV (2012) Selective depletion of rRNA enables whole transcriptome profiling of archival fixed tissue. PLoS One 7:e42882

Chapter 9

Cloning and Profiling of Small RNAs from *Cucumber Mosaic Virus* Satellite RNA

Yuan-Yuan Fang, Neil A. Smith, Jian-Hua Zhao, Joanne R.M. Lee, Hui-Shan Guo, and Ming-Bo Wang

Abstract

RNA silencing is not only a gene regulation mechanism that is conserved in a broad range of eukaryotes but also an adaptive immune response against foreign nucleic acids including viruses in plants. A major feature of RNA silencing is the production of small RNA (sRNA) of 21–24 nucleotides (nt) in length from double-stranded (ds) or hairpin-like (hp) RNA by Dicer-like (DCL) proteins. These sRNAs guide the binding and cleavage of cognate single-stranded (ss) RNA by an RNA silencing complex. Like all plant viruses and subviral agents, replication of viral satellite RNAs (satRNAs) is associated with the accumulation of 21–24 nt viral small interfering RNA (vsiRNA) derived from the whole region of a satRNA genome in both plus and minus-strand polarities. These satRNA-derived siRNAs (satsiRNAs) have recently been shown to play an important role in the trilateral interactions among host plants, helper viruses and satRNAs. Here, we describe the cloning and profile analysis of satsiRNAs from satRNAs of *Cucumber mosaic virus* (CMV). We also describe a method to minimize the strand bias that often occurs during vsiRNA cloning and sequencing.

Key words Small interfering RNA, Virus, CMV, Satellite RNA, satsiRNAs

1 Introduction

Antiviral defense is one of the important functions of RNA silencing in plants. Virus infection is accompanied by the accumulation of vsiRNAs, which guide the destruction of viral genomic RNAs. To survive RNA silencing, viruses have evolved counter-defense mechanisms by expressing RNA silencing suppressor proteins that function primarily through binding and sequestering vsiRNAs [1]. vsiRNAs are processed by DCLs from dsRNA formed between plus and minus strands of viral replicative intermediates, or hp structures formed within single-stranded viral genomic RNAs [2–4]. Plants have four different DCLs, processing 21 nt microRNA (DCL1), and 21 nt (DCL4), 22 nt (DCL2,) and 24 nt (DCL3)

Ichiro Uyeda and Chikara Masuta (eds.), *Plant Virology Protocols*, Methods in Molecular Biology, vol. 1236, DOI 10.1007/978-1-4939-1743-3_9, © Springer Science+Business Media New York 2015

siRNA. vsiRNAs from RNA viruses are processed primarily by DCL4 and DCL2, whereas those from DNA viruses are processed by DCL4, DCL2, and DCL3 [5].

SatRNAs have small RNA genomes that usually do not encode protein, and therefore depend on their helper viruses for replication and spread. For instance, the satRNA of CMV has a 330–400 nt-long linear RNA genome with a highly conserved secondary structure, and requires CMV to support its replication in the cytoplasm of infected plant cells [6]. satRNAs can modulate the symptoms caused by their helper viruses in a nucleotide sequence or RNA structure-dependent manner [7–11]. Recent studies have indicated that satsiRNAs play a key role in this symptom modulation. For instance, the most frequently cloned satsiRNA from the satRNA of the ShanDong strain of CMV (SD-satRNA), satsiR-12 [12], targets the 3′ untranslated region (UTR) of CMV, triggering subsequent host RNA-dependent RNA polymerase 6 (RDR6)-dependent antiviral silencing [13]. Furthermore, one of the satsiRNAs derived from CMV Y-satRNA directs the silencing of a chlorophyll biosynthetic gene *CHLI* in *Nicotiana* species, resulting in the yellowing symptoms observed in CMV Y-satRNA-infected plants [14, 15].

siRNAs derived from a number of CMV satRNAs have recently been reported. siRNAs from SD-satRNA (SD-satsiRNAs) was cloned by both a laboratory small scale cloning method [12] and by deep sequencing using HiSeq2000 (BGI, http://www.genomics.cn/en/index). Bioinformatic analysis showed that SD-satsiRNAs are predominantly 21 and 22 nt in length, corresponding to all regions of the satRNA genome in both sense and antisense orientations (Fig. 1a). However, SD-satsiRNAs are not evenly distributed across either the plus or minus strand of the SD-satRNA genome, with some regions having a higher number of siRNA reads than the other regions (Fig. 1b). The relationship between genome-wide SD-satsiRNA distribution and satRNA secondary structure was recently investigated by matching SD-satsiRNAs to the definite in vivo models of CMV-satRNA secondary structure [16]. This analysis showed that the most frequently cloned positive-strand satsiRNAs corresponded to stem-loop secondary structures, suggesting that these stem-loop structures may be a direct substrate for DCLs to process SD-satsiRNAs [12].

siRNAs from the CMV Y-satRNA was also cloned using Illumina deep sequencing, which again showed a predominant size distribution of 21 and 22 nucleotides, and uneven distribution of siRNAs across the 369 nt Y-satRNA genome [15]. Similar to siRNAs from many RNA viruses obtained using deep sequencing, Y-satsiRNAs showed a strand bias, with plus strand siRNAs more abundant than minus strand siRNAs (Fig. 2). Our previous study using northern blot hybridization suggested that this strand bias is at least partly due to sequestering of minus-strand siRNAs by an

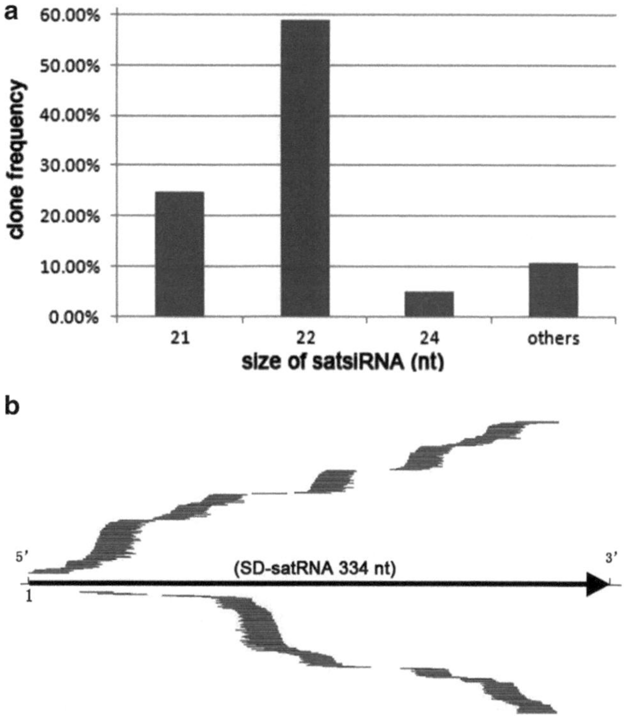

Fig. 1 Size distribution (**a**) and genome wide alignment (**b**) of SD-satsiRNAs obtained by small-scale cloning. *Short lines* above the *arrow headed line* (indicating the 334 nt satRNA genome) represent SD-satsiRNAs derived from the plus strand, and those below represent SD-satsiRNAs derived from the minus strand

excessive amount of plus-strand genomic Y-satRNA during small RNA purification steps, and can be minimized by adding in vitro synthesized minus-strand Y-satRNA transcript to the RNA samples [17]. We further investigated this using deep sequencing by spiking the RNA samples with in vitro minus-strand Y-satRNA transcript. The result showed that strand bias was prevented in the presence of the in vitro transcript, resulting in a relatively equal amount of plus and minus strand Y-satsiRNA reads in the sequencing data (Fig. 2).

Here we describe a step-by-step protocol of a laboratory small scale cloning of SD-satsiRNAs, and the procedure of bioinformatic analysis of sRNA sequencing data. We also describe the method to minimize strand bias of Y-satsiRNAs during deep sequencing.

2 Materials

2.1 Extraction of Small RNAs

1. Extraction Buffer: 0.1 M LiCl, 100 mM Tris–HCl (pH 8.0), 10 mM EDTA, 1 % SDS.
2. Phenol.

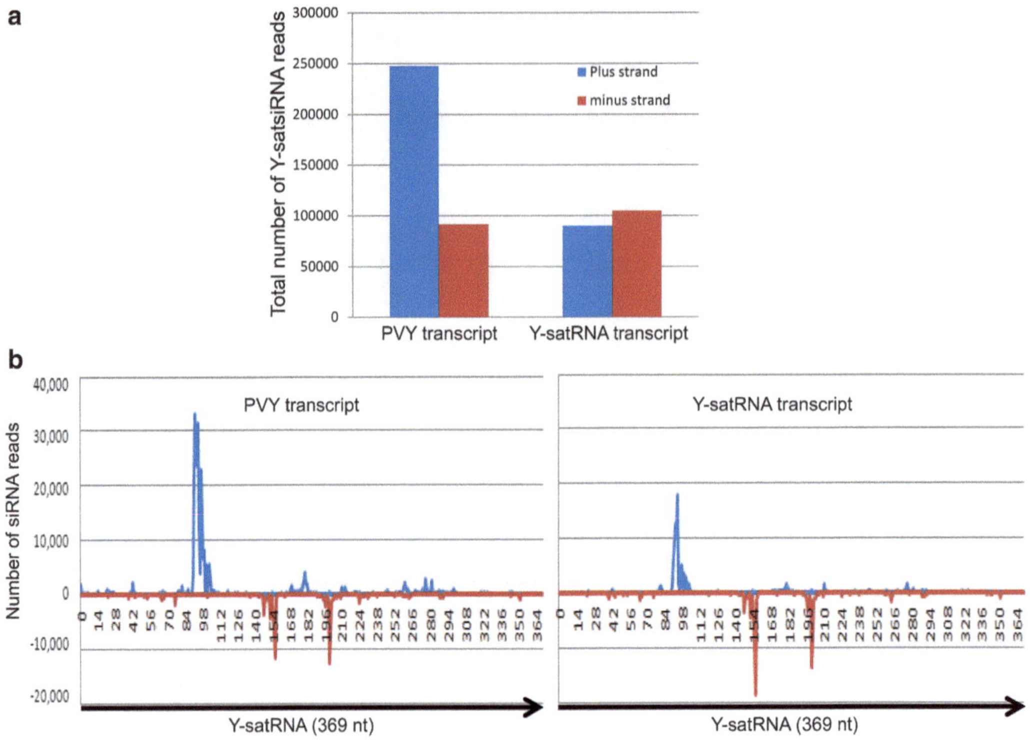

Fig. 2 Spiking RNA samples with minus-strand in vitro Y-satRNA transcript reverses the strand bias of Y-satsiRNAs. (**a**) Total number of plus and minus-strand Y-satsiRNAs in the presence of a control in vitro transcript of *Potato virus Y* (PVY) (*left*) or minus-strand in vitro Y-satRNA transcript (*right*). Note that a strong bias towards plus strand Y-satsiRNA reads occurs in the sample containing the control PVY transcript, but this bias is reversed in the presence of Y-satRNA transcript. (**b**) Distribution of Y-satsiRNA reads along the Y-satRNA genome. This graph was drawn based on the number of Y-satsiRNAs starting from each of the 369 nucleotides along the Y-satRNA genome. The *blue lines* represent the number of plus-strand siRNAs, while the red line minus-strand siRNAs

3. Chloroform.
4. 4 M LiCl.
5. Ethanol.
6. RNase-free water.

2.2 Gel Purification of Small RNAs

1. 15 % polyacrylamide gel (30 ml): Urea 12.6 g, H_2O 3.25 ml, 10× TBE 1.5 ml, 30 % polyacrylamide (acrylamide–bis = 29:1) 15 ml.
2. 10 % APS (0.1 g ammonium persulfate (APS, MW 228.2) in 1 ml of H_2O, stable at 4 °C for up to 1 month).
3. TEMED.
4. Electrophoresis Cell (DYCZ-24A).
5. 0.5× TBE.

6. Deionized formamide.
7. RNA loading buffer.
8. EB (ethidium bromide).
9. 14–30 nt ssRNA Ladder Marker.
10. 0.3 M NaCl.
11. Ethanol.
12. RNase-free water.

2.3 3′ Adapter Ligation and Purification

1. 3′ Adapter: pCTGTAGGCACCATCAAx (x: DMT-O-C3–CPG) (with 5′ phosphorylation and 3′ modification).
2. T4 RNA ligase.

2.4 5′ Adapter Ligation and Purification

1. 5′ Adapter: ATCGTaggcacctgaaa (RNA/DNA mixed oligonucleotide; lowercase letters represent ribonucleotides).
2. T4 RNA ligase.

2.5 RT-PCR of Small RNAs with Adaptors

1. SuperScript™ II Reverse Transcriptase.
2. RT Primer: ATTGATGGTGCCTAC.
3. 5′ DNA Oligo: ATCGTAGGCACCTGAAA.
4. Taq DNA polymerase.

2.6 Cloning RT-PCR Product into TOPO Vector

1. TOPO TA cloning vector.
2. LB medium.
3. X-gal (40 mg/ml).

2.7 Screening and Sequencing

1. M13 Forward primer: GTAAAACGACGGCCAG.
2. M13 Reverse primer: CAGGAAACAGCTATGAC.
3. Taq DNA polymerase.

2.8 Spiking RNA Samples with In Vitro Y-satRNA Transcript

1. DNA of pGEM-based plasmid with a full-length Y-satRNA sequence.
2. *Xho*I restriction enzyme.
3. T7 RNA polymerase.
4. Mixture of ATP, CTP, GTP, and UTP (2.5 mM).
5. 0.1 M DTT.
6. RNase inhibitor.
7. RQ1 RNase-free DNase.
8. Phenol–chloroform–isoamyl alcohol (25:24:1).
9. Chloroform.
10. Ethanol.
11. 3 M Sodium acetate (pH 5.2).

3 Methods

Wild-type Arabidopsis Col-0 were grown in solid MS medium for 2 weeks, transferred to soil, and then grown for 3 weeks under short-day conditions before virus inoculation. Plants were inoculated with fresh sap prepared from SD-CMV-infected tobacco leaves (1 g of ground leaf tissue diluted into 2 ml of phosphate buffer). Plant materials ware collected at 21 days post-inoculation for RNA extraction.

3.1 Extraction of Small RNAs (*see* Note 1)

1. Mix 1 volume of extraction buffer with phenol, and preheat at 80 °C.
2. Grind Arabidopsis leaves in liquid nitrogen to a fine powder.
3. Add 0.2 g of powder per 1 ml extraction buffer mixture, mix well by brief vortexing, then incubate the mixture at 80 °C for 5 min.
4. Add 1 volume of chloroform, vortex briefly.
5. Centrifuge the sample at 15,000×*g* for 5 min at 4 °C.
6. Transfer the supernatant to a new tube, add 1 volume of 4 M LiCl, mix and keep the mixture on ice for at least 2 h or at 4 °C overnight.
7. Centrifuge at 15,000×*g* for 15 min at 4 °C.
8. Transfer the supernatant to a fresh tube, add 3 volumes of ethanol and mix, and precipitate the RNA for 1–2 h at −20 °C.
9. Centrifuge at 1,700×*g* for 30 min at 4 °C.
10. Discard the supernatant and wash the pellet with 1 volume of 80 % ethanol twice.
11. Air-dry the pellet, resuspend in RNase-free water, and store at −80 °C until use.

3.2 Gel Purification of Small RNAs

3.2.1 Preparation of 15 % Polyacrylamide Gel

1. Add urea to the acrylamide–TBE–water mix in a 50 ml bottle and dissolve urea by shaking the bottle in a 50 °C water bath.
2. Add 240 μl of 10 % APS (drop by drop using a pipette) and mix gently.
3. Add 10 μl of TEMED, mix immediately but gently, pour the mix into a gel rig, and leave the gel mix to solidify for 1 h.
4. Assemble gel apparatus and add the running buffer (0.5× TBE), making sure that there are no leaks.

3.2.2 Preparation of Small RNAs and Running Gel

1. Add 1 volume of deionized formamide to small RNA samples.
2. Denature the RNA by heating at 100 °C for 5–10 min, then chill the sample on ice for 5–10 min.
3. Add RNA loading buffer to small RNA samples and mix quickly.
4. Load small RNA samples to the polyacrylamide gel prepared above.
5. Run the gel in 0.5× TBE, at 300 V until the bromophenol blue reaches the bottom of the gel (~3 h).

3.2.3 Purification of Small RNAs from Gel

1. Stain the gel with EB, visualize the RNA band under UV light, and cut out gel slice containing 14–30 nt small RNAs.
2. Put the gel slice in 0.3 M NaCl, and incubate at 4 °C overnight.
3. Transfer the supernatant to a fresh tube, add 3 volumes of ethanol, and precipitate the small RNA at −20 °C for 2 h.
4. Centrifuge at 15,000 × *g* for 15 min at 4 °C.
5. Discard the supernatant from the tube, leaving only the small RNA pellet.
6. Air-dry the pellet, resuspend it in RNase-free water; store at −80 °C.

3.3 3′ Adapter Ligation and Purification

1. Set up 20 μl of ligation reaction containing 6 μl of small RNA, 1 μl of 3′ adapter, 1 μl of 10× ligation buffer, and 2 μl of T4 RNA ligase.
2. Incubate the reaction for 1 h at 37 °C.
3. Gel-purify ligated product (Refer to Subheading 3.2) by cutting gel slice between 30 and 50 nt using siRNA Ladder Marker as a reference (the 3′adapter was 17 nt long, so the ligated products should be about 31–47 nt long).

3.4 5′ Adapter Ligation and Purification

1. Set up 20 μl ligation reaction containing 6 μl of gel-purified RNA-3′adapter ligation product, 1 μl of 5′adapter, 1 μl of 10× ligation buffer, and 2 μl of T4 RNA ligase.
2. Incubate the reaction for 1 h at 37 °C.
3. Gel purify ligated products (Refer to Subheading 3.2) by cutting gel slice between 40 and 70 nt using siRNA Ladder Marker as a reference (the 5′adapter was also 17 nt long, so the ligated products should be about 48–64 nt long).
4. To maximize the yield of RNA, add 1 μl of RT primer to the gel elute before adding the 3 volumes of ethanol for precipitation.

3.5 RT-PCR of Adaptor-Ligated Small RNAs

3.5.1 Reverse Transcription of Adaptor-Ligated Small RNAs

1. Mix 11 μl of gel-purified adaptor-ligation product with 1 μl of dNTPs (10 mM).
2. Incubate the mixture at 65° for 5 min, then chill it immediately on ice. Collect the components of the tube by brief centrifugation and add 4 μl of 5× first-strand buffer; 2 μl of 0.1 M DTT and 1 μl of RNaseOUT (40 units/μl).
3. Mix gently. Incubate at 42 °C for 2 min.
4. Add 1 μl (200 units) of SuperScript™ II reverse transcriptase and mix by pipetting gently up and down.
5. Incubate at 42 °C for 50 min.
6. Inactivate the reaction by heating at 70 °C for 15 min.

3.5.2 PCR

1. Set up 50 μl PCR reaction containing 2 μl of RT product, 5 μl of 10× PCR buffer, 1 μl of dNTPs (10 mM), 1 μl of RT primer, 1 μl of 5′ DNA oligo (forward primer), 1 μl of Taq DNA polymerase, and 39 μl of deionized H_2O.
2. Start the PCR cycling with the following parameters: 94 °C, 2 min; 15 cycles of 94 °C, 45 s; 50 °C, 45 s; and 72 °C, 45 s; 72 °C, 5 min.

3.6 Cloning of PCR Product into TOPO Vector

3.6.1 TOPO Cloning Reaction

1. Mix gently 4 μl of PCR product with 1 μl of Salt solution and 1 μl of TOPO vector.
2. Incubate for 5 min at room temperature (22–23 °C).
3. Place the reaction on ice and proceed to Subheading 3.6.2.

3.6.2 Transforming One Shot TOP10 Competent Cells

1. Thaw One Shot TOP10 competent cells on ice.
2. Add 3 μl of ligation product to competent cell gently.
3. Incubate on ice for 20 min.
4. Heat-shock the cells for 30 s at 42 °C.
5. Immediately transfer the tubes to ice.
6. Add 250 μl of room temperature S.O.C. medium.
7. Shake the tube horizontally (11 × *g*) at 37° for 1 h.
8. In the meantime, pre-warm ampicillin-containing plates at 37° for 30 min; spread 40 μl of X-gal on each plate and incubate at 37° until use.
9. Spread the bacterial culture onto the X-gal plates above. Three different volumes (50, 100 and 150 μl) were plated to ensure that well-spaced colonies are obtained.

3.7 Colony Screening and Sequencing (*see* Note 2)

3.7.1 PCR Screening of TOPO Transformants

1. Set up 10 μl of PCR reactions containing 1 μl of 10× PCR buffer, 0.1 μl of dNTPs (10 mM), 0.1 μl of M13 forward and reverse primers (100 ng/μl), 0.1 μl of Taq DNA polymerase and 8.6 μl of deionized H_2O.
2. Run the PCR reaction with the following cycling: 94 °C, 2 min; 25 cycles of 94 °C, 45 s; 45 °C, 45 s; 72 °C, 2 min; 72 °C, 10 min.
3. Visualize the PCR product using gel electrophoresis in 2 % agarose gel.

3.7.2 Sequencing

1. Submit the DNA from positive TOPO clones to commercial sequencing facilities for direct sequencing using M13F or M13R as primers.
2. Identify satRNA-derived sequences by Blasting against the SD-satRNA sequence.

3.8 Spiking RNA Samples with In Vitro Y-satRNA Transcript (see Note 3)

1. Digest 2 μg of Y-satRNA plasmid DNA with 20 units of *Xho*I in 50 μl of reaction for 2 h or longer at 37 °C to linearize the plasmid.
2. Purify the linearized plasmid with phenol–chloroform and chloroform extractions and ethanol precipitation.
3. Set up 50 μl in vitro transcription reaction containing 1 μg of linearized Y-satRNA plasmid, 10 μl of 5× transcription buffer (Promega), 5 μl of 0.1 M DTT, 2.5 μl of RNase inhibitor, 10 μl of 2.5 mM ATP/CTP/GTP/UTP mix, and 2.5 μl of T7 RNA polymerase.
4. Incubate the transcription reaction at 37 °C for 1 h.
5. Add 5 units of RQ1 RNase-free DNase and incubate the reaction at 37 °C for 10 min.
6. Purify the in vitro transcript with phenol–chloroform and chloroform extractions and ethanol precipitation. Dissolve the pellet in DEPC-treated water.
7. Measure the concentration of in vitro transcript using a NanoDrop spectrophotometer.
8. Add 2 μg of in vitro Y-satRNA transcript (corresponding to the minus strand of Y-satRNA) to 20 μg of total RNA isolated from CMV Y-satRNA-infected *Nicotiana tabacum* leaves.
9. Boil the spiked RNA sample for 5 min, and then chill it on ice for 5 min. The samples are now ready to be sent for deep sequencing.

3.9 Procedure of Bioinformatic Analysis of sRNA Sequencing Data

Software developed by BGI (http://www.genomics.cn/en/index) was used to process the raw data from HiSeq sequencing.

3.9.1 Data Cleaning

1. Remove low quality reads. The criteria for this are described in the Illumina protocol (http://www.illumina.com/).
2. Discard reads with 5′ primer contaminants.
3. Remove reads without 3′primer.
4. Discard reads without the insert tag.
5. Remove reads with poly(A) sequences.
6. Discard reads shorter than 18 nt sequences.

3.9.2 Map sRNA Sequences to SD-satRNA Genome Using BLAST+ (see Note 4)

1. BLAST+ was downloaded from NCBI (http://blast.ncbi.nlm.nih.gov/). Filter sRNA using the parameter *E*-value.
2. Obtain sRNAs of 18–30 nt in size with perfect match to the SD-satRNA genome using Perl script (http://www.perl.org/).

4 Notes

1. Many types of commercial sRNA isolation kits are available now. One can also use these kits for extraction of small RNAs.
2. Today is an era of data explosion. High-throughput sequencing has become more convenient and less costly, and techniques for constructing small RNA cDNA libraries for high-throughput sequencing have been well established [18–22].
3. Strand bias could also occur to endogenous small RNAs including miRNAs if abundant target RNA is present in the RNA samples. Spiking RNA samples with in vitro transcript complementary to the target RNA could also be used to minimize such strand bias in deep sequencing experiments.
4. Since satRNA sequence variants can occur during replication in infected plants, some satsiRNA clones with a few nucleotide mismatch to published satRNA sequences can also be counted as satRNA-derived siRNAs with some caution.

Acknowledgments

This work was supported by an Australian Research Council Future Fellowship (FT0991956) and grants from the Natural Science Foundation of China (31101423).

References

1. Lakatos L, Csorba T, Pantaleo V, Chapman EJ, Carrington JC et al (2006) Small RNA binding is a common strategy to suppress RNA silencing by several viral suppressors. EMBO J 25:2768–2780
2. Molnar A, Csorba T, Lakatos L, Varallyay E, Lacomme C et al (2005) Plant virus-derived small interfering RNAs originate predominantly from highly structured single-stranded viral RNAs. J Virol 79:7812–7818
3. Moissiard G, Voinnet O (2006) RNA silencing of host transcripts by cauliflower mosaic virus requires coordinated action of the four Arabidopsis Dicer-like proteins. Proc Natl Acad Sci U S A 103:19593–19598
4. Szittya G, Molnar A, Silhavy D, Hornyik C, Burgyan J (2002) Short defective interfering RNAs of tombusviruses are not targeted but trigger post-transcriptional gene silencing against their helper virus. Plant Cell 14:359–372
5. Blevins T, Rajeswaran R, Shivaprasad PV, Beknazariants D, Si-Ammour A et al (2006) Four plant Dicers mediate viral small RNA biogenesis and DNA virus induced silencing. Nucleic Acids Res 34:6233–6246
6. Simon AE, Roossinck MJ, Havelda Z (2004) Plant virus satellite and defective interfering RNAs: new paradigms for a new century. Annu Rev Phytopathol 42:415–437
7. Collmer CW, Howell SH (1992) Role of satellite RNA in the expression of symptoms caused by plant viruses. Annu Rev Phytopathol 30:419–442
8. Masuta C, Takanami Y (1989) Determination of sequence and structural requirements for pathogenicity of a cucumber mosaic virus satellite RNA (Y-satRNA). Plant Cell 1:1165–1173
9. Hou WN, Duan CG, Fang RX, Zhou XY, Guo HS (2011) Satellite RNA reduces expression of the 2b suppressor protein resulting in the attenuation of symptoms caused by Cucumber mosaic virus infection. Mol Plant Pathol 12:595–605
10. Sleat DE, Palukaitis P (1992) A single nucleotide change within a plant virus satellite RNA alters the host specificity of disease induction. Plant J 2:43–49

11. Palukaitis P, Roossinck MJ (1996) Spontaneous change of a benign satellite RNA of cucumber mosaic virus to a pathogenic variant. Nat Biotechnol 14:1264–1268
12. Du QS, Duan CG, Zhang ZH, Fang YY, Fang RX et al (2007) DCL4 targets Cucumber mosaic virus satellite RNA at novel secondary structures. J Virol 81:9142–9151
13. Zhu H, Duan CG, Hou WN, Du QS, Lv DQ et al (2011) Satellite RNA-derived small interfering RNA satsiR-12 targeting the 3′ untranslated region of Cucumber mosaic virus triggers viral RNAs for degradation. J Virol 85:13384–13397
14. Shimura H, Pantaleo V, Ishihara T, Myojo N, Inaba J et al (2011) A viral satellite RNA induces yellow symptoms on tobacco by targeting a gene involved in chlorophyll biosynthesis using the RNA silencing machinery. PLoS Pathog 7:e1002021
15. Smith NA, Eamens AL, Wang MB (2011) Viral small interfering RNAs target host genes to mediate disease symptoms in plants. PLoS Pathog 7:e1002022
16. Rodriguez-Alvarado G, Roossinck MJ (1997) Structural analysis of a necrogenic strain of cucumber mosaic cucumovirus satellite RNA in planta. Virology 236:155–166
17. Smith NA, Eamens AL, Wang MB (2010) The presence of high-molecular-weight viral RNAs interferes with the detection of viral small RNAs. RNA 16:1062–1067
18. Lau NC, Lim LP, Weinstein EG, Bartel DP (2001) An abundant class of tiny RNAs with probable regulatory roles in Caenorhabditis elegans. Science 294:858–862
19. Lu C, Meyers BC, Green PJ (2007) Construction of small RNA cDNA libraries for deep sequencing. Methods 43:110–117
20. Lu C, Shedge V (2011) Construction of small RNA cDNA libraries for high-throughput sequencing. Methods Mol Biol 729: 141–152
21. Soares AR, Pereira PM, Santos MA (2012) Next-generation sequencing of miRNAs with Roche 454 GS-FLX technology: steps for a successful application. Methods Mol Biol 822:189–204
22. Thomas MF, Ansel KM (2010) Construction of small RNA cDNA libraries for deep sequencing. Methods Mol Biol 667:93–111

Chapter 10

Drawing siRNAs of Viral Origin Out from Plant siRNAs Libraries

Laura Miozzi and Vitantonio Pantaleo

Abstract

Viruses are obligate intracellular entities that infect all forms of life. In plants, invading viral nucleic acids trigger RNA silencing machinery and it results in the accumulation of viral short interfering RNAs (v-siRNAs). The study of v-siRNAs population in biological samples has become a major part of many research projects aiming to identify viruses infecting them, including unknown viruses, even at extremely low titer. Currently, siRNA populations are investigated by high-throughput sequencing approaches, which generate very large data sets. The major difficulty in these studies is to properly analyze such huge amount of data. In this regard, easy-to-use bioinformatics tools to groom and decipher siRNA libraries and to draw out v-siRNAs are needed. Here we describe a workflow, which permit users with little experience in bioinformatics to draw out v-siRNAs from raw data sequences obtained by Illumina technology. Such pipeline has been released in the context of Galaxy, an open source Web-based platform for bioinformatics analyses.

Key words siRNAs, Plant viruses, v-siRNA population, Next generation sequencing

1 Introduction

RNA silencing was discovered in plants as a mechanism whereby invading nucleic acids, such as transgenes and viruses, are silenced [1]. Indeed, besides being involved in controlling transcriptionally and post-transcriptionally expression of genes (rev. in [2]), nowadays RNA silencing is considered one of the most important natural antiviral strategy in plants, invertebrates (rev. in [2, 3]) and even in specific mammal systems [4].

The key elements of RNA silencing are short interfering RNA molecules (siRNAs) produced by RNase III-like enzymes such as DICERS (DCLs). siRNAs guide the RNA induced silencing complex (RISC) in a sequence-specific manner to DNA or RNA targets [5]. At cellular level, RNA molecules that contain less than 30 nucleotides are usually named small RNAs (sRNAs). Besides several classes of endogenous siRNAs with regulatory functions, in plants other small noncoding RNAs that are not involved in RNA

Ichiro Uyeda and Chikara Masuta (eds.), *Plant Virology Protocols*, Methods in Molecular Biology, vol. 1236,
DOI 10.1007/978-1-4939-1743-3_10,

silencing are also present (e.g., transfer (t)RNAs, small nuclear (sn) RNAs, and small nucleolar (sno)RNAs). As a consequence, the natural composition of cellular sRNA network appears quite complex and even further complexity may be generated as a consequence of viral infections.

Upon infections of plant cells, viral nucleic acids trigger the RNA silencing based defense and massive amount of siRNAs of viral origin (v-siRNAs) are produced. v-siRNAs are processed from double-stranded RNAs or structured single-stranded RNAs DCLs. Genetic evidence indicated that DCL4, DCL2, and at least one additional DCL are involved in the biogenesis of v-siRNAs. Arabidopsis DCL4, DCL2, and DCL3 typically produce sRNAs of 21 nt, 22 nt, and 24 nt, respectively (rev. in [6]).

Over the past years Next Generation Sequencing (NGS) approaches have opened novel doors to study sRNA composition and function in a high-throughput and cost-effective manner. After the standardizations of protocols allowing the generation of sRNA libraries and the improvement of technologies of sequencing, the major limiting factor in these (as in all) NGS approaches continues to be the need for better bioinformatics tools to decipher the large data sets and to dominate the complexity of the cellular sRNA network.

Hence, the use of v-siRNAs for viral genome assembly is a powerful method for identification of novel viruses as well as commonly known viruses that may occur in low virus titers in symptomless plants and animals. A substantial number of studies have been performed which employ v-siRNAs techniques to either analyze (1) known viruses and viroids by means of a reference-guided approach or to (2) discover novel viruses using a de novo-based strategy on siRNA datasets [7–19].

Before starting any viral genome assembly using either a reference-guided or a de novo approach, it is recommended to perform extensive quality filtering of the NGS reads. By trimming low quality bases or even complete reads off, the amount of sequencing errors or artifacts in the data files is reduced, which of course has a direct positive effect on the quality of downstream applications. Raw reads are first preprocessed and then aligned along given reference sequences via mapping tools such as Bowtie [20] or BWA [21]. Output files in "SAM/BAM" format [22] can be managed by other bioinformatics tools in order to perform quantitative and qualitative analysis. To facilitate visualization of the selected reads, output files can be loaded to specific map viewer tools such as Integrative Genomic Viewer [23], Seqmonk (http://www.bioinformatics.babraham.ac.uk/projects/seqmonk/) or MISIS [24].

In this manuscript we show a possible approach for the management of raw sRNA datasets obtained by Illumina platform, from adapter removal, quality evaluation and filtering, until the

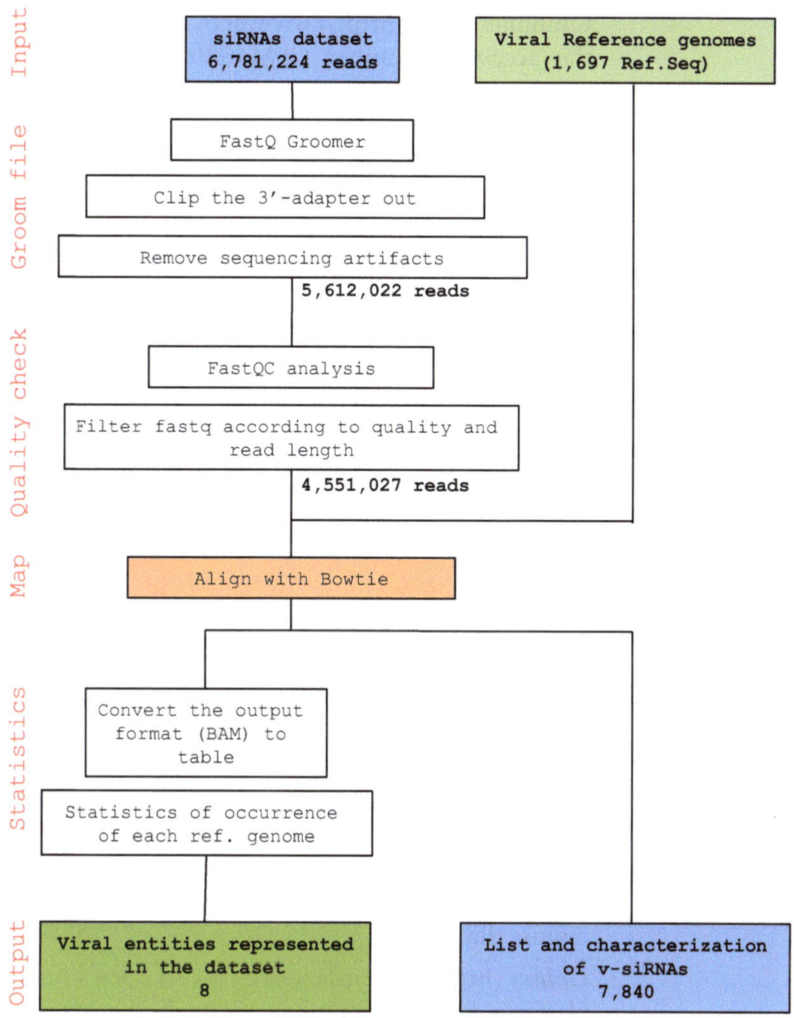

Fig. 1 Schematic representation of the workflow "v-siRNA extractor"

selection of putative v-siRNAs, which could encourage deeper studies on viral genome assembly. The final aim is the profiling of v-siRNAs using an easy-to-use workflow on the Galaxy, an open source Web-based platform for bioinformatics analyses [25–27]. Extensive practical details, description of main problems that can be encountered and how do they can be identified and overcame are here described. A pipeline of the workflow is shown in Fig. 1.

2 Materials

1. *sRNAs library*: short RNA library obtained from berries of *Vitis vinifera* cv. Pinot Noir, clone ENAV115 is used in this study. The library is freely available at Gene Expression

Omnibus site (http://www.ncbi.nlm.nih.gov/geo/) under the accession number GSM458930. The library has been obtained through Illumina technology and it has been previously described in term of both endogenous and exogenous siRNAs (i.e., viral or viroidal siRNAs) [11, 19, 28, 29]. The previous studies performed on this library led us to choose it for using in the Galaxy platform (see below) and for the purposes of the present manuscript. Despite the novel released sRNA Illumina protocols (http://www.illumina.com) the library was obtained with an old generation of adapters (version 2.1), which are detailed in the clipping procedures (*see* Subheading 3.1, **step 5**).

2. *Viral and viroid reference genomes*: here we describe the procedure to select siRNAs derived from viruses and viroids by means of a reference-guided approach, thus excluding any de novo-based strategy. In this context, the definition of a proper reference dataset of viral genomic sequences plays a significative role in the accurate selection of v-siRNAs within the sRNA library. The Viral Genomes section of Entrez Genome database (http://www.ncbi.nlm.nih.gov/genomes/GenomesHome.cgi) is a collection of viruses and viroids genomic sequences, therefore we downloaded all genome sequences of plant viruses (1,634 sequences) and viroids (63) available at the Entrez Genome database at November 2013, and used them as reference genomes. Importantly, the database lacks all tentative viral species and therefore the v-siRNAs originated from such viruses infecting our samples may be missed. In **Note 9** we discuss this case.
3. *Galaxy* (https://usegalaxy.org/) is an open source, Web-based platform for data intensive biomedical research [25–27]. It contains all basilar bioinformatics tools needed for the manipulation of sRNA libraries. We took advantage of such platform in order to summarize and state with care all steps needed to draw out v-siRNAs from raw Illumina sRNA libraries. Therefore, we have generated a workflow named "v-siRNA Extractor" freely available at https://usegalaxy.org/workflow/list_published; here below we follow every step of it providing extensive practical details that, in our knowledge, may encourage the approach of anyone to such analysis.

3 Methods

3.1 Get, Groom and Clip Datasets

1. Use the tool "*Upload the File from your computer*" located in the tool section "*Get Data*" to upload the *.fastq file with the Illumina sequencing raw data (*see* **Note 1**). The dataset we are using as demonstration contains 6,781,224 raw reads.

2. By using the same Galaxy upload tool (see above) get multi-fasta files of reference genomic sequences of plant viruses and viroids. See Subheading 2 for details on reference viral and viroid genomes used.
3. In order to obtain a single file with fasta sequences of plant viruses and viroids reference genomes, go to "*Text Manipulation*" section and use the tool "*Concatenate datasets tail-to-head*": it will merge the files containing reference genomes.
4. Go to "*NGS: QC and manipulation*" section and prepare the *.fastq illumina dataset by using the tool "*FASTQ Groomer*". This tool will convert the quality of the *.fastq file and make it readable for further analysis in Galaxy. For possible options *see* **Note 2**.
5. Go to "*NGS: QC and manipulation*" section and use the tool "*Clip adapter sequences*" to trim 3′-end adapter off. In the menu "*source*" it is possible to insert the sequence of the 3′-adapter to be clipped. In this case we use the 5′-TCGTATG CCGTCTTCTGCTTG-3′. It is recommended to discard sequences with unknown (*N*) bases by checking the appropriate option. In "*output options*", choose to retain only clipped sequences. For details on setting options *see* **Note 3**. A total of 5,612,094 reads were retained after adapter clipping.
6. Go to "*NGS: QC and manipulation*" section and use the tool "*Remove sequencing artifacts*" to filter sequences artifacts out from the dataset (i.e., poly(*N*)s). A total of 5,612,022 reads were retained after this step.

3.2 Dataset Quality Checking

Yet in the workflow described above, tools have been used to perform some quality conversion (i.e., "*FASTQ Groomer*", Subheading 3.1, **step 3**), excluding sRNAs sequences not containing the adapter (i.e., option in "*Clip sequences*", Subheading 3.1, **step 4**) and eliminating artifacts (i.e., "*Remove artifacts*", Subheading 3.1, **step 5**). In plant cells and tissues, sRNAs are generated through enzymatic processes such as dicing thus resulting in specific patterns of siRNAs at least in term of size classes, redundancy and diversity, CG content [30]. Several studies have highlighted biases happening during the creation of sRNA libraries or during sequencing them [31, 32]. The knowledge acquired up-to-date allows us to adopt parameters and criteria to evaluate the dataset with the help of specific bioinformatics tools. Therefore, before proceeding with further extensive studies on endogenous or viral sRNAs, a quality checking is needed.

1. Go to "*NGS: QC and manipulation*" section and run the tool "*FastQC:Read QC*" on your data in order to investigate the quality of your dataset.

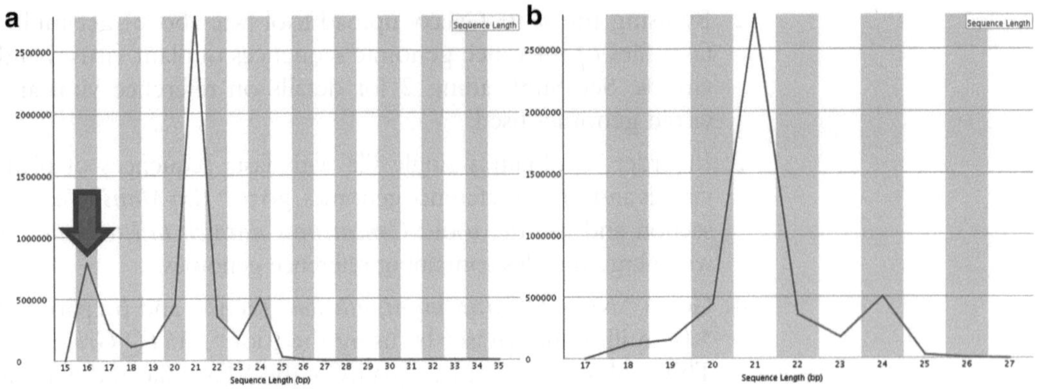

Fig. 2 Sequence length distribution obtained using FastQC before (**a**) and after quality filtering (**b**). *Arrow* indicates the unexpected peak of reads of 16 nt length

2. Download and inspect the "*FastQC:Read QC*" output in order to have a view of your dataset in term of basic statistics, quality score distribution, per base and GC content distribution, reads length distribution, duplication level and overrepresentation of reads (*see* **Note 4**).
3. Go to "*NGS: QC and manipulation*" section and run the tool "*Filter FASTQ reads by quality score and length*" in order to filter reads out according to length and quality parameters. In this example we filtered out reads with a quality score lower than 20 and set the minimum and maximum size equal to 18 and 26, respectively (*see* **Note 5**), in order to remove an unexpected peak at 16 nt length (Fig. 2, panel A). A total of 4,551,027 reads were retained after this step.
4. (optional) Go to "*NGS: QC and manipulation*" section and run again the tool "*FastQC:Read QC*" on your filtered data in order to visualize how your data are changed alter filtering. In Fig. 2 is evident the elimination of the peak at 16 nt length (compare panel A versus panel B). Figure 3 shows the improvement of the dataset quality before (panel A) and after (panel B) filtering. *Y* axis indicates the Illumina quality score: values between 44 and 28 (green area) are good quality score, values between 28 and 20 (orange area) are usually acceptable and values inferior to 20 (pink area) are discarded. As you can see, reads with low quality score (*Y* axis inferior to 20) disappeared after filtering (*see* **Note 6**).

3.3 Align siRNA

1. In order to map short reads onto single viral sequences, go to the "*NGS: Mapping*" section and run the tool "*Map with Bowtie for Illumina*". Select the multi-fasta file with plant viral and viroids genomes available in your history (*see* Subheading 3.1, **step 3**) but leaving the default options for

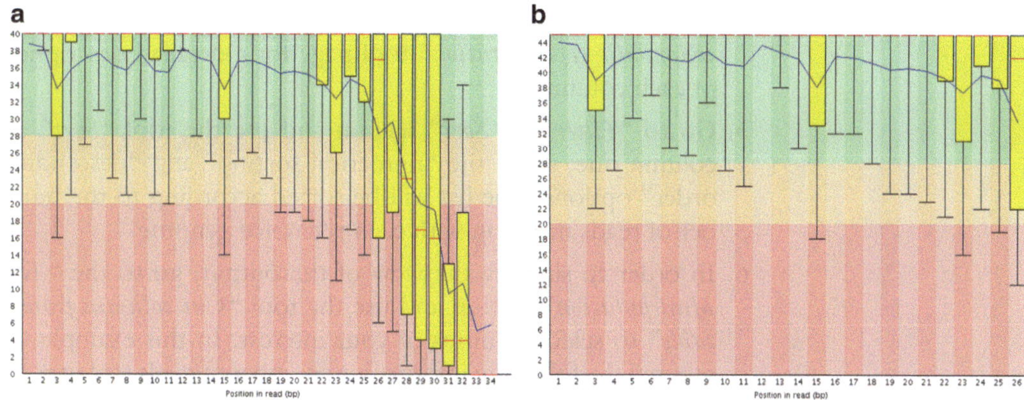

Fig. 3 Sequence quality score histogram obtained using FastQC before (**a**) and after quality filtering (**b**). *Y* axis indicates the Illumina quality score: values between 44 and 28 (*green area*) are good quality score, values between 28 and 20 (*orange area*) are intermediate quality score and values inferior to 20 (*red area*) are low quality score

building indices. Select the option regarding your dataset as single-end or pair-end (single-end in this example). Select the *.fastq file with your reads. Leave all the Bowtie settings as default except for the "*Number of mismatches for SOAP-like alignment policy* (*-v*)" equal to 0 and for the "*Maximum number of mismatches permitted in the seed* (*-n*)" equal to 0 (here we search for siRNAs totally matching to viral genomes, *see* **Note 7**).

3.4 Get the List of Reference Genomes Represented in Your Dataset

1. Go to the "*NGS: SAM Tools*" section and use the tool "*BAM-to-SAM*" to convert the output file of the alignment (Subheading 3.3, **step 1**) into a SAM file. Do not include the header in the output.
2. Go to the "*Text Manipulation*" section and use the tool "*Convert delimiters to TAB*" to substitute pipes with tab-delimiters.
3. Go to the "*Statistics*" section and use the tool "*Count occurrences of each record*" to count the occurrence of each value in column 4 (i.e., GI accession number). This step allows us to count the reads mapping on each viral or viroid reference genome. You will get a file with two columns: the first one contains the number of reads mapping to the reference genome reported in the second column with its GI number (*see* **Note 8**).
4. Go to the "*Metagenomic analyses*" section and use the tool "*Fetch taxonomic representation*" to retrieve taxonomic information of viral genomes on which map the reads. Select column "two" as the column with the GI number and column "one" (containing the number of reads mapping to each virus/viroid) as name column; this allows to retain in the output file the information on read count. Finally, you will get an output file where the first column reports the number of reads mapping to each viral/viroid genome and the other columns report all

retrieved taxonomic information (see the example in the "*Fetch taxonomic representation*" tool page for a detailed description of each column).

5. Go to "*Filter and Sort*" section and use the tool "*Sort*" on column one; select the "numerical sorting" and "descending order" options. The list will be sorted according to the number of reads mapping with each reference genome.
6. In order to simplify the view of the output, go to the "*Text Manipulation*" section and use the tool "*Cut columns from a table*" to select columns of your interest. In this example we selected the following information: number of reads (column 1), taxonomic id (column 2), superkingdom (column 4), family (column 17), genus (column 21), species (column 23), and GI number (column 25). At this point you will get a tab-delimited file reporting the list of viruses and viroids present in your original sample associated with the number of v-siRNAs mapping on each of them (Table 1) (*see* **Notes 9** and **10**).

3.5 Get the Pool of v-siRNAs Matching with a Reference Genome

Fundamental parameters describing v-siRNAs are the sequence, the reference genome to which it matches, the coordinates and the strand polarity. Once investigated the viral diversity in your dataset (Subheading 3.4), it is generally requested to have the list of v-siRNAs mapping on a specific virus/viroid of interest.

1. Go to the "*Filter and Sort*" section and use the tool "*Select lines that match an expression*". As input file use the SAM file obtained at Subheading 3.4, **step 1**. Select rows that match with GI number of interest (i.e., selected from the last column of Table 1). As example, here we selected the 575 redundant v-siRNAs mapping on *Grapevine rupestris stem pitting-associated virus* (GI number: 9630737, Table 1).
2. Download the SAM file you obtained at Subheading 3.5, **step 1**, containing the v-siRNAs mapping on your viral genome of interest. Use this file to visualize the mapping of v-siRNAs along the viral genome. For this purpose we suggest to use the Java standalone program MISIS, freely available at http://www.fasteris.com/apps/ [24] (*see* **Note 11**) (Fig. 4).
3. In order to obtain a tab-delimited file reporting information on coordinates of alignment in the reference genome, strand polarity and read sequence, go to section "*NGS SAM Tools*" and use the tool "*Convert SAM to interval*" to convert positional information from the SAM file with all alignments obtained at Subheading 3.4, **step 1** into interval format with 0-based start and 1-based end.
4. Go to the "*Filter and Sort*" section and use the tool "*Select lines that match an expression*" to select lines containing the GI number of interest (i.e., *Rupestris stem pitting virus*) As input file use the

Table 1
List of viruses and viroids reference genomes and the number of mapped v-siRNAs

No. of reads	Taxonomy ID	Superkingdom	Family	Genus	Species	GI number
3,419	12893	Viroids	Pospiviroidae	Hostuviroid	Hop stunt viroid	11497495
2,859	12904	Viroids	Pospiviroidae	Apscaviroid	Grapevine yellow speckle viroid 1	11496576
648	46342	Viroids	Pospiviroidae	Apscaviroid	Grapevine yellow speckle viroid 2	20153376
575	196400	Viruses	Betaflexiviridae	Foveavirus	Grapevine rupestris stem pitting-associated virus	9630737
245	103722	Viruses	Tymoviridae	Maculavirus	Grapevine fleck virus	18138525
34	47985	Viruses	Closteroviridae	Ampelovirus	Grapevine leafroll-associated virus 1	366898511
30	1051792	Viruses	Betaflexiviridae	Trichovirus	Grapevine Pinot gris virus	339906182
30	12154	Viruses	Tymoviridae	Tymovirus	Turnip yellow mosaic virus	21686950

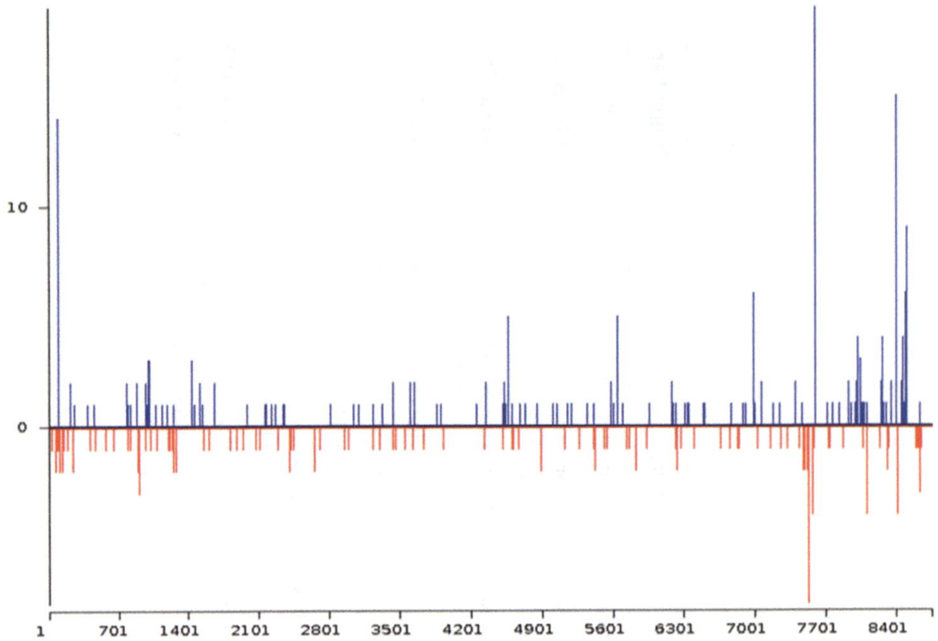

Fig. 4 Visualization of v-sRNAs mapping on *Grapevine rupestris stem pitting-associated virus* reference genome obtained by MISIS

converted file obtained at Subheading 3.5, **step 3**. Go to section "*Text Manipulation*" and use the tool "*Cut columns from a table*" to select columns with the following information: reference genome (column 1), start (column 2), end (column 3), strand (column 4), and read sequence (column 14).

4 Notes

1. The Galaxy tool used to get data is generally able to autodetect the file format for such datasets and also it accepts zipped files.
2. When using *Basic* options, the output will be *sanger* formatted, which is suggested for such kind of analysis.
3. Here we have set the minimum sequence length at 16 nt instead of the default of 15 nt. However, the minimal sequence length acceptable can depend on the dataset. Checking the read length distribution using the "*FastQC*" tool available in Galaxy at the "*NGS: QC and manipulation*" section and descriwbed in Subheading 3.2 can be useful to better set this parameter.

4. The inspection of the output file is mandatory. Here you can be sure whether your dataset contains sufficient information in term of quality and quantity allowing us to go ahead with the analysis. Therefore, consider performing in depth quality inspection before going on with further analyses.
5. Use the FastQC analysis to set the parameters for your dataset. In this example, we select reads with a length above 18 nt because the FastQC analysis highlighted an unexpected peak of reads of 16 nt-length (Fig. 1a). The quality cutoff was kept as default because this setting was in line with the quality value distribution among the reads.
6. Perform a new FastQC analysis after filtering is really useful to inspect how your dataset has changes.
7. The –v option is used to report alignments with a given number of mismatches. You can easily check how many v-sRNAs map onto your reference genome with one or more mismatches, simply changing the "*Number of mismatches for SOAP-like alignment policy* (*–v*)" value. Once "–v" option is defined, other alignment parameters (i.e., "–e" and "–l") will be ignored.
8. The GI number is the sequence identifier used by the National Center for Biotechnology Information.
9. The number of v-siRNAs mapping on each genome may widely vary depending on the biological case of study. Only one or just few v-siRNAs are generally not significative in term of the effective presence of this virus/viroid in the sample. This is particularly true if selected v-siRNAs do not overlap and, therefore, do not generate a contig enough representative of the reference genome. Indeed, Table 1 has been subjected to a cutoff for the number of reads below 30. In any case is strongly suggested a careful inspection of the outputs of the analyses accordingly to the biological information available.
10. During the inspection of the list of viruses/viroids present in the sample it should be considered that a reference-guided approach has been performed to select v-siRNAs. Therefore you may have missed v-siRNAs related to viruses/viroids not present in the list of reference genomes used. The example here reported is emblematic; biological assays carried out on grapevine samples used for sRNA libraries revealed the presence of *grapevine asteroid mosaic associated virus* (GAMaV) and *Grapevine red globe virus* (GRGV) [28] while bioinformatics analysis failed to identify v-siRNAs derived from these viruses. This is due to the fact that the two viruses were not present in the reference genome dataset. In similar cases, it is useful to integrate the reference genome dataset with missed specific genomic references [28]. Molecular approaches (i.e., PCR) must be applied to the sample to validate the presence of the viral entities not yet having a reference genome in public dataset.

11. We suggest to use MISIS to visualize v-siRNAs along the viral genome, because it is a freely available, easy-to-use tool requiring low or no experience in bioinformatics. However, other tools are available for the same purpose, such as Integrative Genomic Viewer [23] or Seqmonk (http://www.bioinformatics.babraham.ac.uk/projects/seqmonk/).

Acknowledgements

This work was supported by a grant from Italian Ministry of Economy and Finance dedicated to CNR (C.I.S.I.A. Legge 191/2009) and by a DAA-CNR grant (award prize 2010 for advanced research in "Animal and Plant Genomics" to VP). COST FA0806 supported L.M. for attending a T.S. in Applied Bioinformatics in Plant Science.

References

1. Fire A (1999) RNA-triggered gene silencing. Trends Genet 15:358–363
2. Vaucheret H, Beclin C, Fagard M (2001) Post-transcriptional gene silencing in plants. J Cell Sci 114:3083–3091
3. Ding SW, Voinnet O (2007) Antiviral immunity directed by small RNAs. Cell 130:413–426
4. Maillard PV, Ciaudo C, Marchais A, Li Y, Jay F et al (2013) Antiviral RNA interference in mammalian cells. Science 342:235–238
5. Baulcombe DC (2000) Molecular biology. Unwinding RNA silencing. Science 290: 1108–1109
6. Shimura H, Pantaleo V (2011) Viral induction and suppression of RNA silencing in plants. Biochim Biophys Acta 1809:601
7. Giampetruzzi A, Roumi V, Roberto R, Malossini U, Yoshikawa N et al (2012) A new grapevine virus discovered by deep sequencing of virus- and viroid-derived small RNAs in Cv Pinot gris. Virus Res 163:262–268
8. Kreuze JF, Perez A, Untiveros M, Quispe D, Fuentes S et al (2009) Complete viral genome sequence and discovery of novel viruses by deep sequencing of small RNAs: a generic method for diagnosis, discovery and sequencing of viruses. Virology 388:1–7
9. Wu Q, Luo Y, Lu R, Lau N, Lai EC et al (2010) Virus discovery by deep sequencing and assembly of virus-derived small silencing RNAs. Proc Natl Acad Sci U S A 107:1606–1611
10. Loconsole G, Saldarelli P, Doddapaneni H, Savino V, Martelli GP et al (2012) Identification of a single-stranded DNA virus associated with citrus chlorotic dwarf disease, a new member in the family Geminiviridae. Virology 432: 162–172
11. Navarro B, Pantaleo V, Gisel A, Moxon S, Dalmay T et al (2009) Deep sequencing of viroid-derived small RNAs from grapevine provides new insights on the role of RNA silencing in plant-viroid interaction. PLoS One 4:e7686
12. Szittya G, Moxon S, Pantaleo V, Toth G, Rusholme Pilcher RL et al (2010) Structural and functional analysis of viral siRNAs. PLoS Pathog 6:e1000838
13. Kreuze J, Koenig R, De Souza J, Vetten HJ, Muller G et al (2013) The complete genome sequences of a Peruvian and a Colombian isolate of Andean potato latent virus and partial sequences of further isolates suggest the existence of two distinct potato-infecting tymovirus species. Virus Res 173:431–435
14. De Souza J, Fuentes S, Savenkov EI, Cuellar W, Kreuze JF (2013) The complete nucleotide sequence of sweet potato C6 virus: a carlavirus lacking a cysteine-rich protein. Arch Virol 158:1393–1396
15. Fuentes S, Heider B, Tasso RC, Romero E, Zum Felde T et al (2012) Complete genome sequence of a potyvirus infecting yam beans (Pachyrhizus spp.) in Peru. Arch Virol 157:773–776
16. Cuellar WJ, Cruzado RK, Fuentes S, Untiveros M, Soto M et al (2011) Sequence characterization of a Peruvian isolate of Sweet potato chlorotic stunt virus: further variability and a model for p22 acquisition. Virus Res 157:111–115
17. Aregger M, Borah BK, Seguin J, Rajeswaran R, Gubaeva EG et al (2012) Primary and secondary

siRNAs in geminivirus-induced gene silencing. PLoS Pathog 8:e1002941
18. Blevins T, Rajeswaran R, Aregger M, Borah BK, Schepetilnikov M et al (2011) Massive production of small RNAs from a non-coding region of Cauliflower mosaic virus in plant defense and viral counter-defense. Nucleic Acids Res 39:5003
19. Wu Q, Wang Y, Cao M, Pantaleo V, Burgyan J et al (2012) Homology-independent discovery of replicating pathogenic circular RNAs by deep sequencing and a new computational algorithm. Proc Natl Acad Sci U S A 109:3938–3943
20. Langmead B, Trapnell C, Pop M, Salzberg SL (2009) Ultrafast and memory-efficient alignment of short DNA sequences to the human genome. Genome Biol 10:R25
21. Li H, Durbin R (2009) Fast and accurate short read alignment with Burrows-Wheeler transform. Bioinformatics 25:1754–1760
22. Li H, Handsaker B, Wysoker A, Fennell T, Ruan J et al (2009) The sequence alignment/map format and SAMtools. Bioinformatics 25:2078–2079
23. Robinson JT, Thorvaldsdottir H, Winckler W, Guttman M, Lander ES et al (2011) Integrative genomics viewer. Nat Biotechnol 29:24–26
24. Seguin J, Otten P, Baerlocher L, Farinelli L, Pooggin MM (2013) MISIS: a bioinformatics tool to view and analyze maps of small RNAs derived from viruses and genomic loci generating multiple small RNAs. J Virol Methods 195C:120–122
25. Giardine B, Riemer C, Hardison RC, Burhans R, Elnitski L et al (2005) Galaxy: a platform for interactive large-scale genome analysis. Genome Res 15:1451–1455
26. Blankenberg D, Von Kuster G, Coraor N, Ananda G, Lazarus R, et al. (2010) Galaxy: a web-based genome analysis tool for experimentalists. Curr Protoc Mol Biol Chapter 19: Unit 19 10 11–21
27. Goecks J, Nekrutenko A, Taylor J (2010) Galaxy: a comprehensive approach for supporting accessible, reproducible, and transparent computational research in the life sciences. Genome Biol 11:R86
28. Pantaleo V, Saldarelli P, Miozzi L, Giampetruzzi A, Gisel A et al (2010) Deep sequencing analysis of viral short RNAs from an infected Pinot Noir grapevine. Virology 408:49–56
29. Pantaleo V, Szittya G, Moxon S, Miozzi L, Moulton V et al (2010) Identification of grapevine microRNAs and their targets using high-throughput sequencing and degradome analysis. Plant J 62:960–976
30. Schwach F, Moxon S, Moulton V, Dalmay T (2009) Deciphering the diversity of small RNAs in plants: the long and short of it. Brief Funct Genomic Proteomic 8: 472–481
31. Linsen SE, de Wit E, Janssens G, Heater S, Chapman L et al (2009) Limitations and possibilities of small RNA digital gene expression profiling. Nat Methods 6:474–476
32. Sorefan K, Pais H, Hall AE, Kozomara A, Griffiths-Jones S et al (2012) Reducing ligation bias of small RNAs in libraries for next generation sequencing. Silence 3:4

Chapter 11

Viral Detection by High-Throughput Sequencing

Daisuke Motooka, Shota Nakamura, Katsuro Hagiwara, and Takaaki Nakaya

Abstract

We applied a high-throughput sequencing platform, Ion PGM, for viral detection in fecal samples from adult cows collected in Hokkaido, Japan. Random RT-PCR was performed to amplify RNA extracted from 0.25 ml of fecal specimens ($N=8$), and more than 5 μg of cDNA was synthesized. Unbiased high-throughput sequencing using the 318 v2 semiconductor chip of these eight samples yielded 57–580 K (average: 270 K, after data analysis) reads in a single run. As a result, viral genome sequences were detected in each specimen. In addition to bacteriophage, mammal- and insect-derived viruses, partial genome sequences of plant, algal, and protozoal viruses were detected. Thus, this metagenomic analysis of fecal specimens could be useful to comprehensively understand viral populations of the intestine and food sources in animals.

Key words Plant virus, Metagenome, High-throughput sequencing, Fecal specimen, Random RT-PCR, NGS

1 Introduction

Newly developed next-generation sequencing (NGS) technologies have facilitated the obtaining of large amounts of unbiased data (gigabases) in a single round of operation [1]. Among these sequencing technologies, benchtop sequencers, such as 454 GS-Junior from Roche; GA, MiSeq from Illumina, and Ion PGM from Life Technologies, are faster NGS platforms with lower costs and are useful for microbiological research, including amplicon (bacterial 16S rRNA gene) sequencing and metagenomics [1].

We previously conducted a metagenomic diagnosis of virus infections in human clinical specimens directly by NGS [2, 3]. Feces of norovirus-positive patients ($N=5$) were chosen as model samples. cDNA, as a template for the GS FLX platform, was prepared by random RT-PCR using the RNA extracted from the clinical samples. A high-throughput approach using the 454 GS FLX platform detected the targeted norovirus sequences in all five

Ichiro Uyeda and Chikara Masuta (eds.), *Plant Virology Protocols*, Methods in Molecular Biology, vol. 1236,
DOI 10.1007/978-1-4939-1743-3_11,

specimens [3]. In addition to the human virus, pepper mild mottle virus (PMMV) was found in two of the five specimens, and the total number of PMMV reads was more than that of norovirus reads in one specimen [3]. In addition, Kyuri green mottle mosaic virus (KGMMV) was detected in one specimen and was also detected from the precipitated fraction of ultracentrifugation, suggesting that viral particles of KGMMV are present in the human gut [3]. Another research group reported that PMMV was detected at 10^9 virions per gram dry weight of fecal matter and was also detected in 12 (66.7 %) of 18 fecal samples collected from healthy humans, and that fecal PMMV was infectious to host plants [4]. Therefore, these plant-derived viruses may show infectivity in feces and use mammals as a vector.

Here, we present our current metagenomic analysis of bovine fecal samples using the Ion PGM benchtop NGS platform with a larger scale and lower costs.

2 Materials

2.1 RNA Isolation from Fecal Samples

1. Fecal samples: Stool samples of adult Holstein cattle ($n=8$) were collected in Hokkaido, Japan. Each collected fecal sample was suspended with 9 volumes of PBS and was homogenized with zirconia beads by TissueLyser (Qiagen). The feces suspensions were centrifuged at $9{,}100\times g$ for 10 min. The supernatants (0.25 ml) were used for RNA isolation.
2. RNA isolation: TRizol-LS (Life Technologies) is a ready-to-use reagent for the isolation of total RNA from liquid samples, including fecal specimens.
3. RNA purification: The Purelink RNA Mini kit (Life Technologies) was used to purify total RNA from TRizol-treated solution.

2.2 Random RT-PCR (cDNA) Amplification

1. Whole transcriptome amplification: WTA-2 (Sigma).
2. cDNA purification: Ampure XP (Agencourt).
3. Sizing and quality control of cDNA: 2100 Bioanalyzer (Agillent).
4. Reagent kit of 2100 Bioanalyzer: High Sensitivity DNA kit (Agillent).
5. Vortex Mixer: IKA-Model MS3.
6. cDNA quantification: NanoDrop (ND-1000) (Thermo Scientific).

2.3 High-Throughput Sequencing by NGS

1. Library preparation for Ion PGM sequencing: Ion Plus Fragment Library kit, Ion Xpress Barcode Adapters 1-16 kit, and Ion Library Quantitation kit (Life Technologies).
2. Emulsion PCR: Ion OneTouch 2 and Ion OneTouch ES (Life Technologies).

3. NGS benchtop sequencer: Ion PGM™ (Life Technologies).
4. Sequencing: Ion PGM Sequencing OT2 400 kit and Ion 314 v2 or 318 v2 semiconductor chip.

3 Methods

3.1 Preparation of Total RNA from Fecal Samples

1. Suspend a collected stool (~1 g) with 9 volumes of PBS and homogenize with zirconia beads by TissueLyser (Qiagen). The feces suspensions were centrifuged at 9,100 × *g* for 10 min. Filtrate the supernatant (0.25 ml) with a 0.45-μm filter to remove cellular and bacterial components (*see* **Note 1**).
2. Mix the supernatant (0.25 ml) with TRizol LS reagent (0.75 ml) and incubate for 10 min at room temperature to allow the complete dissociation of nucleoprotein complexes (*see* **Note 2**).
3. Add 0.2 ml of chloroform to the solution and shake the tube vigorously by hand for 15 s.
4. Incubate at room temperature for 2–15 min, and centrifuge the sample at 12,000 × *g* for 15 min at 4 °C.
5. Transfer ~400 μl of the colorless upper phase containing the RNA to a fresh RNase-free tube.
6. Add an equal volume of 70 % ethanol to obtain a final ethanol concentration of 35 %. Vortex to mix well.
7. Transfer ≤700 μl of the solution to a Purelink spin cartridge (with a collection tube).
8. Further steps including binding, washing, and elution of RNA are performed according to the manufacturer's protocol (http://www.lifetechnologies.com). The volume of RNA elution is 30 μl.

3.2 Random Amplification (WTA 2)

Library synthesis and amplification reactions are basically performed according to the manufacturer's protocol (http://www.sigmaaldrich.com), with some modifications (*see* **Note 3**).

Library Synthesis Reaction

1. Add 5 μl of the RNA solution to 2.5 μl of Library Synthesis Solution and 9.1 μl of nuclease-free water, giving a total of 16.6 μl.
2. Mix and incubate the solution at 70 °C for 5 min, and then cool to 18 °C. To the cooled-primed RNA, immediately add the following (individually or premixed)
   (a) 2.5 μl of Library Synthesis Buffer
   (b) 3.9 μl of Water
   (c) 2 μl of Library Synthesis Enzyme

3. Incubate the mixed solution (a total of 25 μl) in a thermal cycler with the following schedule:

   18 °C for 10 min

   25 °C for 10 min

   37 °C for 30 min

   42 °C for 10 min

   70 °C for 20 min

   4 °C

Amplification Reaction

4. Prepare the following master mix
   (a) 60.25 μl of Nuclease-Free Water
   (b) 7.5 μl of Amplification Mix
   (c) 1.5 μl of WTA dNTP Mix
   (d) 0.75 μl of Amplification Enzyme
5. Add the 5 μl of the Library Synthesis Reaction from **step 3** to the above (**step 4**) master mix solution (70 μl) and mix.
6. Incubate the solution in a thermal cycler using the following schedule:

   94 °C for 2 min.

   17–29 cycles × (94 °C for 30 s, 70 °C for 5 min) (*see* **Note 4**)

   4 °C.
7. For the removal of residual primers and nucleotides, use the QIAquick PCR purification kit (Qiagen)

3.3 Purify WTA DNA Using Agencourt AMPure XP (Beckman Coulter)

1. Add 90 μl of Agencourt AMPure XP (×1.8) to the 50 μl of WTA DNA solution in a 1.5-ml tube (*see* **Note 5**).
2. Pipette-mix 10 times.
3. Incubate at room temperature for 5–10 min.
4. Place the reaction tube onto a magnet stand for 2 min to separate beads from solution.
5. Aspirate the supernatant from the reaction plate and discard.
6. Dispense 200 μl of 70 % ethanol and incubate at room temperature for at least 30 s. Aspirate off the ethanol and discard. Repeat for a total of two washes.
7. Add 40 μl of elution buffer and pipette-mix 10 times.
8. Incubate at room temperature for 2 min.
9. Place the reaction tube onto a magnet stand for 1 min to separate beads from solution.
10. Transfer the purified product to a new 1.5-ml tube.

3.4 Analysis of WTA DNA Using 2100 Bioanalyzer (Agilent Technologies)

1. The purified WTA DNA is analyzed regarding its sizing and quantify with the High-sensitivity DNA kit using the 2100 Bioanalyzer (or 2200 TapeStation) according to the manufacturer's protocol.
2. The DNA concentration can also be measured using NanoDrop such as ND-1000 or ND-2000 (Thermo Scientific).

3.5 Library Preparation for Ion PGM Sequencing (Life Technologies)

Because most of the WTA DNA is shorter than 400 bp (Fig. 1a), no further fragmentation is required.

Endrepair

1. A total of 1 μg of WTA DNA (*see* **Note 6**) is treated with end repair enzyme (Ion Plus Fragment Library kit) and incubated for 20 min at room temperature.
2. The DNA is purified with AMPure XP, as described previously.

Adapter Ligation

3. Ion P1 Adapter and Ion Xpress Barcode X are added to the DNA solution and incubated with DNA ligase according to the manufacturer's protocol (*see* **Note 7**).
4. The adapter-ligated DNA is purified with AMPure XP.

Amplify the Library

5. The adapter-ligated DNA is amplified with a thermal cycler using Library Amplification Primer Mix (Ion Plus Fragment Library kit) according to the manufacturer's protocol (*see* **Note 8**).
6. The amplified DNA is purified with AMPure XP (Fig. 1b).

Library Quantitation

7. In order to quantify the library DNA, quantitative real-time PCR is performed using the Ion Library Quantification kit according to the manufacturer's protocol. It yields an optimized dilution (concentration) of the library DNA for further template amplification (emulsion PCR) onto Ion Sphere Particles (ISP) (*see* **Note 9**).

3.6 Emulsion PCR by Ion OneTouch 2 and ES (Life Technologies)

1. The diluted library DNA (*see* **Note 10**) is mixed with ISP, and clonal amplification (emulsion PCR) by Ion OneTouch 2 is performed according to the manufacturer's protocol for the Ion PGM Template OT2 400 kit.
2. Template-positive ISP is recovered and enriched with Ion OneTouch ES.

3.7 High-Throughput Sequencing by Ion PGM (Life Technologies)

1. High-throughput sequencing is performed using the Ion PGM sequencer to detect hydrogen ions produced during DNA replication with semiconductor chips such as 314 v2 or 318 v2 with a range of 100 K to 5 M reads (clones).

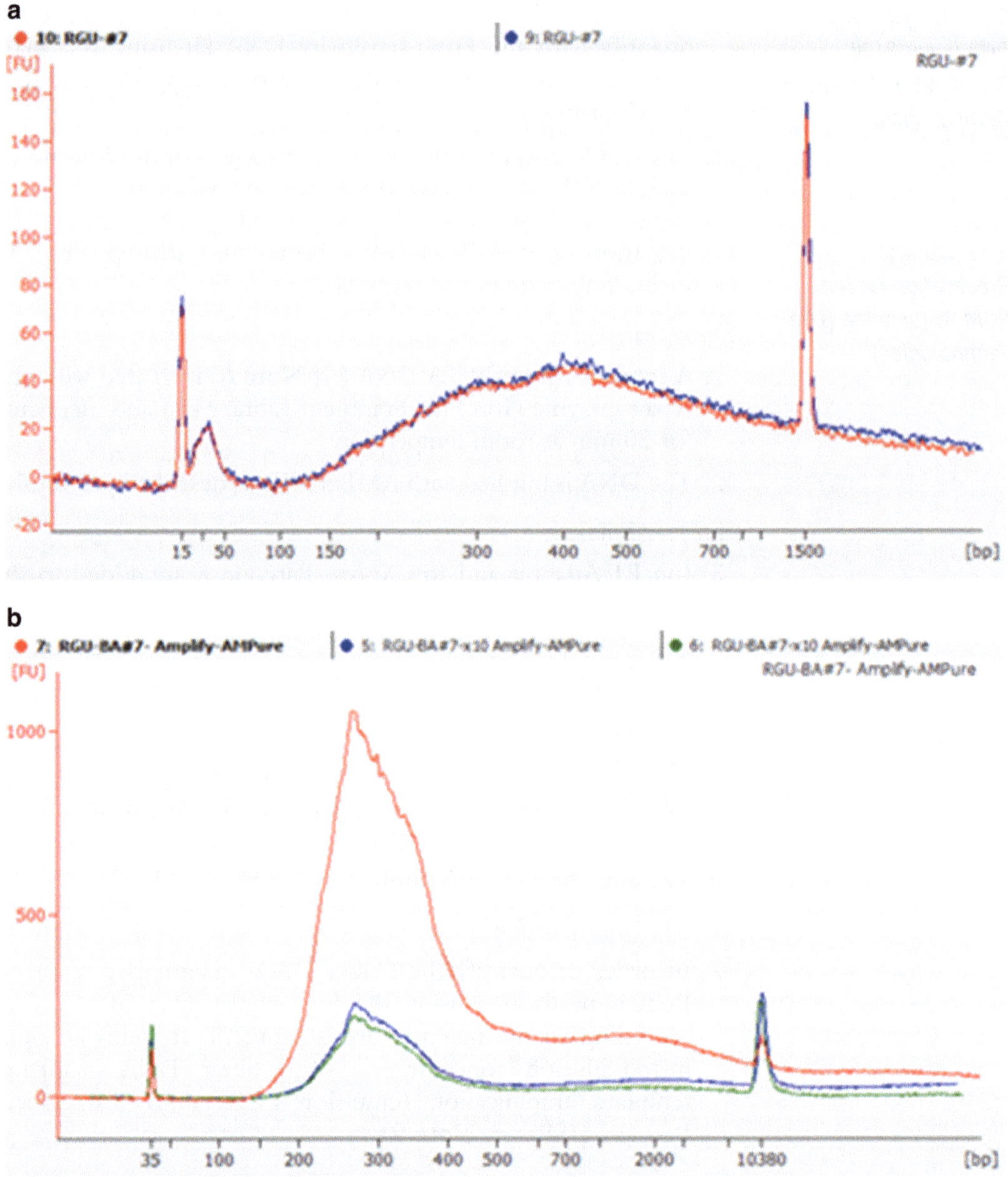

Fig. 1 Bioanalyzer instrument analysis of WTA DNA and amplified library. (**a**) Bioanalyzer instrument analysis of WTA DNA. Fragment profile of RGU-BA-4 WTA DNA (duplicated). (**b**) Bioanalyzer instrument analysis of the amplified library. Analysis was conducted with the Agilent High Sensitivity DNA Kit. Peaks at 35 and 10,380 bp represent low- and high-molecular-weight markers, respectively. Sizing and semi-quantification of undiluted (*red*) and tenfold diluted (*blue* and *green*, duplicated) library DNA

2. We usually mix 3–5 samples (with different barcode sequences) per run using 314 chip v2 (*see* **Note 11**) and the Ion PGM Sequencing OT2 400 kit at first according to the manufacturer's protocol.

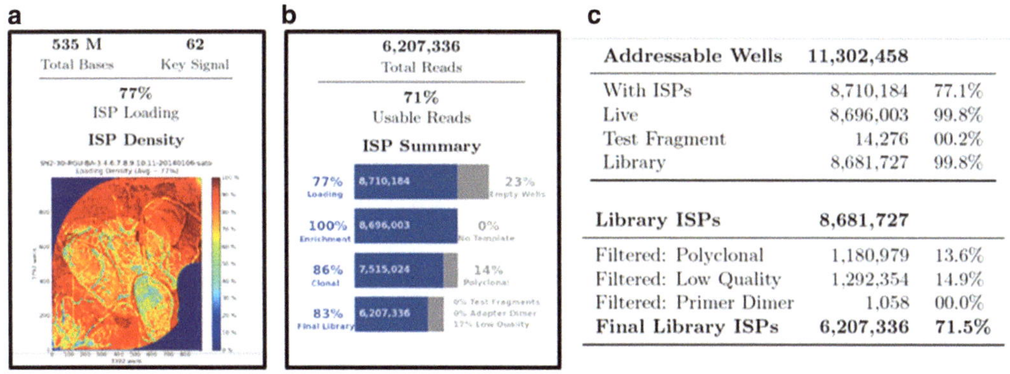

Fig. 2 Run report of Ion PGM sequencing. Ion sphere particle (ISP) identification summary. (**a**) Review of loading density (*heat map*), ISP summary (*bar graph*: **b**), and detailed data (**c**)

3. If more sequencing information is required, 318 chip v2 is used (*see* **Note 12**).

A Torrent Browser run report contains statistics and quality metrics for the sequencing run (Fig. 2).

3.8 Data Analysis

Data analysis (Fig. 3) was performed on each read sequence using computational tools, as developed previously [5], with some modifications. The analysis steps were:

1. Remove adapter sequences and sequences shorter than 50 bp using Cutadapt version 1.2.1 [6].
2. Map the trimmed sequences to the host genome (GenBank code: 313728) using Stampy [7] and extract the unmapped sequences.
3. Execute a BLAST [8] search of the extracted sequences against the NCBI BLAST database.
4. Identify the scientific name for each read by text mining of the BLAST output. This text mining was carried out by utilizing a BLAST parser of BioRuby [9].

4 Notes

1. Similar procedures are available for nasal swabs, bronchoalveolar lavage, and body fluids including blood (plasma and serum) and urine. Tissue and cell analysis are difficult because of a large amount of host-derived (cellular) DNA and RNA contamination.
2. TRizol LS (TRizol) contains phenol and destroys any microbes, so there is zero infectivity. It is suitable for further experiments regarding biosafety concerns. It can be stored for several months in a −80 °C freezer.

a

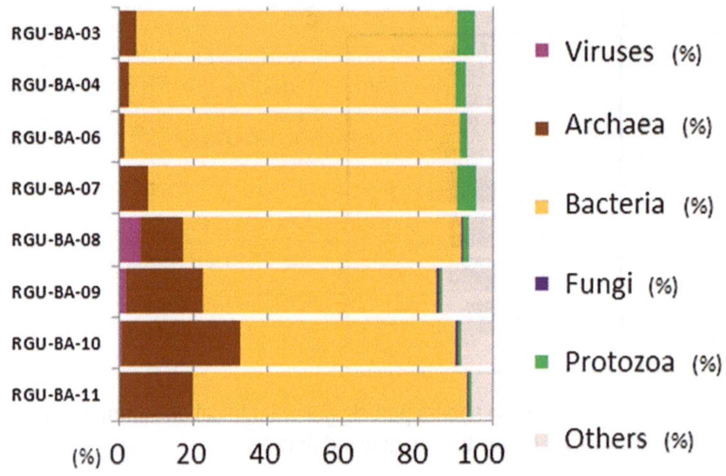

b

	Viruses (%)	Detected viruses (best match)
RGU-BA-03	0.01	Red clover vein mosaic virus, Paramecium bursaria Chlorella virus, Megavirus courdo7, Choristoneura biennis entomopoxvirus 'L'
RGU-BA-04	0.01	Montastraea cavernosa colony-associated virus
RGU-BA-06	0.01	Red clover vein mosaic virus, Paramecium bursaria Chlorella virus NE-JV-4, Moumouvirus Monve
RGU-BA-07	0.02	Montastraea cavernosa colony-associated virus
RGU-BA-08	0.12	Acanthamoeba polyphaga mimivirus, Moumouvirus Monve, Acanthamoeba castellanii mamavirus
RGU-BA-09	0.05	Montastraea cavernosa colony-associated virus
RGU-BA-10	0.04	Organic Lake phycodnavirus 1
RGU-BA-11	0.03	Acanthamoeba castellanii mamavirus

Fig. 3 (**a**) Comparison of the organisms showing the best matches for the sequences. (**b**) Summary of detected viruses without bacteriophages or mammalian viruses. Virus-derived sequences were detected in 0.01–0.12 % of the total reads. Because *Montastraea cavernosa* colony-associated virus was detected in all samples, it is listed only in RGU-BA-04, -07, and -09

3. The SeqPlex RNA amplification kit is also available in place of WTA. For DNA (DNA virus) amplification, the WGA or SeqPlex DNA amplification kit can be used. However, these RNA purification kits can also be used to isolate transcripts (mRNA) and are useful for detection of DNA viruses.

4. The optimal number of amplification cycles varies with the amount and quality of the template. Seventeen cycles are recommended for 5 ng of high-quality RNA or 50 ng of FFPE RNA, according to the manufacturer's protocol. However, isolated RNA is usually under the detection limit; thus, we selected 29 cycles. According to the guide of Sigma, an optimal cycle number is achieved by proceeding for a further 2–3 cycles beyond the amplification "plateau," as observed with real-time quantitative PCR.

5. Because AMPure XP selectively binds to DNA fragments larger than 100 bp, we do not use any size-purification system such as agarose gel electrophoresis.
6. In place of 1 μg, 50–100 μg of WTA DNA is also available for adapter ligation (according to the manufacturer's protocol).
7. The barcode adaptor sequence is 10 nucleotides with 96 types. The Ion Xpress™ Barcode Adapters 1-16 kit provides a set of 16 unique barcode adaptors.
8. Minimize the number of cycles to avoid over-amplification, the production of concatemers, or introduction of PCR-induced errors (manufacturer's protocol). We usually perform 5 cycles for 1 μg of input DNA.
9. Real-time PCR shows "relative qPCR quantity" for each library, and the "Template Dilution Factor" is determined as follows: Template Dilution Factor = [(relative qPCR quantity) × (sample library fold-dilution)]/0.32. In our experiments, the "Template Dilution Factor" for an undiluted sample library is usually around 1,000, indicating that the sample library has been diluted 1:1,000 for the Template Preparation Reaction.
10. The volume of the "diluted library" is 25 μl. When five libraries are analyzed in a single sequencing run, add 5 μl of each of the (diluted) libraries to the Ion OneTouch 2 instrument.
11. Use half the volume of enriched, template-positive ISPs in a single run with 314 chip v2 according to the manufacturer's protocol.
12. The remaining ISPs (*see* **Note 11**) can be used in another run using 318 chip v2 with a larger scale.

Acknowledgement

We thank Ms. Yuka Tanaka and Ms. Misa Sato (Kyoto Prefectural University of Medicine, Kyoto, Japan) for technical assistance.

References

1. El-Metwally S, Hamza T, Zakaria M, Helmy M (2013) Next-generation sequence assembly: four stages of data processing and computational challenges. PLoS Comput Biol 9:e1003345
2. Nakamura S, Nakaya T, Iida T (2011) Metagenomic analysis of bacterial infections by means of high-throughput DNA sequencing. Exp Biol Med 236:968–971
3. Nakamura S, Yang CS, Sakon N, Ueda M, Tougan T, Yamashita A et al (2009) Direct metagenomic detection of viral pathogens in nasal and fecal specimens using an unbiased high-throughput sequencing approach. PLoS One 4:e4219
4. Zhang T, Breitbart M, Lee WH, Run JQ, Wei CL et al (2006) RNA viral community in human feces: prevalence of plant pathogenic viruses. PLoS Biol 4:e3
5. Nakamura S, Maeda N, Miron IM, Yoh M, Izutsu K, Kataoka C et al (2008) Metagenomic diagnosis of bacterial infections. Emerg Infect Dis 14:1784–1786

6. Martin M (2011) Cutadapt removes adapter sequences from high-throughput sequencing reads. EMBnetjournal 17:3
7. Lunter G, Goodson M (2011) Stampy: a statistical algorithm for sensitive and fast mapping of Illumina sequence reads. Genome Res 21: 936–939
8. Altschul SF, Gish W, Miller W, Myers EW, Lipman DJ (1990) Basic local alignment search tool. J Mol Biol 215:403–410
9. Goto N, Nakao MC, Kawashima S, Katayama T, Kanehisa M (2003) BioRuby: open-source bioinformatics library. Genome Inform 14: 629–630

Chapter 12

Analysis and Application of Viroid-Specific Small RNAs Generated by Viroid-Inducing RNA Silencing

Charith Raj Adkar-Purushothama, Zhixiang Zhang, Shifang Li, and Teruo Sano

Abstract

Viroids are noncoding RNA pathogens inducing severe to mild disease symptoms on agriculturally important crop plants. Viroid replication is entirely dependent on host transcription machinery, and their replication/accumulation in the infected cells can activate RNA silencing—a host defense mechanism that targets the viroid itself. RNA silencing produces in the cell large amounts of viroid-specific small RNAs of 21–24-nucleotides by cleaving (or "dicing") entire molecules of viroid RNA. However, viroid replication is resistant to the effects of RNA silencing and disrupts the normal regulation of host gene expression, finally resulting in the development of disease symptoms on infected plant.

The molecular mechanisms of biological processes involving RNA silencing and underlying various aspects of viroid–host interaction, such as symptom expression, are of special interests to both basic and applied areas of viroid research. Here we present a method to create infectious viroid cDNA clones and RNA transcripts, the starting material for such analyses, using *Hop stunt viroid* as an example. Next we describe methods for the preparation and analysis of viroid-specific small RNAs by deep sequencing using tomato plants infected with *Potato spindle tuber viroid* as an example. Finally we introduce bioinformatics tools and methods necessary to process, analyze, and characterize these viroid-specific small RNAs. These bioinformatic methods provide a powerful new tool for the detection and discovery of both known and new viroid species.

Key words Viroids, RNA silencing, Host interaction, Infectious cDNA clones, Mechanical inoculation, Viroid-specific small RNAs, RNA gel-blot hybridization, Next-generation sequencing, Bioinformatics

1 Introduction

As self-replicating low molecular RNA pathogens, viroids have long attracted researchers intrigued by their unique molecular structures in which undiscovered RNA functions are believed to reside. Although viroids are noncoding RNAs which do not encode any genetic information translatable into protein, their replication in certain host species can induce severe to mild disease symptoms [1]. The molecular mechanisms underlying their biological functions, namely viroid–host interactions, have been studied using a variety

Ichiro Uyeda and Chikara Masuta (eds.), *Plant Virology Protocols*, Methods in Molecular Biology, vol. 1236, DOI 10.1007/978-1-4939-1743-3_12, of different approaches—molecular genetics, plant physiology, cell biology and histochemical studies, and evolutionary biology [2–4].

Viroid replication is completely dependent on the transcriptional machinery of the host and can evolve more rapidly than other known infectious agents [5]. Survival is severely restricted by structural constraints, however [6]. The fact that viroid-infected plants accumulate diverse mixtures of viroid sequence variants forming quasi-species make it difficult to assign biological functions to specific molecular structures. Therefore, infectious cDNA clones and/or their corresponding in vitro RNA transcripts are essential tools for sequence- and structure-based molecular biological analysis of viroid replication, movement, pathogenicity, and evolution. A cDNA-based experimental strategy is also useful to test whether or not a newly discovered single-stranded RNA molecule with a viroid-like structure is able to replicate autonomously.

Although viroids are single-stranded circular RNAs that replicate via an RNA–RNA pathway in the nucleus or chloroplast of host cells, it was shown in mid-1980s that certain full-length double-stranded cDNA copy of viroids were also infectious and can produce single-stranded circular RNA progeny [7–10]. Analysis of viroid structures necessary for replication has shown that the upper portion of the "central conserved region" (CCR) of viroids in the family *Pospiviroidae* plays a critical role as the processing site for producing monomeric progeny [8, 10, 11]. In fact, infection assays using monomeric, dimeric, or multimeric viroid cDNAs and their respective in vitro RNA transcripts indicate that (1) dimers and multimers are more infectious than the monomeric forms and (2) infectivity of monomer is limited and site-dependent. Tandem dimers of full-length viroids cDNAs are now routinely used to create infectious cDNA clones. A more sophisticated ribozyme-based expression system to produce highly infectious transcripts in vitro was devised by Feldstein et al. [12], in which full-length viroid cDNA was embedded in between hammerhead and paperclip ribozyme cassettes to produce precisely full-length RNA whose termini are capable of undergoing facile circularization in vitro.

RNA silencing is a sequence-specific transcriptional and posttranscriptional gene-regulation system triggered in response to double-stranded or hairpin RNAs. It acts as a cell level host defense mechanism against the invasion of viruses and viroids [13]. As expected from their highly base-paired stem-loop structure, viroids were shown to trigger RNA silencing and the accumulation of abundant viroid-specific small RNAs in infected plants [14–16]. The intimate connections between RNA silencing pathways that regulate host gene expression and those responsible for host defense mechanism targeting viroid invasion suggest that viroid-inducing RNA silencing could negatively effects normal homology-dependent regulation of host gene expression, thereby resulting in expression of disease symptoms [17]. Support for this hypothesis was provided by studies of peach "calico" (albino symptoms) disease involving a specific variant of *Peach latent mosaic viroid* (PLMVd) [18]. In addition, several other studies have used next generation sequencing to examine the biogenesis of viroid-specific small RNAs and their possible role(s) in regulating host gene expression, with special emphasis on host micro-RNA pathways [19–24]. Here in this chapter, we focus on the fundamental and applied methods necessary to analyze the role of RNA silencing in mediating viroid–host interaction, i.e., (1) preparation of infectious cDNA clones and RNA transcripts, (2) infectivity assays, (3) preparation and detection of viroid-specific small RNAs, (4) preparation of viroid-specific small RNAs for deep sequencing, and (5) bioinformatic analysis of viroid-specific small RNAs.

2 Materials (Use Autoclaved, Deionized Water in All Recipes and Protocol Steps)

2.1 Construction of Infectious cDNA Clone for Infection Assay

1. ATP, CTP, GTP, UTP (Roche Applied Science).
2. Agarose S (Wako).
3. Agarose HS (Wako).
4. *Cfr*9I (5 U/μl, Fermentas).
5. dATP, dCTP, dGTP, dTTP (TaKaRa Bio).
6. DNA ligation kit (Takara Bio).
7. *Eco*RV (5 U/μl, Fermentas).
8. Ethidium Bromide (Amresco, Catalog code E406-5ML).
9. Gel elution buffer: 0.5 M ammonium acetate, 1 mM EDTA (pH 8.0), 0.1 % (w/v) SDS.
10. HpSVd 83M: 5′-AA<u>CCCGGG</u>GCTCCTTTCTCA-3′ (underline shows *Cfr*9I site).
11. HpSVd 78P: 5′-AA<u>CCCGGG</u>GCAACTCTTCTC-3′ (underline shows *Cfr*9I site).
12. LA *Taq* polymerase (TaKaRa Bio).
13. Lithium chloride (Wako).
14. Marker dye solution: 0.02 % bromophenol blue, 0.02 % xylene cyanol, 50 % glycerol.
15. M-MuLV reverse transcriptase (200 U/μl, Invitrogen).
16. pBluescript II(SK-) (Stratagene).
17. Phenol–chloroform solution: Mix equal volume of water-saturated phenol and chloroform.
18. RNase inhibitor (Wako).
19. RQ1 RNase (Promega)

20. T3 RNA polymerase (10–20 U/μl, Promega, Catalog No. P2083).
21. TEMED (*N,N,N′,N′*-Tetramethylethylenediamine, Wako).
22. Trizol reagent (Invitrogen).
23. TAKARA suprec-01 (TaKaRa Bio).
24. 2× CTAB (cetyltrimethylammonium bromide) buffer: Combine 10 ml 1 M Tris–HCl (pH 9.5), 46.6 ml 3 M NaCl, 10 ml 0.2 M disodium EDTA (pH 7.0), 2 g CTAB powder, 0.5 ml 2-mercaptoethanol, and 28.4 ml distilled water (final volume = 100 ml).
25. 50× TAE buffer: 2 M Tris-acetate, 50 mM EDTA, pH 8.0.
26. 10× TBE buffer: 0.89 M Tris base, 0.89 M boric acid, 25 mM EDTA, pH 8.3.
27. 10 % (w/v) ammonium persulfate (APS, Wako).
28. 12 % polyacrylamide gel (PAGE) containing 8 M urea in 1× TBE buffer: Combine 6 ml 40 % acrylamide (acrylamide:bisac rylamide = 19:1 w/w) stock solution, 10 g urea, 6 ml 10× TBE buffer, and sufficient distilled water to yield a final volume of 20 ml. Incubate with gentle shaking in hot water bath until urea dissolves completely. Cool down to room temperature, then add 0.2 ml of 10 % ammonium persulfate solution and 0.02 ml of TEMED. Mix gently and pour immediately into the gel apparatus.

2.2 Northern Hybridization

1. Pre-Hybridization and Hybridization solution: Mix 25 ml 20× SSC, 2 ml 50× Denhardt's reagent (Wako), 5 ml 20 % SDS, 0.25 ml 100 mg/ml yeast tRNA (Wako), and 67.75 ml of distilled water.
2. 2× SSC (3 M NaCl, 0.3 M sodium citrate, adjust to pH 7.0 with NaOH).
3. 20 % sodium dodecyl sulfate (SDS) (Wako).
4. 0.5 M sodium phosphate buffer (pH 7.5).
5. Blocking solution: Combine 100 ml of 0.1 M maleic acid–0.15 M NaCl (pH 7.5) and 1.0 g Blocking reagent (Roche diagnostic).
6. Anti-DIG solution: Mix 1.0 μl Anti-DIG Fab fragment (Roche diagnostic) in 10 ml of Blocking solution.
7. Washing buffer: Dissolve 0.3 ml Tween 20 (Wako) in 100 ml 0.1 M maleic acid–0.15 M NaCl (pH 7.5).
8. Buffer 3: Mix 3.0 ml 1 M Tris–HCl (pH 9.5), 3.0 ml 1 M NaCl, 1.5 ml 1 M $MgCl_2$, and 22.5 ml distilled water.
9. CSPD, ready-to-use (Roche diagnostic, Catalog No.11 755 633 001).

2.3 RT-PCR, Cloning and Sequence Analysis of Known and New Viroid Species

1. M-MuLV Reverse Transcriptase (200 U/μl), 5× buffer, and 50 mM DTT (Promega, Catalog No. M1701).
2. RNasin Ribonuclease Inhibitor, 40 U/μl (Promega, Catalog No. N2111S).
3. dNTP mix (10 mM) (Promega, Catalog No.U1511).
4. PCR Master Mix (Promega, Catalog No. M7502).
5. Viroid-specific synthetic oligodeoxynucleotide primers.
6. Axygen AxyPrep DNA Gel Extraction Kit (Corning/Axygen, Product #AP-GX-4).
7. 100 bp DNA ladders (Promega, Catalog No. G2101).
8. pGEM-T Vector Systems I (Promega, Catalog No. A3600).
9. Biowest Regular Agarose G-10 (Spain, Lot No. 111860).
10. *E. coli* DH5α Competent Cells (TaKaRa Bio., Catalog No.9057).
11. Luria broth (LB).
12. LB agar plates (with 50 μg/ml Ampicillin/L).
13. M13 forward and M13 reverse PCR primers (Fasmac).
14. RNase-free water (TaKaRa Bio, Catalog No.9012).

3 Methods

3.1 Preparation of Monomeric and Dimeric Full-Length Viroid cDNAs and RNA Transcripts

3.1.1 RT-PCR Amplification of Unit-Length Viroid cDNAs

1. *Hop stunt viroid* (HpSVd) contains a unique *Cfr*9I recognition site (CCCGGG) in the upper portion of the CCR; similarly, *Potato spindle tuber viroid* (PSTVd) has a unique *Bam*HI recognition site (GGATCC) in the upper portion of the CCR. These restriction enzyme recognition sites, if they are unique, can be used to create a unit-length viroid cDNA. Sets of primers such as HpSVd-78P (for PCR) and HpSVd-83M (for RT and PCR) can be designed to overlap a unique restriction site and used for RT-PCR amplification (*see* **Notes 1** and **2**).

Reaction mixture for reverse transcription	
Low molecular RNA containing HpSVd [25]	2.0 μl
5× Reverse transcription buffer	5.0 μl
Nucleotide stock solution containing 2.5 mM each dNTP	2.0 μl
Oligodeoxynucleotide primer (HpSVd 83M; 20 μM)	1.0 μl
M-MuLV reverse transcriptase	1.0 μl
Distilled water	14.0 μl
Total volume	25.0 μl

2. Incubate the mixture at 37 °C for 45–60 min to reverse transcribe HpSVd cDNA, then heat at 70 °C for 10 min to inactivate reverse transcriptase.

Reaction mixture for Polymerase chain reaction (PCR)	
Reverse transcription solution	2.0 µl
10× PCR buffer (with 25 mM $MgCl_2$)	2.5 µl
Nucleotide stock solution (2.5 mM each dNTP)	2.0 µl
Primer (HpSVd 83M; 20 µM)	1.0 µl
Primer (HpSVd 78P; 20 µM)	1.0 µl
LA *Taq* polymerase	0.5 µl
Distilled water	16.0 µl
Total volume	25.0 µl

3. The reaction mixture is heat denatured at 95 °C for 4 min, then incubated for a 30–35 cycles of heat denaturation at 94 °C for 1 min, primer annealing at 55 °C for 1 min, and primer extension at 72 °C for 1 min to synthesize full-length double-stranded HpSVd cDNA, and followed by final primer extension at 72 °C for 7 min.
4. The HpSVd cDNA amplicon is precipitated by addition of 2.5 volumes of 99.5 % ethanol to the PCR reaction, storage at −30 °C for 1 h, and centrifugation at 13,000×*g* for 10 min at 4 °C. The resulting pellet is dissolved in 20 µl distilled water and digested with *Cfr*9I.

Reaction mixture for *Cfr*9I restriction enzyme digestion	
HpSVd cDNA amplicon	20.0 µl
10× reaction buffer (with 0.1 % BSA)	5.0 µl
*Cfr*9I (5 U/µl)	2.0 µl
Distilled water	23.0 µl
Total volume	50.0 µl

5. Incubate the reaction at 37 °C overnight and check the extent of digestion using agarose gel electrophoresis. Remove 2 µl of digestion reaction, mix with an equal amount of marker dye solution, and run at 50 V for 30 min in a 1.5 % agarose gel submerged in 1× TAE buffer. After electrophoresis, stain the gel with ethidium bromide and place it on trans-illuminator to detect unit-length double-stranded HpSVd cDNA band.
6. If sufficient unit-length HpSVd cDNA is detected, the remaining solution (48 µl) is added to 52 µl distilled water, mixed with an equal volume of phenol–chloroform, shaken vigorously

for 1 min using a vortex mixer, and centrifuged at 13,000 × *g* for 1 min. The upper aqueous phase is transferred to a new sampling tube (1.5 ml), and 2.5 volumes of 99.5 % ethanol are added. After storage at −30 °C for 1 h, the unit-length double-stranded HpSVd cDNA is recovered by centrifugation at 13,000 × *g* for 10 min at 4 °C. Dry the resulting precipitate for 10 min at room temperature, dissolve in 10 μl distilled water, mix with 2 μl marker dye, and separate in a 7.5 % polyacrylamide gel in 1× TAE buffer at 250 constant voltage for 90 min.

7. After electrophoresis, stain the gel with ethidium bromide, visualize the band on a trans-illuminator, and excise a gel slice containing the band of interest (ca. 300-bp) using a clean razor blade. The gel slice is further cut into small pieces (ca. 1 mm^3), submerged in 500 μl of gel elution buffer, and incubated at 37 °C with gentle shaking (120 rpm) overnight. The unit-length double-stranded HpSVd cDNA is recovered by ethanol precipitation, air-dried at room temperature for 10 min, and dissolved in 10 μl distilled water.

3.1.2 Construction of Dimeric Viroid cDNA Clone with the Same Restriction Enzyme Recognition Sites in Both Termini

1. Aliquots (3–5 μl) of a unit-length HpSVd cDNA with *Cfr*9I termini are combined with an equal volume of Solution I from the DNA ligation kit and incubated at 16 °C for 2–5 h to make self-ligated tandem dimers of HpSVd cDNA. The ligation mixture is then added to 1 μl (0.1 μg/μl) of dephosphorylated plasmid DNA such as pBluescript II (SK-) digested with *Cfr*9I (the same enzyme used for digestion of HpSVd cDNA) and 1 μl of Solution I from the DNA ligation kit, incubated at 16 °C overnight, and used for transformation of *E. coli* [26] (*see* **Note 3**).
2. After transformation, select white colonies, extract recombinant plasmid DNA using either the alkaline lysis with SDS method [27] or a commercially available plasmid DNA extraction kit, and check the size by electrophoresis using 1.5 % agarose gel (Fig. 1a) (*see* **Note 4**).

3.1.3 Preparation of Dimeric Transcripts

1. Plasmid DNAs containing dimeric viroid cDNAs are infectious and can be used for infectivity assays, but in vitro RNA transcripts, especially those of positive-(i.e., genomic) sense, are more infectious [7–10]. Therefore, in vitro plus-sense RNA transcripts are desirable for infectivity assays. Depending on plasmid vectors used and the orientation of viroid cDNA, either of T7-, T3-, or SP6-RNA polymerase can be used for transcription.
2. After consulting the sequencing data, select the RNA polymerase required to produce positive-sense transcript. Plasmid DNA containing dimeric viroid cDNA is then digested with a restriction enzyme site cleaving downstream from the 3′-end

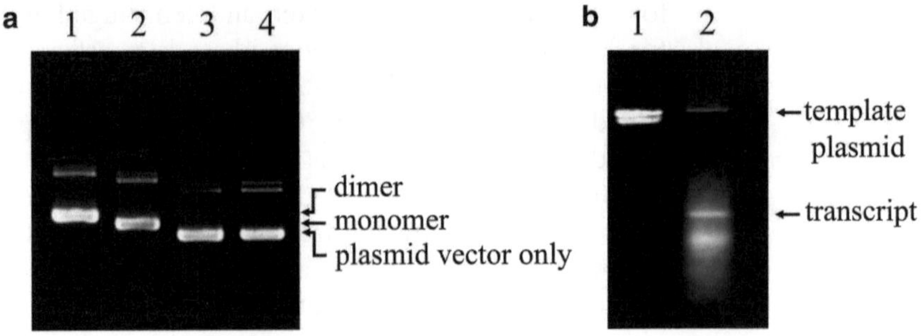

Fig. 1 Agarose gel (1.5 %) electrophoresis of recombinant plasmid DNA containing monomeric and dimeric HpSVd cDNA insert (**a**), and in vitro transcript of HpSVd (**b**)

of the viroid cDNA insert. Take care to check for the presence of additional restriction enzyme recognition site(s) in the viroid cDNA. Here we present an example of dimeric PSTVd cDNA in pBluescript II SK(-) plasmid vector.

Reaction mixture for *Not* I restriction enzyme digestion	
Plasmid DNA (ca. 0.1 μg/μl)	25 μl
10× reaction buffer	5 μl
Distilled water	18 μl
Not I (5 U/μl)	2 μl
Total volume	50 μl

3. After incubation at 37 °C overnight, the reaction mixture is extracted with phenol–chloroform as described above. The linearized plasmid DNA is recovered by ethanol precipitation and dissolved in 16 μl of distilled water for later use in transcription reactions.

Reaction mixture for T3 transcription	
Plasmid DNA/*Not* I	8 μl
10× T3 reaction buffer	2 μl
T3 RNA polymerase (10–20 U/μl)	1 μl
Ribonucleotide stock (10 mM each NTP)	2 μl
RNase inhibitor (40 U/μl)	1 μl
Distilled water	6 μl
Total volume	20 μl

4. After transcription at 37 °C for 2 h, the reaction is stopped by adding 2 μl 0.1 M EDTA pH 7.0, 2.5 μl 4 M LiCl, and 75 μl 99.5 % ethanol, and stored for 1 h at −30 °C. RNA transcripts

are recovered by centrifugation at 12,000 × *g* for 5 min at 4 °C, air dried for 1 min, and dissolved in 50–100 μl of distilled water. Transcripts can be quantified by UV spectrophotometry and their integrity confirmed by gel electrophoresis (Fig. 1b). The reaction shown here produces ca. 200 μg of transcript per μg of DNA template.

3.2 Infectivity Assay

3.2.1 Mechanical Inoculation

1. Inocula are prepared at concentrations of 100–200 ng RNA transcripts/μl in 50 mM sodium phosphate buffer (pH 7.0). For mechanical inoculation, an aliquot (e.g., 10 μl) is placed on a leaf dusted with carborundum (600-mesh) and gently rubbed against the leaf 10–20 times using a sterile glass-bar [28]. Plants are rinsed immediately with tap water and incubated in a growth chamber controlled at 25–30 °C with high light intensity (fluorescent, 40 W × 4, ca. 60 cm distance) (*see* **Notes 5** and **6**).

3.3 Detection and Preparation of Viroid-Specific Small RNAs

Viroid-specific small RNAs can be extracted using the same methods used to extract viroid genomic RNA. Several different extraction methods have been reported to be well-suited for specific viroid–host combinations. Therefore, it may be necessary to compare different extraction methods when beginning studies with a new viroid–host combination. It is often true that methods that give high yields of viroid-RNA yield higher amounts of contaminants and degradation products derived from host nucleic acids or even from viroid-RNA itself that can disturb the subsequent analysis of viroid-specific small RNAs. Here we describe a protocol for enrichment of viroid-specific small RNAs consisting of (1) extraction of total nucleic acids, (2) enrichment of viroid and viroid-specific small RNAs by 2 M LiCl fractionation, (3) elimination of DNA by RNase-free DNase treatment, and (4) PAGE-fractionation of viroid-specific small RNAs.

3.3.1 Extraction of Total Nucleic Acids

The method described below follows that originally described by Doyle and Doyle [29] with some modifications for use with viroids [23]. A protocol using Trizol reagent is also suitable to prepare high quality total RNA extracts from tomato tissues [24].

1. Harvest an appropriate amount of viroid-infected tissue at appropriate time points from desired part of individual plants.
2. Weigh the fresh tissue, immediately transfer to a porcelain mortar containing liquid nitrogen, and carefully grind until the tissue has been reduced to a fine powder and the liquid nitrogen has evaporated.
3. Add an appropriate volume (e.g., 1 ml/100 mg tissue) of 2× CTAB buffer (*see* Subheading 2), mix quickly to form a thick paste, and allow slurry to thaw.

4. Add the same volume (e.g., 1 ml/100 mg tissue) of phenol–chloroform (1:1, v/v), shake vigorously for 1 min to extract nucleic acids, and centrifuge at 13,000 × *g* for 5 min at 4 °C.
5. Collect aqueous phase, add 2.5 volumes (v/v) of 99.5 % ethanol, mix gently, and stand for 1 h at −30 °C.
6. Precipitate total nucleic acid by centrifugation at 13,000 × *g* for 10 min at 4 °C, air-dry for 10 min at room temperature.

3.3.2 Enrichment of Viroid and Viroid-Specific Small RNAs by 2 M LiCl Fractionation

1. Dissolve the nucleic acid pellet completely in 200 μl distilled water, slowly add an equal volume (200 μl) of 4 M LiCl with gentle shaking, and allow to stand on ice for 4 h to precipitate high molecular weight RNAs.
2. Collect the supernatant (i.e., 2 M LiCl soluble nucleic acids fraction) by centrifugation at 13,000 × *g* for 5 min, add 2.5 volumes (1 ml) of 99.5 % ethanol, mix gently, and allow to stand overnight at −30 °C.
3. Recover the 2 M LiCl soluble nucleic acids by centrifugation at 13,000 × *g* for 10 min at 4 °C, and air-dry the resulting pellet for 10 min at room temperature (*see* **Note 7**).

3.3.3 DNase I Treatment

1. Dissolve the precipitate in 160 μl distilled water, add 40 μl 5× *DNase* I buffer and 1 μl *DNase* I (10–20 U/μl), and incubate for 30 min at 37 °C (*see* **Note 8**).
2. Extract the *DNase* I digest with an equal volume (200 μl) of phenol–chloroform (1:1), and centrifuge at 12,000 × *g* for 5 min at 4 °C.
3. Collect aqueous phase, add 2.5 volumes of 99.5 % ethanol, and stand for 1 h at −30 °C.
4. Recover the DNA-free LMW-RNA by centrifugation at 13,000 × *g* for 10 min at 4 °C, air-dry the resulting pellet for 10 min at room temperature.
5. Dissolve pellet in an appropriate amount (e.g., 100 μl) of *RNase* free distilled water (*see* **Note 9**).

3.3.4 PAGE-Fractionation of Viroid-Specific Small RNAs

1. Low molecular weight RNA (e.g., 200 μg/100 μl) is mixed with 0.9 volume of 50 % urea solution (90 μl) and 0.1 volume of marker dye solution (10 μl), and heat-denatured at 68 °C for 15 min.
2. Apply the sample to a 12 % polyacrylamide gel (acrylamide:bisacrylamide = 19:1) containing 1× TBE buffer–8 M urea and run at 450 V (constant voltage) until the bromophenol blue marker dye migrates 8 cm from the origin (Fig. 2a) (*see* **Note 10**).
3. After electrophoresis, remove one of gel plates, place a dry nylon membrane (13 cm × 10 cm, Biodyne plus, Poll) directly on the gel surface, followed by three sheets of Whatman 3MM filter (13 × 9 cm), and a clean gel plate (14 cm × 14 cm × 1 mm).

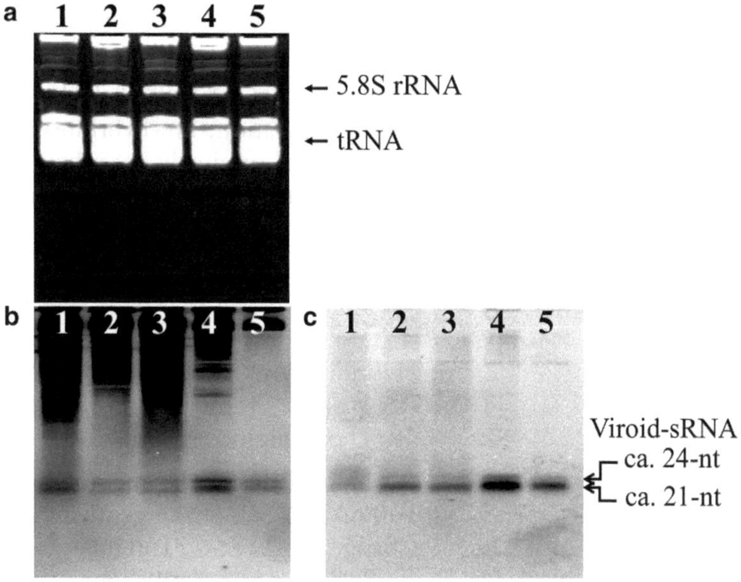

Fig. 2 RNA gel-blot analysis of viroid-specific small RNAs. (**a**) 8 M urea–12 % polyacrylamide gel stained with ethidium bromide, (**b**) PSTVd-specific small RNA with plus orientation, (**c**) PSTVd-specific small RNA with minus orientation. *Lane 1–5*; PSTVd-infected tomato samples #1–5. *Arrows* indicate PSTVd-specific small RNA containing.24-nt and 21-nt

4. Wrap the assembly with vinyl film to prevent drying out, place a bottle containing ca. 1–2 L of water on top, and leave several hours—over night at *ca.* 25 °C for contact-blotting of RNAs to nylon membrane.
5. Remove the membrane and use for RNA gel-blot analysis of viroid-specific small RNA as described in the next section.
6. After removing the membrane, excise a gel slice from 3 to 5 cm below from the meniscus, i.e., the region containing small RNAs with the size 15–35 nucleotides.
7. Crush the gel into small pieces, add 1 ml RNA gel elution buffer, and shake gently overnight at 37 °C.
8. Remove gel debris by passage through a TaKaRa suprec-01 spin column according to the manufacturer's instruction and recover the eluate.
9. Small RNAs are precipitated by adding 2.5 volumes of 99.5 % ethanol, overnight storage at −30 °C, and centrifugation at 13,000 × *g* for 15 min at 4 °C.
10. Air-dry the resulting pellet for 10 min at room temperature and dissolve in 10 μl *RNase*-free distilled water.
11. Quantify the amount of small RNAs by UV absorption (*see* **Note 11**).

*3.3.5 Detection of Viroid-Specific Small RNAs by RNA-Gel Blot Analysis: Northern Hybridization for Small RNA Detection [Continues from Subheading 3.3.4, **step 5**]*

1. UV crosslinking: Place the moist nylon membrane in UV crosslinker (Stratagene) with the gel contact side on top, and crosslink using a setting of 1,200 (×100 μJ/cm^2).
2. Pre-hybridization: Pre-hybridization and hybridization follow the protocol originally described by Papaefthimiou et al. [15] with modifications. Place the membrane in a hybridization bottle with the gel contact side on top, add pre-hybridization solution (5 ml for 13.5 cm×9.5 cm membrane), and remove any air bubbles between membrane and bottle. Incubate at 58 °C for at least 1 h.
3. Hybridization: Add 1 ml of hybridization solution (1 μl DIG-RNA probe/1 ml of pre-hybridization buffer). Incubate at 58 °C overnight.
4. Washing: Transfer the membrane to a plastic box filled with 100 ml 2× SSC. Shake gently at room temperature for 5 min. Change the 2× SSC (100 ml) and shake for 5 min. Discard the 2× SSC and transfer the membrane to a second plastic box containing 100 ml 2× SSC–0.5 % SDS (preheated to 58 °C). Shake gently at 58 °C for 30 min. Change 2× SSC–0.5 % SDS (100 ml), and shake gently for an additional 30 min at 58 °C.
5. DIG ELISA: Place the membrane in a plastic bag add 5 ml of blocking solution, and incubate at room temperature for 30 min. Discard the blocking solution, add 5 ml anti-DIG solution (1 μl anti-DIG Fab fragment in 5 ml of blocking solution) to the same bag, and incubate at room temperature for 30 min. Discard the solution and transfer the membrane to a plastic box containing 75 ml of washing buffer. Shake gently for 10 min at room temperature, change the washing buffer (75 ml), and shake gently for 10 min at room temperature. Discard the solution and submerge the membrane in a plastic box containing 30 ml Buffer 3 for ca. 5 min. Transfer the membrane to a plastic bag containing 500 μl diluted CSPD (dilute 1/2–1/5× in Buffer 3), and carefully remove any air bubbles before sealing the bag completely. Place the bag in contact with Hyper film (Amersham) in the dark for 1–3 h at room temperature or process as follows.
6. Chemiluminescent Development: Place the membrane in Chemidoc (XRS, Bio-Rad), expose for 10 min–2 h to detect viroid-specific small RNAs (Fig. 2b, c).

3.4 Analysis of Viroid-Specific Small RNAs by Deep Sequencing

3.4.1 Preparation and Construction of cDNA Libraries for Deep Sequencing

Total RNA or small RNA, the starting material, can be extracted using the methods described in this chapter (Subheading 3.3). Integrity of the resulting RNA preparations must be checked before preparing cDNA libraries for deep sequence analysis. If an Agilent Technologies 2100 Bioanalyzer is used to check RNA integrity, the RNA Integrity Number (RIN) value should be greater than 8. Alternatively, the integrity of total RNA preparations can be assessed by electrophoresis on a formaldehyde–1 % agarose gel and ethidium

a

Sample	A230	A260	A280	A260/A280	A260/A230	ng/ul	volume(ul)	total (ug)
PSTVd-tomato	0.595	0.877	0.442	1.99	1.46	35.1	17	0.597

b

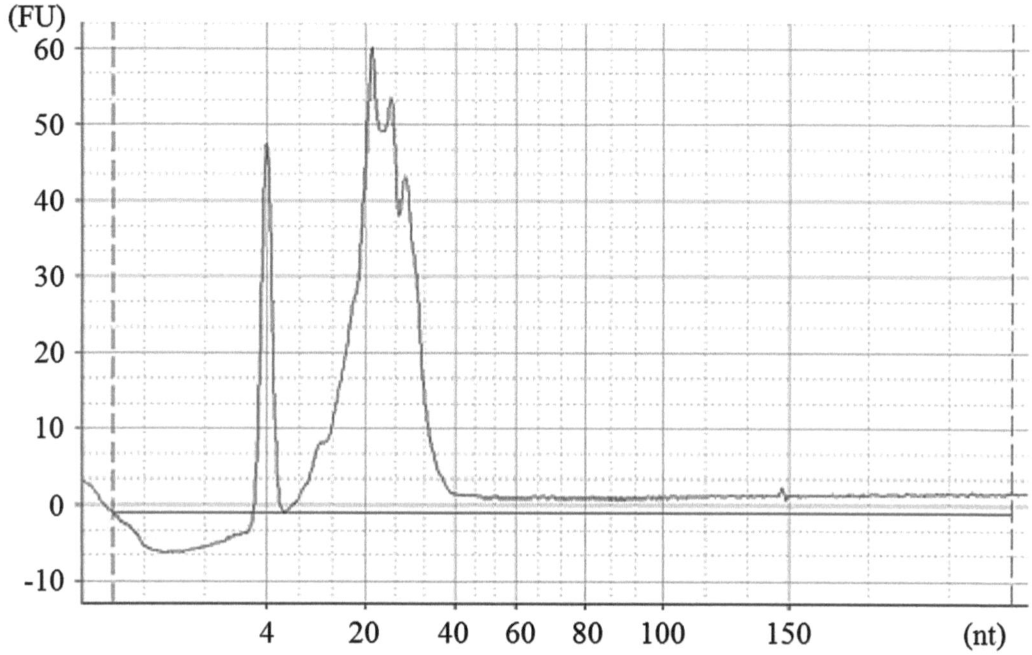

Fig. 3 Quality assessment of a gel-purified small (15–30-nt) RNA preparation (*see* Subheading 3.3.4) extracted from PSTVd-infected tomato by (**a**) spectrophotometer and (**b**) Agilent Technologies 2100 Bioanalyzer. *Horizontal line* indicates length of RNA and *vertical line* indicates quantity

bromide staining. High quality total RNA will show two clear bands of 28S and 18S rRNA, in which the amount of 28S rRNA should be twice as much as 18S rRNA and 5.8S RNA will not be specifically visible. The concentration of total RNA can be measured using a spectrophotometer such as the NanoDrop (Thermo Scientific) (Fig. 3a).

Approximately 10 μg of total RNA is necessary to construct a small RNA library. Methods to create cDNA library suitable for deep sequence analysis starting with small RNA extracted from virus-infected plants are described in detail elsewhere [29–32]. The pipeline to construct a viroid-specific small RNA library is essentially the same as that used for viruses. Total RNA preparations can be sent directly to a commercial company for further processing to produce small RNA. In case of gel-purified small RNA ranging in size from 15 to 30-nt, ~100–300 ng is sufficient to produce ~20 millions reads [23]. When shipping preparations of either total or small RNA preparation to a company for analysis, samples must be preserved in 70 % ethanol (ethanol precipitation) and kept on dry ice during transportation. For preparation of cDNA libraries from small RNA, Illumina's small RNA sample preparation Kit can be used [19, 21].

1. Isolate small RNA by electrophoresis of total RNA in 12 % PAGE containing 8 M urea (*see* Subheadings 3.3.1–3.3.4).
2. Ligate 5′-adapter (e.g., GTTCAGAGTTCTACAGTCCGAC GATC) to 5′-end of small RNA, then purify by PAGE.
3. Ligate 3′-adapter (e.g., ATCTCGTATGCCGTCTTCTGCT TG) to 3′-end of small RNA containing the 5′-adapter and purify by PAGE.
4. Synthesize and amplify double-stranded cDNAs from small RNA by reverse transcription and PCR; purify the resulting PCR products by PAGE.
5. High-throughput DNA sequencing on the Illumina Genome Analyzer [33–36] (*see* **Notes 12** and **13**).

3.4.2 Outline of Viroid-Specific Small RNA Analysis

The analysis of viroid-specific small RNA involves the following procedures [37] (*see* **Note 14**).

1. Validate quality of the trimmed small RNAs based on Q20 value (>99 %).
2. Filter the obtained trimmed small RNAs and extract 18–30-nt reads.
3. Calculate statistics for size and depth distribution.
4. Align extracted viroid-specific small RNAs against the full genome of viroid variants without any mismatch.
5. Calculate additional statistics for viroid-derived small RNAs, including amount, size distribution with different polarity, composition of the 5′ nucleotide, and profile of the location on viroid genome.
6. Visualize statistics of viroid-specific small RNA using software such as Excel, R language, and Origin (*see* **Notes 15** and **16**).

3.4.3 Processing of Raw Sequence Data: Adapter Trimming

Most modern sequencers produce their output as FASTQ files, which are modified versions of FASTA formatted files. FASTQ files encode both nucleotide calls as well as quality information about each nucleotide. FASTQ files are normally given the extension ".fq" or ".fastq". A number of free software programs are available to retrieve these data, but their use requires a basic knowledge of "Linux" and "C" language. There are also a few commercial software packages like CLC genomic workbench (http://www.clcbio.com/products/clc-genomics-workbench/) and Geneious software (http://www.geneious.com/) which are more convenient for biologists.

Hence, from the heading 3.4.3 to 3.4.7, we will show an example of the analysis of viroid-specific small RNA data according to the protocol (descriptions) of CLC genomic workbench.

Once the data are retrieved, the first step is to remove the adapters from each sequence read, a step known as "adapter trimming". Adapter sequences are removed from the ends of the resulting raw short-read data based on the presence of an exact 10-nt match with the termini of the respective adapters. This can be done using number of computer programs. Alternatively, a set of custom Perl programming language scripts can be written to carry out this analysis.

FastqMcf (http://code.google.com/p/ea-utils/wiki/FastqMcf).

Trimmomatic (http://www.usadellab.org/cms/index.php?page=trimmomatic).

HTSeq (http://www-huber.embl.de/users/anders/HTSeq/doc/overview.html).

The following example uses the Cutadapt program as explained in ARK-genomics (http://www.ark-genomics.org/events-online-training-eu-training-course/adapter-and-quality-trimming-illumina-data):

1. Cutadapt is a tool specifically designed to remove adapter sequences from sequencing data by next generation sequencer. To see the full list of help and options, type "cutadapt –h".
2. Because our data contains many adapter sequences, we are going to run Cutadapt with all of them. The command is:

```
cutadapt  -a CGACAGGTTCAGAGTTCTACAGTCCGACGATC \
          -a TACAGTCCGACGATC \
          -a ATCTCGTATGCCGTCTTCTGCTTG \
          -e 0.1 -O 5 -m 15 \
          -o SRR026762_adaprm.fastq SRR026762.fastq.gz
```

3. The options mean:

   -a CGACAGGTTCAGAGTTCTACAGTCCGACGATC: the first adapter to remove (the full Small RNA primer)

   -a TACAGTCCGACGATC: the second adapter to remove (the portion of the Small RNA primer that we have observed)

   -a ATCTCGTATGCCGTCTTCTGCTTG: the third adapter to remove (the full Small RNA 5′ adapter)

   -e 0.1: allow a mismatch rate of one mismatch in ten bases between the read and the adapters

   -O 5: The overlap must be at least five base-pairs

   -m 15: after trimming, reads less than 15-bp are discarded

   -o SRR026762_adaprm.fastq: put the trimmer data in this file

4. The output of the command will look something like (*see* **Note 17**):

Maximum error rate:	10.00 %
Processed reads:	5,677,631
Trimmed reads:	1,927,223 (33.9 %)
Total base pairs:	204,394,716 (204.4 Mbp)
Trimmed base pairs:	41,594,144 (41.6 Mbp) (20.35 % of total)
Too short reads:	1,016,544 (17.9 % of processed reads)
Too long reads:	0 (0.0 % of processed reads)
Total time:	287.36 s
Time per read:	0.05 ms

=== Adapter 1 ===
Adapter 'CGACAGGTTCAGAGTTCTACAGTCCGACGATC', length 32, was trimmed 333 times.
Lengths of removed sequences

Length	Count	Expected
5	331	5544.6
7	1	346.5
10	1	5.4

=== Adapter 2 ===
Adapter 'TACAGTCCGACGATC', length 15, was trimmed 52,873 times.
Lengths of removed sequences

Length	Count	Expected
5	93	5544.6
6	1,121	1386.1
7	1,082	346.5
8	7	86.6
9	2	21.7
10	3	5.4
11	2	1.4
>=17	50,563	0.1

=== Adapter 3 ===
Adapter 'ATCTCGTATGCCGTCTTCTGCTTG', length 24, was trimmed 1,874,017 times.
Lengths of removed sequences

Length	Count	Expected
5	322	5544.6
6	90	1386.1
7	162	346.5
8	20	86.6
9	132	21.7
10	384	5.4
11	1,257	1.4
12	4,026	0.3
13	10,585	0.1
14	33,836	0.0
15	84,695	0.0
16	104,889	0.0
17	86,715	0.0
18	77,590	0.0
19	121,253	0.0
20	199,881	0.0
21	182,110	0.0
22	205,659	0.0
23	184,010	0.0
24	232,069	0.0
25	137,080	0.0
>=26	207,252	0.0

3.4.4 Adapter Trimming Using Commercial Analyzer: CLC Genomic Work Bench

1. To start trimming:

   Toolbox > NGS Core Tools > Trim Sequences
2. Add the sequences to be trimmed.
3. Click "Next" to display a dialogue box where the parameters for quality trimming can be entered (*see* **Note 18**).
4. Click "Next" to specify adapter trimming.
5. Go to "Trim sequences", then set the parameters and select the appropriate adapter (adapters), to be deleted from the data.
6. Go to "Trim adopter list", then check "Search on both strands" which will search both the minus and plus strand for the adapter sequence.

7. Click "Next" to get new dialogue box. At the top we can choose to "Trim bases" by specifying a number of bases to be removed from either the 3′ or the 5′ end of the reads. Below we can choose to "Discard reads below length and Discard reads above length". For example, if you want to retain only 21–24-nt, then select 21 for "Discard reads below length" and 24 for "Discard reads above length".
8. Clicking "Next" will ask us about "result handling". In this we can choose from a number of options like "Save discarded sequences", "Create report", "Save results", "Open results" etc.
9. "Create report" includes the following (*see* **Note 19**):
   Trim summary.

   Name: The name of the sequence list used as input.

   Number of reads: Number of reads in the input file.

   Avg. length: Average length of the reads in the input file.

   Number of reads after trim: The number of reads retained after trimming.

   Percentage trimmed: The percentage of the input reads that are retained.

   Avg. length after trim: The average length of the retained sequences.

   Read length before/after trimming: This is a graph showing the number of reads of various lengths. The numbers before and after are overlaid so that we can easily see how the trimming has affected the read lengths (right-click the graph to open it in a new view).

   Trim settings: A summary of the settings used for trimming.

   Detailed trim results: A table with one row for each type of trimming.

   Input reads: The number of reads used as input. Since the trimming is done sequentially, the number of retained reads from the first type of trim is also the number of input reads for the next type of trimming.

   No trim: The number of reads that have been retained, unaffected by the trimming.

   Trimmed: The number of reads that have been partially trimmed. This number plus the number from "No trim" is the total number of retained reads.

   Nothing left or discarded: The number of reads that have been discarded either because the full read was trimmed off or because they did not pass the length trim (e.g., too short) or adapter trim (e.g., if "Discard when not found" was chosen for the adapter trimming).

10. Select the appropriate adapter(s) to be deleted from your data.
11. In the dialogue box select "Remove adopter" under "Action" and then specify parameters for "Alignment scores cost" and "Match thresholds" according to our requirements.
12. The advantage of this method is single step processing that simultaneously trims primer and segregate the trimmed sequences according to length (*see* **Note 20**).

3.4.5 Mapping of Viroid-Specific Small RNAs on the Viroid Genome

All the small RNA profiling tools available are programmed to carry out the mapping of small RNA reads on a linear reference sequence. Therefore, it is necessary to modify the target sequence slightly in order to carry out mapping using a circular viroid molecule. For instance, when mapping small RNAs against the PSTVd genome, two 30-nt sequences containing positions 329–359 and 1–30 in the PSTVd genome (accession AY937179) were added to the 5′- and 3′-ends of the unit-length linear PSTVd molecule, respectively. Thus, the reference sequence contains PSTVd nucleotide numbers 5′-329–359–1–359–1–30-3′, i.e., a total 419-nt. Once this has been done, mapping of PSTVd-specific small RNA reads against the modified sequence can begin.

1. Toolbox/NGS Core Tools/Map Reads to Reference: In the dialog box, select the sequences or sequence lists containing the sequencing data. Then click "Next" and select the "reference" sequence against which these sequences are to be mapped. At the top we select one or more reference sequences by clicking "Browse" and then selecting the "Element" button. We can select either single sequences, a list of sequences, or a sequence track as reference (*see* **Note 21**).
2. Click "Next" to allow the parameters for read mapping to be entered.
   (a) Mismatch cost: number of mismatches allowed between the read and the reference sequence.
   (b) Insertion cost: Number of insertions in the read.
   (c) Deletion cost: Number of gaps in the read.
   (d) Length fraction: Set minimum length fraction of a read that must match the reference sequence.
   (e) Similarity: Set minimum fraction of identity between the read and the reference sequence (*see* **Note 22**).
3. Select the type of alignment to be performed, i.e., either "local alignment" or "global alignment". By default, mapping is done with local alignment of the reads to the reference.
4. Clicking "Next" lets choice of how the results of the mapping will be reported. After specifying the desired parameters, click "Finish" to begin mapping (*see* **Note 23**).

a

Number of reads	Ave. length	Number of reads after trim	Percentage trimmed	Ave. length after trim
5,189,398	23.5	4,681,117	90.21%	23.4

b

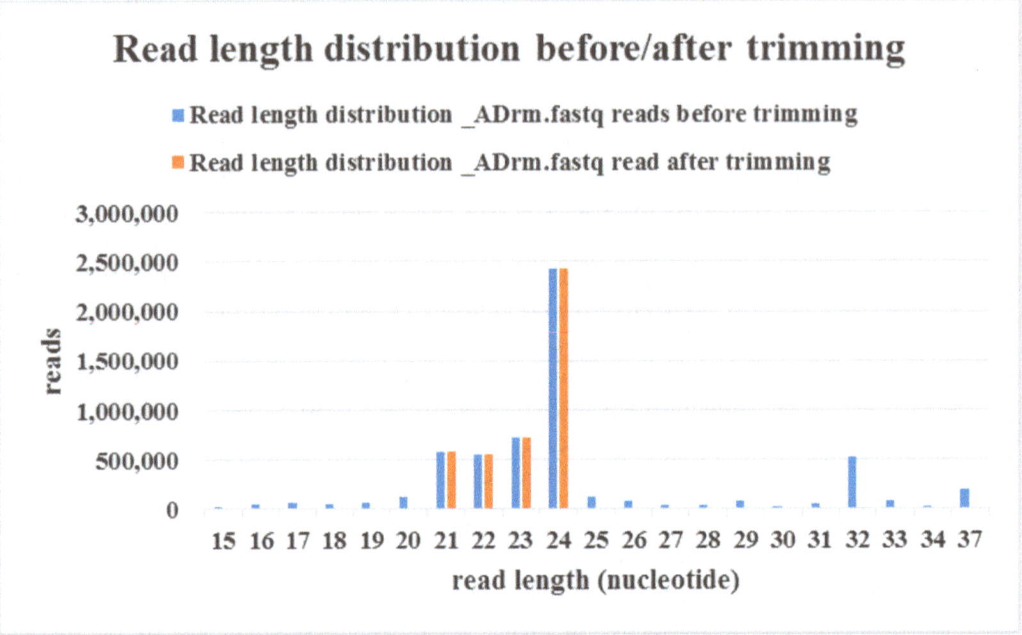

Fig. 4 Read length distribution after trimming. (**a**) Trim summary. (**b**) Read length before/after trimming. Trim settings are "Removal of ambiguous nucleotide: No ambiguous nucleotides allowed" and "Removal of sequences on length: minimum length 21-nt and maximum length 24-nt"

5. Final result will appear as shown in Fig. 4a, b. In this example, we have analyzed small RNA data obtained from a tomato plant infected with PSTVd-Intermediate strain. A total 5,189,398 reads of small RNA data was subjected to adapter trimming and trimmed sequences of 21–24-nt were analyzed. The results were as follows:
6. Visualization of the data: Alignments will be shown as below (Fig. 5).
7. This data can be exported to .bam or .sam files. For "compact" viewing of the entire PSTVd genome, either the free software program "Tablet" (http://bioinf.scri.ac.uk/tablet/) or the commercial "Genious" package can be used as shown below (Fig. 6) (*see* **Note 24**).

```
TTCGGCTACTACCCGGTGGAAACAACTGAAG          reference sequence

TTCGGCTACTACCCGGTG-AA                      mapped sequence
TTCGGCTACTACCCGGTGGAA
TTCGGCTACTACCCGGTGGAA
TTCGGCTACTACCCGGTGGAA
TTCGGCTACTACCCGGTGGAA
TTCGGCTACTACCCGGTGGAAA
TTCGGCTACTACCCGGTGGAAA
TTCGGCTACTACCCGGTGGAAA
TTCGGCTACTACCCGGTGGAAAC
TTCGGCTACTACCCGGTGGAAAC
TTCGGCTACTACCCGG-GGAAAC
TTCGGCTACTACCCGGTGGAAAC
TTCGGCTACTACCCGGTGGAAACA
TTCGGCTACTACCCGG-GGAAACA
-TCGGCTACTACCCGGTGGAAACA
--CGGCTACTACCCGGTGGAAACA
--CGGCTACTACCCGGTGGAAACA
--CGGCTACTACCCGGTGGAAACAA
---GGCTACTACCCGGTGGAAACAA
----GCTACTACCCGGTGGAAACAAC
----GCTACTACCCGGTGGAAACAAC
----GCTACTACCCGGTGGAAACAACT
-----CTACTACCCGGTGGAAACAACT
----GCTACTACCCGGTGG-AACAACTGA
------TACTACCCGGTGGAAACAACTGAA
-------ACTACCCGGTGGAAACAACTGAAG
```

Fig. 5 Mapping of PSTVd-specific small RNAs on the PSTVd genomic strand. In this case, only the plus strand was used as reference, and PSTVd-specific small RNA reads of both plus and minus polarity are used for mapping. Black codes (*top*) are the reference PSTVd-Intermediate sequences. The second law or below denote PSTVd-specific small RNA derived from plus strand and *underlined* denotes those from minus strand. One mismatch was allowed, and (*dash*) indicates a gap

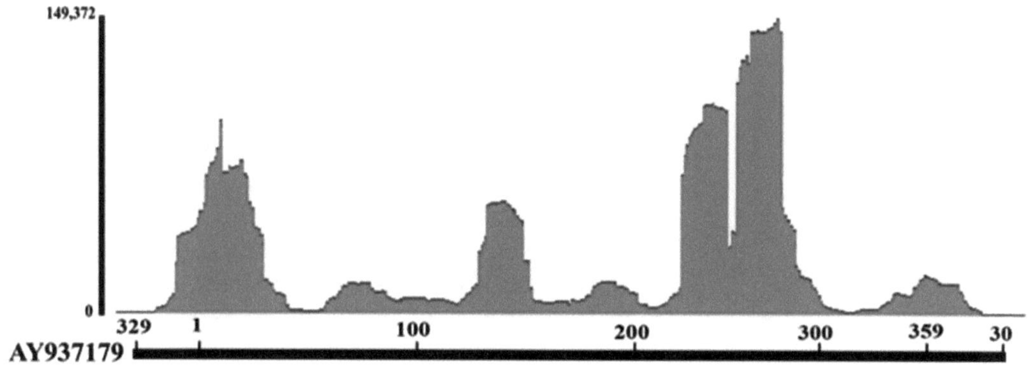

Fig. 6 Mapping profile of PSTVd-specific small RNA reads on PSTVd-intermediate genome (AY937179). *Horizontal line* indicates the modified PSTVd reference sequence. *Numbers* indicate genomic positions from 329–1–359–30. *Vertical axis* indicates the number of reads

3.4.6 *In Silico Analysis of Host Gene Expression Based on Deep Sequencing Data: (1) Analysis of Host MicroRNA Expression and Detection of Viroid-Specific Small RNAs Potentially Targeting Host MicroRNA in Viroid-Infected Plants*

MicroRNAs (miRNA) play a vital role in the plant life cycle. Recent studies have shown that relative expression levels of certain host miRNAs can be changed in viroid-infected plant [24, 38]. Hence, it is quite important to examine sequence data sets for potential differences in miRNA expression levels in healthy and viroid-infected host plants. Furthermore, it is also of interest to determine whether or not any of the viroid-specific small RNA can target mature miRNA or stem-loop sequence of miRNA precursors. In this section we describe an *in silico* analysis of PSTVd-specific small RNAs that has identified such an RNA.

In the previous section we explained how to trim the adapters and how to segregate small RNAs based on length. Once this is done, the small RNAs need to be subjected to "Extract and count".

1. Toolbox/Transcriptomics Analysis/Small RNA Analysis/ Extract and Count: Follow the instructions in dialogue box and extract the small RNA. As part of the extraction process we can specifically select only 21-nt small RNA from the total set of 21–24-nt viroid-specific small RNAs. Once the samples are extracted, save the file.
2. Download miRBase: In order to make use of the additional information about mature regions on the precursor miRNAs in miRBase, we need to use another tool integrated into the CLC Genomics Workbanch to download miRBase rather than downloading it from http://www.mirbase.org/:
3. Toolbox/Transcriptomics Analysis/Small RNA Analysis/ Download miRBase: This will download a list of sequences that includes precursor miRNAs together with annotations for mature regions. This list can then be selected when annotating the samples.
4. Creating own miRBase file: To construct a file to be used as a miRBase file for annotation, the file must be formatted in the same way as the miRBase data file. In particular, the following need to be in place:
   (a) The sequences need "miRNA" annotation on the precursor sequences. In the Workbench, we can add a miRNA annotation by selecting a region and right click to "Add Annotation". A maximum of two miRNA annotations should be add per precursor sequences. Matches to first miRNA annotation are counted in the 5′ column; matches to second miRNA annotation are counted as 3′ matches.
   (b) If the sequence list contains sequences from multiple species, the "Latin name" of the sequences should be set. This is used in the "Annotate and merge small RNA samples" dialog where we can select specific species. If the Latin name is not set, the dialog will show "N/A".

5. Once this file has been created, it has to be imported as described above (*see* **Note 25**).
6. Annotating and merging small RNA samples: The small RNA data set produced by counting the tags can be enriched using "CLC Genomics Workbench" by comparing the tag sequences with annotation resources such as miRBase and other small RNA annotation sources. Note that the annotation can also be performed on an "experiment", created from a number of small RNA data sets (see Setting up an experiment).
7. Besides adding annotations to known small RNAs in the sample, it is also possible to merge variants of the same small RNA to produce a cumulative count. When initially counting the tags, the Workbench requires that the trimmed reads are identical for them to be counted as the same tag. However, we will often see different variants of the same miRNA in a sample, and it is useful to be able to count these together. This is also possible using the "annotate and merge" tool.
8. Toolbox/Transcriptomics Analysis/Small RNA Analysis/ Annotate and Merge Counts: This will open a dialog box where we select the "small RNA samples" to be annotated. Note that if we have included several samples, they will be processed separately but summarized in one report providing a good overview of all samples. You can also input "Experiments" (see Setting up an experiment) created from small RNA samples. Click "Next" when the data is listed in the right-hand side of the dialog.
9. After each round of mapping, the tags that are mapped will be removed from the list of tags that continue to the next round. This means that a tag mapping with perfect match in the first round will not be considered for the subsequent one-mismatch round of mapping. Following the mapping, the tags are classified into the following categories according to where they match; namely, Mature 5′ exact, Mature 5′ super, Mature 5′ sub, Mature 5′ sub/super, Mature 3′ exact, Mature 3′ super, Mature 3′ sub, Mature 3′ sub/super, Precursor, and Other.
10. All categories except "Other" refer to hits in miRBase. For hits on mirBase sequences we distinguish between where on the sequences the tags match. The mirBase sequences may have up to two mature micro RNAs annotated. We refer to a mature miRNA that is located closer (or equally close) to the 5′-end than to the 3′-end as "Mature 5"". A mature miRNA that is located closer to the 3′-end is referred to as "Mature 3′". Exact means that the tag matches exactly to the annotated mature 5′- or 3′-region; "Sub" means that the observed tag is shorter than the annotated mature 5′ or mature 3′; "super" means that the observed tag is longer than the annotated mature 5′ or mature 3′.

Table 1
Total tomato small RNA counts annotated to miRNA in healthy and PSTVd-infected tomato leaf by using CLC genomic workbench

Name	Total small RNA reads	Annotated to miRNA	Percentage (%)	Ambiguously annotated to miRNA	Percentage (%)*
Small RNA from healthy leaf	2,345,032	2,762	0.1	384	0
Small RNA from PSTVd-infected leaf	2,187,096	2,556	0.1	337	0

*ambiguously annotated to miRNA

Table 2
Detection of miRNA-related sequences among PSTVd-specific small RNA using CLC genomic workbench

Name	PSTVd-sRNAs	Annotated to miRNA	Percentage (%)	Ambiguously annotated to miRNA	Percentage (%)*
PSTVd-specific small RNA (PSTVd-sRNAs)	27,906	2	0	1	0

*ambiguously annotated to miRNA

The combination "sub/super" means that the observed tag extends the annotation in one end and is shorter at the other end. "Precursor" means that the tag matches a mirBase sequence, but outside the annotated mature region(s). The "Other" category is for hits in the other resources (the information about resource is also shown in the output). The "Other" category is for hits elsewhere on mirBase sequences (that is, outside any annotated mature regions) or hits in the other resources (information about resource is also shown in the output).

11. The miRNA analysis shown below was performed on small RNA data obtained from healthy and PSTVd-infected tomato plants.
12. Note that the miRNA counts for healthy and PSTVd-infected tomato samples are very similar, i.e., 2,762 in healthy versus 2,556 in PSTVd-infected sample (Table 1).
13. Next, all 27,906 PSTVd-specific small RNAs 21-nt in length and in the PSTVd-infected sample were examined for the presence of sequences similar to those of tomato miRNAs. Two PSTVd-specific small RNAs which exhibit the similarity with tomato miRNA were identified (Table 2).

14. To check the specificity and identify a possible target miRNA in tomato, BLAST analysis was performed on the two PSTVd-specific small RNAs in miRBase http://www.mirbase.org/search.shtml. One of the PSTVd-specific small RNAs having the sequence 5′-GGAAACAACTGAAGCTCCCGA-3′ can bind to miR167b as shown below.

```
                 1                     19
PSTVd sRNA        GGAAACAACUGAAGCUCCC
                  || ||||| ||||||| ||
ath-MIR167b       GGGAACAAGUGAAGCUGCC
                 1                     19
```

3.4.7 In Silico Analysis of Host Gene Expression Based on Deep Sequencing Data: (2) Use of Plant Genomic cDNA Against Viroid-Specific Small RNA to Predict Possible Target Genes

Identification of host genes whose expression is affected by viroid infection/replication and, thus, are potentially responsible for the appearance of viroid-specific disease symptoms such as stunting and epinasty, is a major interest for several viroid research groups. Because the tomato genome contains more than 33,000 genes, *in silico* analysis has many advantages for this purpose. Every step in the analysis must be carried out as explained above in "Mapping against viroid genome" except for the choice of reference sequence. Results from an analysis carried out on PSTVd-infected tomato plants are presented below.

1. Down-load tomato genome data from NCBI and export the tomato genome coding sequences (CDs) to CLC work-bench.
2. Toolbox/NGS Core Tools/Map Reads to Reference: In the dialog box, select the sequences or sequence lists containing the sequencing data. Preferably, this is a data set containing PSTVd-specific small RNAs selected by after mapping against PSTVd genome.
3. Then Click "Next" and select the "reference" sequence against which to map. Here, we need to select all coding sequences (CDs) in the tomato genome if we are interested in looking for PSTVd-specific small RNA targets in whole genome. If the focus is on particular gene, then select only those CDs as reference.
4. Follow steps as explained above in "Mapping against viroid genome" and wait for the analysis to finish (*see* **Note 26**).
5. For example, when "Tomato cDNA contig, Solyc02g065770.1.1 which codes for COBRA-like (LOC101251844), mRNA (GenBank Acc. No. NM_001278954)" is aligned with PSTVd-specific small RNAs, the nucleotides 533–553 of Solyc02g 065770.1.1 exhibited a marked similarity to several PSTVd-specific small RNAs (Fig. 7).

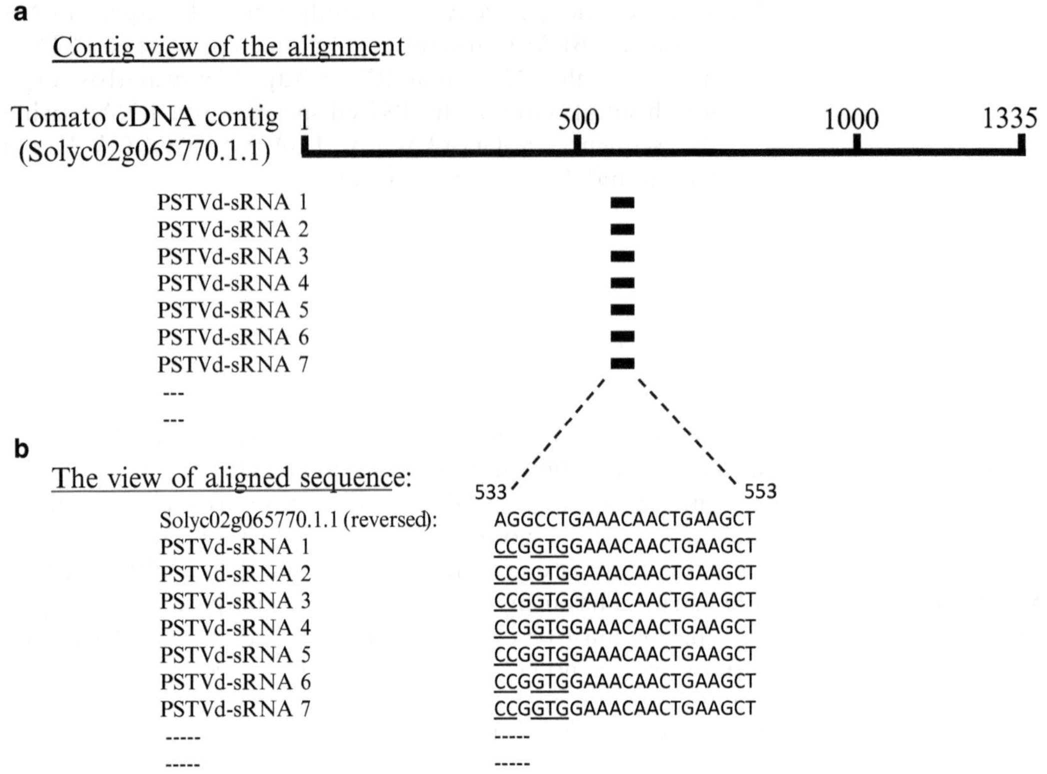

Fig. 7 Alignment of PSTVd-specific small RNA with tomato cDNA contig (Solyc02g065770.1.1). *Underlined letters* indicate mismatches

3.5 Detection of Known Viroid Species by Deep Sequencing Analysis: Detection of Partial Known Sequence, Primer Design, and RT-PCR Amplification of Complete Sequence

Databases and software required for this analysis are: (1) Sequence reads lacking their 5′ and 3′ adapters, (2) Viroid reference sequences which can be downloaded from NCBI Genome Resources: http://www.ncbi.nlm.nih.gov/genomes/GenomesGroup.cgi?taxid=12884&opt=Viroid, (3) Host genome sequence. If the genome of the plant species being analyzed for the presence of viroids has been sequenced, the sequence can be recovered from either the NCBI database or special Web sites cited in published papers, and (4) the Velvet program for assembly of small RNA sequences [39].

1. Identify small RNAs derived from highly abundant host RNA species (e.g., rRNA and tRNA) by aligning the trimmed reads against the host reference genome [34, 40].
2. Assembly of the subtracted reads with Velvet using a k-mer value of 17 to generate contigs (*see* **Note 27**).
3. Identify sequences of known viroids by BLAST analysis.
4. Based on the viroid sequence(s) detected, design one or more pairs of adjacent primers of opposite polarity to amplify the full-length viroid cDNA.

5. Prepare RNA templates for reverse transcription. Add the following reagents to a sterile, RNase-free, 200 μl microcentrifuge tube.

Total RNA extract	2 μl
RT-primer (20 μM)	1 μl
10 mM each dNTPs	2 μl
Total volume	5 μl

6. Incubate the mixture at 65 °C on a thermal cycler for 10 min. Place the tube on ice and centrifuge briefly to collect contents.
7. Using the following individual reagent volumes per sample assemble a reverse transcription master mix. Add 10 % extra volume of each reagent.

RNase-free distilled water	7 μl
0.1 M DTT	2 μl
5× first strand buffer	4 μl
RNasin	1 μl
M-MuLV reverse transcriptase (200 U/μl)	1 μl
Total volume	15 μl

8. Add 15 μl master mix to the cooled tube containing the annealed primer-RNA templates. Mix and incubate at 37 °C for 1 h. Heat 3 min at 70 °C to inactivate reverse transcriptase, chill on ice for 2 min, and briefly centrifuge to collect the tube contents.
9. Premix, in a separate tube, the following reagents in the order listed. Multiply each volume by the number of samples to be processed and prepare an additional 10 % reagent mixture.

RNase-free distilled water	17 μl
2× PCR Master Mix	25 μl
Reverse-primer (20 μM)	2 μl
Forward-primer (20 μM)	2 μl
Total volume:	46 μl

10. Add 46 μl of prepared PCR master mix to a sterile, nuclease-free 200 μl PCR tube containing 4 μl of reverse transcription reaction mixture. Run the following PCR program; one cycle of 94 °C for 5 min, 30 cycles of 94 °C—30 s, 55 °C—30 s, and 72 °C—1 min, followed by a final 7 min incubation at 72 °C (*see* **Note 28**).

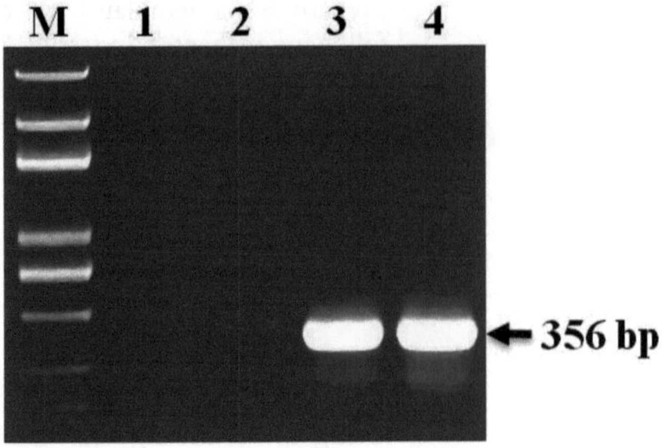

Fig. 8 RT-PCR detection of *Chrysanthemum stunt viroid* (CSVd). *M*: DNA marker, *1*: distilled water; *2*: healthy plant; *3*: unknown sample; *4*: positive control

11. Load 10 μl aliquots of the PCR reactions and 5 μl aliquots of the 100 bp DNA ladder (Promega) on a 2 % agarose gel. Run the gel for about 30 min at 100 V. Stain the gel with ethidium bromide for 5–10 min and view the gel on a UV transilluminator to detect cDNA amplification products of the predicted size (Fig. 8).
12. Excise a gel slice containing the cDNA band of the predicted size under UV illumination using a clean scalpel blade. Purify the cDNA using AxyPrep™ DNA Gel Extraction Kit according to the manufacturer's instruction.
13. Ligate the purified cDNA to pGEM®-T Vector using the following reaction mixture:

2× ligation buffer	5 μl
pGEM-T vector	1 μl
PCR product	2 μl
T4 DNA ligase	1 μl
DEPC-treated water	1 μl
Total volume	10 μl

14. Incubate the mixture at room temperature for 2 h or overnight at 4 °C (to increase the yield of recombinant plasmid).
15. Add 2 μl aliquot of the ligation reaction to 50 μl of *E. coli* DH5α competent cells. Incubate on ice for 30 min and then heat-shock at 42 °C for 45 s. Chill on ice for 3–5 min, add 1 ml of LB, and incubate cells for 1 h at 37 °C. Spread 100 μl aliquot of the culture on an LB-Agar plate containing 50 μg/mL ampicillin and X-gal/IPTG, then incubate the plate overnight at 37 °C.

16. Randomly select white colonies using a micropipette tip, resuspend the bacterial cells in 1 ml of LB, and incubate in a shaker at 37 °C for 3–5 h. Remove a 1 μl aliquot for PCR analysis using M13 forward and M13 reverse primers. The PCR reaction is carried out in a 20 μl volume containing 10 μl of 2× PCR Master Mix, 0.5 μl each of M13 forward and M13 reverse primers (20 μM) using the following PCR program; one cycle of 94 °C for 5 min, 25 cycles of 94 °C—30 s, 55 °C—30 s, and 72 °C—30 s, and incubation for 7 min at 72 °C.
17. Analyze 5 μl aliquots of the PCR reaction of each picked clone on a 1.5 % agarose–1× TAE gel. Confirm that the PCR product has the expected size and select positive clones.
18. Use the remaining LB cultures of positive clones for sequence analysis (*see* **Note 29**).

3.6 Detection of New Viroid and Viroid-Like Circular RNA Species by Deep Sequencing Analysis

The databases and software required for the analysis are: (1) Clean reads that have been removed 5′ and 3′ adapters, (2) Host genome sequence information. If the genome sequence of the host species used for the analysis is available, we can find the database for NCBI or special Web site supplied in the published papers, and (4) PFOR program [41].

1. Run the PFOR program with default k-mer = 17.
2. Change variant k-mers (14–17) to obtain more contigs.
3. BLAST analysis of obtained contigs with nucleotide databases of GenBank (*see* **Note 30**).
4. Separate contigs of 200–500-nt length, and validate new circular RNAs from these contigs (*see* **Note 31**).
5. Calculate the secondary structure of the possible new viroids using Mfold [42].
6. Search for conserved motifs such as the central conserved region (CCR) for family *Pospiviroidae* or hammerhead ribozyme for family *Avsunviroidae.*
7. Design a pair of adjacent primers of opposite polarity to validate circularity of the predicted new viroids or viroid-like satellite RNAs using RT-PCR (Fig. 9) (*see* **Note 32**).

4 Notes

1. Low molecular RNA at a concentration of ca.100 ng/μl is normally appropriate. Several protocols and reagents commercially available can be used for RNA extraction; for more details, *see* Owens et al. [25].
2. When an appropriate restriction enzyme recognition site does not exist in the upper strand of central conserved region, unique restriction enzyme recognition sites locating elsewhere

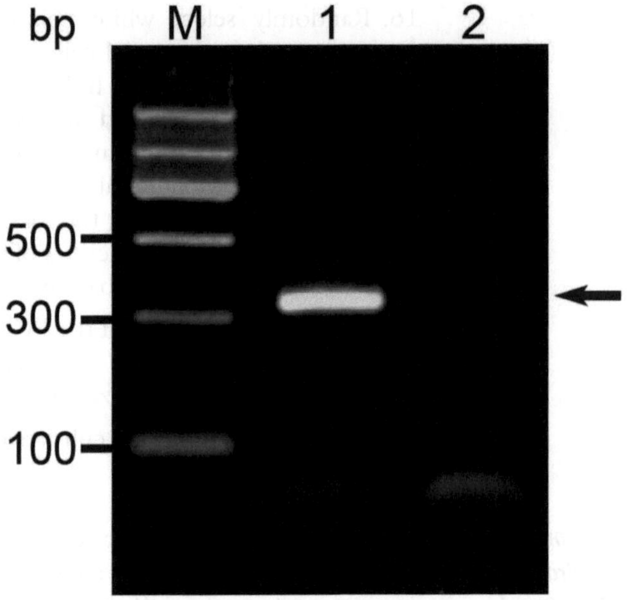

Fig. 9 Validation of circularity of a new viroid-like RNA from grapevine by RT-PCR. *M*: DNA marker; *lane 1*: grapevine sample; *lane 2*: water control

in the molecule can be utilized. In cases where an appropriate unique restriction enzyme recognition site is not present, it is necessary to use another method such as In-Fusion cloning technology; for details see In-Fusion HD cloning kit user manual (Clone Laboratories, Inc. TaKaRa Bio Company; www.clontech.com).

3. Alternatively, after self-ligation the resulting dimeric molecules can be first separated by electrophoresis using polyacrylamide or agarose before ligation to dephosphorylated plasmid DNA.
4. After colony selection, analysis by nucleotide sequencing should be performed to ensure that the dimeric HpSVd cDNA selected does not contain any unwanted mutations. Sequence analysis will also provide information on the orientation of the insert, information which is essential to prepare the transcripts with desired orientation, i.e., plus or minus strand.
5. We normally add purified bentonite (final concentration = 1 mg/ml) to the inoculum in order to inhibit *RNase* activity. Commercially available *RNase* inhibitors can also be used.

Incubation conditions are very important, because viroids in general replicate more actively and show clearer symptoms under higher temperature (e.g., 25–30 °C) and stronger light conditions (>20,000 lx).

For herbaceous indicator plants such as tomato (*Solanum lycopersicom*) or cucumber (*Cucumis sativus*) for infection assay, it is desirable to inoculate the fully expanded cotyledons.

When we infect a perennial woody species, it is desirable to inoculate actively growing young seedlings or shoots by razor slashing, i.e., put 10–20 μl inoculum on the shoot and make 30–50 shallow cuts with a razor blade through the droplet to introduce the inoculum into the vascular tissue.

6. Inoculation by particle bombardment is another very useful choice. Parameters for biolistic transfer of viroid nucleic acids using a Helios Gene Gun device were examined in detail by Matoušek et al. [28]. The method is highly efficient for inoculation of linear monomeric PSTVd cDNAs and RNA transcript, requiring only 50 ng and 200 pg per tomato plant, respectively. Infection is readily achieved with exact length monomeric RNA transcripts having 5′-triphosphate and 3′-OH termini in amounts ranging from 2 to 20 ng per plant, suggesting no need for any supplementary modifications of ends or RNA circularization.
7. When total RNA is extracted using Trizol extraction protocol, the *DNase* I treatment described below is unnecessary, and the final precipitate can be dissolved in an appropriate amount (e.g., 100 μl) of distilled water.
8. *RNase*-free *DNase* I such as RQ1 *DNase* should be used for DNA digestion.
9. In experiments followed this protocol, average ca. 200 μg of low molecular weight RNAs can be obtained from 1 g of tomato leaf tissues.
10. The amount of nucleic acid to be applied to the gel is critical. When low molecular weight RNA preparations from HpSVd-infected cucumber or PSTVd-infected tomato are fractionated on a normal size gel (14 cm × 14 cm × 1 mm), 10–20 μg RNA (i.e., 10–20 μl heat-denatured sample) per well (width × depth = ca. 8 mm × 8 mm) allows reliable detection of viroid-specific small RNAs.
11. In our experience, 1 g of tomato leaf tissue yields an average of ca. 150–300 ng of small RNAs ranging in size from ca. 15–35 nucleotides.
12. It is strongly advised to follow Illumina's method strictly or, at the very least, make only minor adjustments.
13. The scale of sequencing can be adjusted to fit the requirements of each study. It is our choice how many reads of small RNA sequence will be necessary for the research purpose. To analyze the profile of viroid-specific small RNAs in plants infected with known viroid species, ~2 million reads for herbaceous plants [33] and ~5 million reads for woody species [19] are sufficient. In the case of PSTVd-infected tomato (cv. Rutgers) plants, ~21 million reads derived from total small RNA contained ~2 million

(~10 %) PSTVd-specific reads [23]. An experimental strategy known as called "Tagged", "Multiplex", or "Bar-code" sequencing is essential to compare the small RNA profiles in different samples [23]. If the aim is to detect known or unknown viroids or viroid-like RNAs in the samples collected in the field, a higher number of small RNA reads may be required. However, there is no established criterion for the number of reads required, because the concentration of viroid small RNAs varies greatly with the viroid–host combination. It is safe to say that higher resolution requires more small RNA reads. Higher resolution means higher cost, however, and the following examples provide useful guidance. For analysis of herbaceous host such as tomatoes grown in the fields, 5–7 million reads will be suitable [34]. In case of woody species such as peach [35] or grapevine [21] growing in the field, more than 15 million reads may need to be examined. Furthermore, when multiple samples are to be pooled for analysis by deep sequencing [36], more reads are required than for a single sample.

14. Bowtie, an ultrafast, memory-efficient short read aligner, is a popular tool for alignment of small RNAs [37].

15. Although these programs to analyze our data can easily be run without any knowledge of how they work, it is enormously beneficial to understand the theory of ligation and trimming adapters and what the reads file contain. First-time users are strongly recommended to read the description available at http://tucf-genomics.tufts.edu/documents/protocols/TUCF_Understanding_Illumina_TruSeq_Adapters.pdf. If you have any questions about adapter trimming, you can visit the Web site: http://seqanswers.com/, and ask for help.

16. It is necessary to obtain, by cloning and sequencing, information on possible viroid sequence variants that can emerge from the inoculum source during replication in a particular host. If sequence variation is detected, one must consider which sequence variant(s) should be used for alignment. Moreover, the circularity of the viroid genome also needs to be taken into consideration during alignment. There are many protocols that work fine with these statistics, but the one covering this circular nature of viroid genome has not yet available. We need to make minor adjustments when we use these programs (*see* Subheading 3.4.5).

17. We can see from this output that >33 % of reads contained some form of adapter. Also, > 17 % of the trimmed reads were shorter than 15-bp and therefore discarded. The results are contained in a file named SRR026762_adaprm.fastq.

18. Generally, companies offering next generation sequencing services remove adapter sequences before returning sequence data to the investigator. Hence, the next several steps may not be relevant.

19. To trim the adapters, it is first necessary to create an adapter list to be supplied to the trim tool in this step: File > New > Trim Adapter List
20. From two examples described here, it is quite clear that commercial software is more user-friendly for biologists lacking a background in computational biology or computer programming. Hence, hereafter we describe small RNA profiling using CLC genomic workbench.
21. In this step we can select two references, one for the plus strand and a second sequence as minus strand. Thus, we can map small RNAs to both the plus and minus strand at the same time.
22. If we want the reads to have for example at least 90 % identity with the reference sequence in order to be included in the final mapping, set this value to 0.9. Note that the similarity fraction does not apply to the whole read; it relates to the "Length fraction".
23. Here we have the option to collect the data about un-mapped sequences. These are sometimes helpful if we are looking for differences between samples from healthy and infected tissue.
24. To check the sequence of any region, just select that region and right-click to extract its sequence.
25. In this chapter, the directory was constructed from the plant miRNA database, because we were going to analyze the changes in miRNA expression level in tomato.
26. Because the reference sequences are longer, this analysis takes more time than "Mapping against viroid genome".
27. It is always necessary to try different parameters in order to obtain more and longer contigs.
28. Prepare a 2 % agarose–1× TAE buffer gel while the PCR reaction is running.
29. Even after trying many different assembly parameters with the Velvet program, no contigs with homology to known viroids or only a few, very short contigs may be obtained. As a possible alternative strategy, either the subtracted reads or the trimmed reads can be directly aligned against all known viroid reference genomes. One or two mismatches are allowed.
30. Most of the assembly contigs are from host genome and ribosome. If the host genome is available, BLAST analysis can remove most of the "contaminants".
31. Separation of 200–500-nt length contigs can save many following works of validation, when we obtained hundreds of contigs.
32. The circularity of contigs can be confirmed by RT-PCR using two adjacent primers of opposite polarity. If the new potential viroid-like RNA can be folded into rod-like or branched

rod-like secondary structure typical of viroids in the families *Pospiviroidae* and *Avsunviroidae* it can be considered as a possible new viroid species. Further direct evidence can be obtained by PAGE, northern-blot analysis, and finally bioassay to show the autonomous replication in a certain host species. The methods described in Subheadings 3.1 and 3.2 of this chapter are useful for this purpose.

Acknowledgment

We thank Dr. Robert A Owens (USDA/ARS, MPPL, USA) for critical reading and valuable suggestions on the manuscript.

References

1. Diener TO (1979) Viroids and viroid diseases. Wiley, New York, 252 pages
2. Flores R, Hernandez C, de Alba AEM, Daros JA, Di Serio F (2005) Viroids and viroid-host interactions. Annu Rev Phytopathol 43:117–139
3. Ding B (2009) The biology of viroid-host interactions. Annu Rev Phytopathol 47:105–131
4. Owens RA, Hammond RW (2009) Viroid pathogenicity: one process, many faces. Viruses 1:298–316
5. Gago S, Elena SF, Flores R, Sanjuan R (2009) Extremely high mutation rate of a hammerhead viroid. Science 323:1308
6. Elena SF, Gomez G, Daros JA (2009) Evolutionary constraints to viroid evolution. Viruses 1:241–254
7. Cress DE, Kiefer MC, Owens RA (1983) Construction of infectious potato spindle tuber viroid cDNA clones. Nucleic Acids Res 11: 6821–6835
8. Meshi T, Ishikawa M, Ohno T, Okada Y, Sano T, Ueda I, Shikata E (1984) Double-stranded cDNAs of hop stunt viroid are infectious. J Biochem 95:1521–1524
9. Tabler M, Sänger HL (1985) Infectivity studies on different potato spindle tuber viroid (PSTV) RNAs synthesized in vitro with the SP6 transcription system. EMBO J 4:2191–2199
10. Visvader JE, Forster AC, Symons RH (1985) Infectivity and in vitro mutagenesis of monomeric cDNA clones of citrus exocortis viroid indicates the site of processing of viroid precursors. Nucleic Acids Res 13:5843–5856
11. Diener TO (1986) Viroid processing: a model involving the central conserved region and hairpin I. Proc Natl Acad Sci U S A 83:58–62
12. Feldstein PA, Hu Y, Owens RA (1998) Precisely full length, circularizable, complementary RNA: an infectious form of potato spindle tuber viroid. Proc Natl Acad Sci U S A 95:6560–6565
13. Ding SW, Voinnet O (2007) Antiviral immunity directed by small RNAs. Cell 130:413–426
14. Itaya A, Folimonov A, Matsuda Y, Nelson RS, Ding B (2001) Potato spindle tuber viroid as inducer of RNA silencing in infected tomato. Mol Plant Microbe Interact 14:1332–1334
15. Papaefthimiou I, Hamilton AJ, Denti MA, Baulcombe DC, Tsagris M, Tabler M (2001) Replicating potato spindle tuber viroid RNA is accompanied by short RNA fragments that are characteristic of post-transcriptional gene silencing. Nucleic Acids Res 29:2395–2400
16. Itaya A, Zhong X-H, Bundschuh R, Qi Y-J, Wang Y, Takeda R, Harris AR, Molina C, Nelson RS, Ding B (2007) A structured viroid RNA serves as a substrate for dicer-like cleavage to produce biologically active small RNAs but is resistant to RNA-induced silencing complex-mediated degradation. J Virol 81:2980–2994
17. Wang MB, Bian XY, Wu LM, Liu LX, Smith NA, Isenegger D, Wu RM, Masuta C, Vance VB, Watson JM, Rezaian A, Dennis ES, Waterhouse PM (2004) On the role of RNA silencing in the pathogenicity and evolution of viroids and viral satellites. Proc Natl Acad Sci U S A 101:3275–3280
18. Navarro B, Gisel A, Rodio ME, Delgado S, Flores R, Di Serio F (2012) Small RNAs containing the pathogenic determinant of a chloroplast-replicating viroid guide degradation of a host mRNA as predicted by RNA silencing. Plant J 70:991–1003

19. Di Serio F, Gisel A, Navarro B, Delgado S, de Alba AEM, Donvito G et al (2009) Deep sequencing of the small RNAs derived from two symptomatic variants of a chloroplastic viroid:implications for their genesis and for pathogenesis. PLoS One 4(10):e7539. doi: 10.1371/journal.pone.0007539
20. St-Pierre PI, Hassen F, Thompson D, Perreault JP (2009) Characterization of the siRNAs associated with peach latent mosaic viroid infection. Virology 383:178–182
21. Navarro B, Pantaleo V, Gisel A, Moxon S, Dalmay T, Bisztray G et al (2009) Deep sequencing of viroid-derived small RNAs from grapevine provides new insights on the role of RNA silencing in plant-viroid interaction. PLoS One 4:e7686
22. Martinez G, Donaire L, Llave C, Pallas V, Gomez G (2010) High-throughput sequencing of *Hop stunt viroid*-derived small RNAs from cucumber leaves and phloem. Mol Plant Pathol 11:347–359
23. Wang Y, Shibuya M, Taneda A, Kurauchi T, Senda M, Owens RA, Sano T (2011) Accumulation of Potato spindle tuber viroid-specific small RNAs is accompanied by specific changes in gene expression in two tomato cultivars. Virology 413:72–83
24. Owens RA, Tech KB, Shao JY, Sano T, Baker CJ (2012) Global analysis of tomato gene expression during potato spindle tuber viroid infection reveals a complex array of changes affecting hormone signaling. Mol Plant Microbe Interact 25:582–598
25. Owens RA, Sano T, Duran-Vila N (2012) Plant viroids: isolation, characterization/detection, and analysis. Methods Mol Biol 894:253–271
26. Matoušek J, Orctová L, Steger L, Riesner D (2004) Biolistic inoculation of plants with viroid nucleic acids. J Virol Methods 122:153–164
27. Sumbrook J, Russell DW (2001) Preparation of plasmid DNA by alkaline lysis with SDS. In: Molecular cloning. A laboratory manual, vol 1, 3rd edn. Cold Spring Harbor Laboratory Press, Cold Spring Harbor, pp 1.31–1.34
28. Sumbrook J, Russell DW (2001) The Hanahan method for preparation and transformation of competent *E. coli*: high-efficiency transformation. In: Molecular cloning. A laboratory manual, vol 1, 3rd edn. Cold Spring Harbor Laboratory Press, Cold Spring Harbor, pp 1.105–1.111
29. Doyle JJ, Doyle JL (1987) A rapid DNA isolation procedure for small quantities of flesh leaf tissue. Phytochem Bull 19:11–15
30. Curtin S, Zsögön A, Watson J, Waterhouse P (2012) Isolation and analysis of small RNAs from virus-infected plants. In: Watson JM, Wang M-B (eds) Antiviral resistance in plants, vol 894, Methods in molecular biology. Humana Press, New York, pp 173–189
31. Smith N, Eamens A (2012) Isolation and detection of small RNAs from plant tissues. In: Watson JM, Wang M-B (eds) Methods in molecular biology, vol 894. Humana Press, New York, pp 155–172
32. Zhu Q-H, Helliwell C (2012) Generation of plant small RNA cDNA libraries for high-throughput sequencing. In: Watson JM, Wang M-B (eds) Antiviral resistance in plants, vol 894, Methods in molecular biology. Humana Press, New York, pp 123–137
33. Di Serio F, de Alba AEM, Navarro B, Gisel A, Flores R (2010) RNA-dependent RNA polymerase 6 delays accumulation and precludes meristem invasion of a viroid that replicates in the nucleus. J Virol 84:2477–2489
34. Li R, Gao S, Hernandez AG, Wechter WP, Fei Z, Ling KS (2012) Deep sequencing of small RNAs in tomato for virus and viroid identification and strain differentiation. PLoS One 7:e37127
35. Bolduc F, Hoareau C, St-Pierre P, Perreault JP (2010) In-depth sequencing of the siRNAs associated with peach latent mosaic viroid infection. BMC Mol Biol 11:16
36. Coetzee B, Freeborough MJ, Maree HJ, Celton JM, Rees DJG, Burger JT (2010) Deep sequencing analysis of viruses infecting grapevines: virome of a vineyard. Virology 400: 157–163
37. Langmead B, Trapnell C, Pop M, Salzberg SL (2009) Ultrafast and memory-efficient alignment of short DNA sequences to the human genome. Genome Biol 10:R25. doi:10.1186/gb-2009-10-3-r25
38. Diermann N, Matoušek J, Junge M, Riesner D, Steger G (2010) Characterization of plant miRNAs and small RNAs derived from potato spindle tuber viroid (PSTVd) in infected tomato. Biol Chem 391:1370–1390
39. Zerbino DR, Birney E (2008) Velvet: algorithms for de novo short read assembly using de Bruijn graphs. Genome Res 18:821–829
40. Isakov O, Modai S, Shomron N (2011) Pathogen detection using short-RNA deep sequencing subtraction and assembly. Bioinformatics 27: 2027–2030
41. Wu QF, Wang Y, Cao MJ, Pantaleo V, Burgyan J, Li W-X et al (2012) Homology-independent discovery of replicating pathogenic circular RNAs by deep sequencing and a new computational algorithm. Proc Natl Acad Sci U S A 109:3938–3943
42. Zuker M (2003) Mfold web server for nucleic acid folding and hybridization prediction. Nucleic Acids Res 31:3406–3415

Web Links

http://www.ark-genomics.org/events-online-training-eu-training-course/adapter-and-quality-trimming-illumina-data
http://bioinf.scri.ac.uk/tablet
http://www.clcbio.com/products/clc-genomics-workbench
http://www.geneious.com
http://www.mirbase.org

Chapter 13

Detection and Characterization of Mycoviruses in Arbuscular Mycorrhizal Fungi by Deep-Sequencing

Tatsuhiro Ezawa, Yoji Ikeda, Hanako Shimura, and Chikara Masuta

Abstract

Fungal viruses (mycoviruses) often have a significant impact not only on phenotypic expression of the host fungus but also on higher order biological interactions, e.g., conferring plant stress tolerance via an endophytic host fungus. Arbuscular mycorrhizal (AM) fungi in the phylum Glomeromycota associate with most land plants and supply mineral nutrients to the host plants. So far, little information about mycoviruses has been obtained in the fungi due to their obligate biotrophic nature. Here we provide a technical breakthrough, "two-step strategy" in combination with deep-sequencing, for virological study in AM fungi; dsRNA is first extracted and sequenced using material obtained from highly productive open pot culture, and then the presence of viruses is verified using pure material produced in the in vitro monoxenic culture. This approach enabled us to demonstrate the presence of several viruses for the first time from a glomeromycotan fungus.

Key words Arbuscular mycorrhizal fungi, Deep-sequencing, dsRNA, Mycoviruses, Obligate biotroph

1 Introduction

Fungal viruses (mycoviruses) have been found in a variety of fungi, including the ascomycetes, basidiomycetes, and deuteromycetes. Their genomes are composed of double-stranded RNA (dsRNA) in most cases and unexceptionally possess an RNA-dependent RNA polymerase (RdRp) gene in the genomes [1]. Infections of mycoviruses are asymptomatic in many cases, but often have a significant impact not only on phenotypic expression of the host fungus but also on higher order biological interactions, e.g., attenuation of virulence of a plant pathogenic fungus [2] and conferring plant thermal tolerance via an endophytic host fungus [3].

Arbuscular mycorrhizal (AM) fungi that belong to the phylum Glomeromycota associate with most land plants and supply mineral nutrients, in particular phosphorus, to the host plants through extensive hyphal networks constructed in the soil [4]. The AM

Ichiro Uyeda and Chikara Masuta (eds.), *Plant Virology Protocols*, Methods in Molecular Biology, vol. 1236, DOI 10.1007/978-1-4939-1743-3_13, association with plant roots occurred more than 400 million years ago, and the coincidence of the appearances of early land plants and AM associations suggests that the plant–AM fungal symbiosis were instrumental in the colonization toward land for primitive plants [5]. Although distribution of mycoviruses in AM fungi and their impacts on the plant–fungal interactions are of interest, little information has been obtained due to their obligate biotrophic nature.

One breakthrough for virological study in AM fungi is the establishment of a highly productive open pot culture system for AM fungal mycelia [6, 7]. In this system about 1 g of mycelia could be obtained from 70 to 80 plants grown in growth chambers, which is insufficient for purification of viral particles, but may be sufficient for detection and amplification of viral genomes for sequencing. Purity of fungal material, however, is best in the in vitro monoxenic culture [8], although fungal growth is generally poorer than in open culture. To overcome these problems, the "two-step strategy" in combination with the next-generation sequencing technology has been developed recently [9]; dsRNA is first extracted from fungal material produced in the open pot culture and subjected to deep-sequencing, and the presence of viral genomes is subsequently verified by RT-PCR using dsRNA extracted from material produced in the in vitro monoxenic culture. This approach enabled us to demonstrate for the first time that glomeromycotan fungi harbor diverse mycoviruses and, further, that a new class of virus found in an AM fungus has a significant impact on spore productivity of the host fungus [9].

2 Materials

2.1 Mass Production of Fungal Material in Open Pot Culture

1. Fungal isolates established from either single or multiple spores (*see* **Note 1**).
2. Plant host. Small plants that grow well under low-light conditions (e.g., fluorescent light) are suitable. Dwarf marigold (*Tagetes patula*) and *Lotus japonicus* are ideal plants for mass production of hyphal material in a growth chamber (*see* **Note 2**).
3. Washed river sand, autoclaved. A mixture of fine (<1 mm in diam) and coarse (≥1 mm) particles at ratios between 30:70 and 70:30 (v:v), preferably 50:50, is required (*see* **Note 3**).
4. Plastic pots with drainage holes (6 cm in diam, 120 ml in vol).
5. Cone-shaped nylon-mesh bags for separation of hyphal and root compartments (Fig. 1). Pore-size of the nylon mesh is 37 μm that is small enough to prevent roots from passing, but large enough to allow AM fungal hyphae to pass through. Cut a mesh sheet in a semicircular shape (7 cm in radius for 6-cm pot), fold in the middle, and seal the margin by a heat-sealer (about 26 ml in vol).

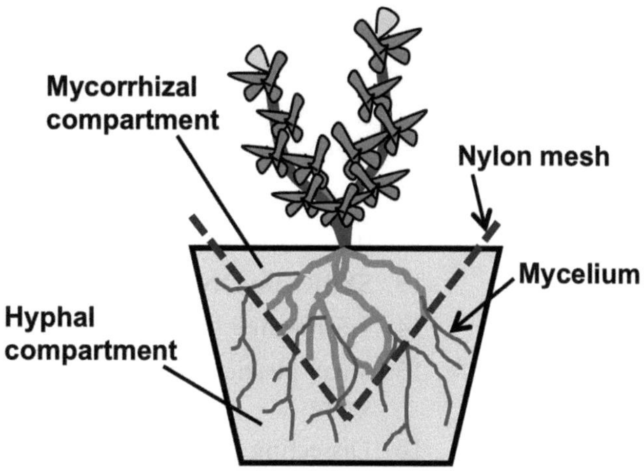

Fig. 1 Mesh bag-separated open pot culture system. The *dotted line* represents the 37-μm nylon mesh bag that separates the mycorrhizal and hyphal compartments. The pore size of the mesh is small enough to prevent roots from passing through, but large enough to allow passage of AM fungal hyphae

6. Low-P nutrient solution: 4 mM NH_4NO_3, 1 mM K_2SO_4, 75 μM $MgSO_4$, 2 mM $CaCl_2$, 50 μM Fe-EDTA, and 50 μM KH_2PO_4. The working solution may be made from concentrated (×100–1,000) stock solutions for each salt. Microelements may not be necessary in sand culture.

2.2 In Vitro Monoxenic Culture

1. Minimal medium [8] solidified with 0.3 % gellan gum (w/v) in petri dishes.
2. Ri T-DNA transformed root organ culture. Carrot or chicory hairy roots have widely been used (*see* **Note 4**).
3. Chloramine-T solution: 20 g/l chloramine-T.
4. Antibiotic solution: 0.2 g/l streptomycin and 0.1 g/l gentamycin.
5. Melting solution: 0.5 M EDTA.

2.3 dsRNA Extraction and Electrophoresis

1. Multi-beads shocker (vibration grinder).
2. 3-ml O-ring sealed plastic tubes fitted to the grinder, liquid nitrogen-resistant.
3. Metal cone fitted to the tube.
4. Extraction buffer: 100 mM Tris–HCl at pH 8.0, 200 mM NaCl, 2 mM EDTA, 0.1 % sodium dodecyl sulfate (SDS) (w/v), and 0.1 % 2-mercaptoethanol (v/v).
5. TE (100 mM Tris–HCl at pH 8.0, 2 mM EDTA).
6. PCI (TE-saturated phenol–chloroform–isoamyl alcohol = 25:24:1).

7. CIA (chloroform–isoamyl alcohol = 24:1).
8. Absolute ethanol.
9. 3 M sodium acetate (pH 5.2) for ethanol precipitation.
10. DNase I (RNase-free) and S1 nuclease.
11. Autoclaved deionized water.
12. 0.8 % agarose gel for electrophoresis.
13. SYBR safe DNA gel stain for dsRNA staining.
14. 1.5-ml tube (for extraction from spores).
15. Small plastic pestle fitted to 1.5-ml tube (for extraction from spores).
16. Dissecting microscope (for extraction from spores).

2.4 cDNA Amplification and Deep-Sequencing

1. Dialysis tube, 25 kDa cutoff.
2. First strand cDNA synthesis kit.
3. Primers for cDNA amplification [3]: 6N-anchor primer, CCTGAATTCGGATCCTCC-NNNNNN; anchor primer, CCTGAATTCGGATCCTCC.
4. *Taq* polymerase.
5. Thermal cycler.
6. DNA cleanup spin columns.
7. Next-generation sequencer (*see* **Note 5**).

2.5 Rapid Amplification of cDNA End (RACE)

1. 5′/3′ RACE kit, including appropriate primers for cDNA synthesis and amplification.
2. First, second, and third specific primers designed based on the 5′- and 3′-end sequences of target dsRNA.
3. TA-cloning vector.
4. Ligase and buffer.
5. *E. coli* JM109 competent cell.
6. Medium for blue-white selection: LB medium with 50 mg/l ampicillin solidified by 1.5 % agar to which 20 μl of 100 μM isopropylthio-β-D-galactoside and 35 μl of 50 mg/ml 5-bromo-4-chloro-3-indolyl-β-D-garactoside are applied before use.
7. Sequence primers: U-19mer primer (GTTTTCCCAGTCACGACGT) and T7 promoter primer (CTAATACGACTCACTATAGGG).
8. Dideoxy-cycle sequencing kit.
9. Capillary sequencer.

2.6 Sequence Analysis and Phylogeny

1. BLAST program (http://www.ncbi.nlm.nih.gov/).
2. Pfam database (http://pfam.sanger.ac.uk/).
3. MEGA 5 software for alignment and phylogeny (http://www.megasoftware.net/) [10].

2.7 Detection of Mycoviruses from In Vitro Culture

1. A primer for first strand cDNA synthesis (virus-specific or 6N random primer).
2. Virus-specific primer pairs for the first and second (nested) PCR.

2.8 Establishment of Single Spore-Culture Lines for Virus Elimination

1. *L. japonicus* or marigold as host plant.
2. Washed river sand, autoclaved.
3. Small (6 cm in diam, 120 ml in vol) and medium (9 cm in diam, 350 ml in vol) plastic pots with a drainage hole.
4. Low-P nutrient solution.
5. Dissecting microscope

3 Methods

A key step for successful detection of dsRNA in AM fungi is mass production of fungal material; at least 0.5 g of fresh mycelia is necessary for detection of major dsRNA segments in electrophoresis. Representative results are shown in Fig. 2. Deep-sequencing is much more powerful than electrophoresis for detection of dsRNA, and, in fact, we unexpectedly found a viral segment from a fungus in which no dsRNA was observed in electrophoresis during transcriptome analysis by the Illumina RNA-seq (Kikuchi and Ezawa, unpublished observation). It is also true, however, that deep-sequencing is still costly at present, and thus one might not apply the method unless dsRNA segments are detected in electrophoresis. Therefore, electrophoresis is the first key analysis in the virological study of the fungi.

Among factors that influence productivity of mycelium, fungal species is most critical; some species are highly productive in sand culture, but some are not. Water regimes also critically affect mycelial

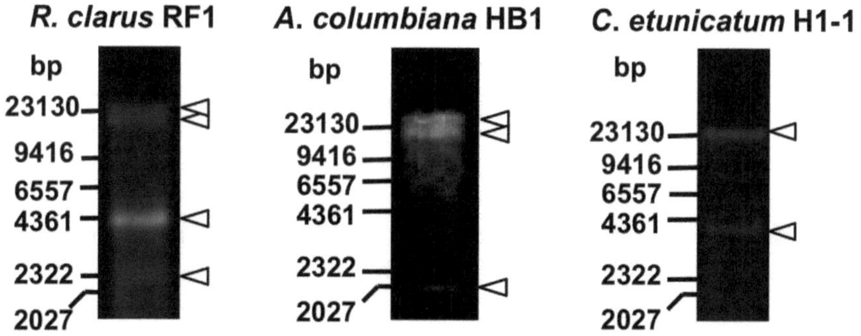

Fig. 2 Electrophoretic analysis of dsRNA fractions prepared from extraradical mycelia of *Rhizophagus clarus* strain RF1, *Acaulospora colombiana* strain HB1, and *Claroideoglomus etunicatum* strain H1-1. Total nucleic acid was extracted from 0.5 to 1.0 g mycelia by the SDS-phenol method, digested by DNase I and S1 nuclease, and subjected to electrophoresis. *Arrowheads* indicate dsRNA segments

yield. Tips for optimizing the growth conditions of the fungi are suggested in this section. After successful detection and sequencing of viral segments, it is highly recommended to validate their presence in the fungus grown in vitro as long as the segments are detected in the material obtained from open culture. RT-(nested) PCR using viral sequence-specific primers is sensitive enough to detect viral segments from a small amount of fungal material.

3.1 Mass Production of Fungal Material in Open Pot Culture

1. About 100 mL sand is put into the pots, watered lightly, and the surface is shaped to fit to the cone-shaped mesh bag. After fitting the mesh bag, the bag is filled with sand (*see* **Note 6**). Plant seeds are sowed in the middle of the bag, inoculated with AM fungal spores (e.g., 500–1,000 spores per pot for *Rhizophagus* spp.), and grown for 6–8 weeks (*see* **Note 7**) in a growth chamber at a photosynthetic photon flux of 150 μmol/m^2 s (20–25 °C; 16-h day length; 60–70 % relative humidity).
2. A sufficient amount of the low-P nutrient solution is given until the solution flowed out from the drainage at appropriate intervals (*see* **Note 8**). Accumulated salts in the medium may be washed out with deionized water once a week.
3. At harvest, the mycorrhizal compartment in which the plants are growing is removed together with the mesh bag, and mycelia in the hyphal compartment are collected on a stainless steel mesh (50 μm opening) by wet sieving as quickly as possible, blotted on a paper towel, transferred into a plastic tube, frozen in liquid nitrogen, and stored at −80 °C. Three to five pots could be processed together, which will save time and also improve recovery of mycelia. In the case of *Rhizophagus clarus* strain RF1 (=*Glomus* sp. strain RF1), 10–30 mg FW mycelium per pot could be obtained under the conditions.

3.2 In Vitro Monoxenic Culture

1. Spores are surface sterilized with the chloramine-T solution for 10 min, followed by the antibiotic solution for 10 min, and then washed with deionized water.
2. Ri T-DNA transformed roots maintained on the Minimal medium are inoculated at five spores per plate and incubated in the dark at 27 °C for 10 weeks.
3. Spores produced in the medium are collected on a 53-μm stainless mesh after melting the gel by shaking in the melting solution and subjected to dsRNA extraction.

3.3 dsRNA Extraction and Electrophoresis

1. The frozen mycelia (0.5–1.0 g) are ground in the presence of liquid nitrogen in the 3-mL tube with the metal cone using the Multi-beads shocker at 3 × 15 s at 2,000 rpm, suspended in 4 ml of the extraction buffer, extracted twice with an equal volume of PCI and once with CIA, precipitated with 2

volumes of ethanol and 1/10 volume of 3 M sodium acetate, and then dissolved in 50–100 μl of TE.

2. In the case of dsRNA extraction from spores, approx. 200 spores are crushed in 100 μl of the extraction buffer in a 1.5-ml tube with the small pestle under a dissecting microscope and then the slurry is subjected to PCI/CAI extractions.
3. To digest genomic DNA and single-stranded RNA, the fraction is incubated twice with 0.4 U/μl DNase I and 3.4 U/μl S1 nuclease at 37 °C overnight, precipitated by ethanol after PCI/CAI extractions, dissolved in 25 μl deionized water.
4. The dsRNA fraction is subject to electrophoresis and visualized with the SYBR Safe.

3.4 cDNA Amplification and Deep-Sequencing

1. dsRNA segments are eluted from the gel pieces in the dialysis tube, digested again by overnight incubation with DNase I and S1 nuclease, extracted with PCI/CAI, precipitated with ethanol, and then dissolved in 25 μl deionized water.
2. First strand cDNA is synthesized with 2 ng dsRNA and 800 ng 6N-anchored primer in a final volume of 20 μl according to the instruction of the kit.
3. The cDNA is randomly amplified with *Taq* polymerase using 1 μl cDNA solution and 5 μM anchor primer in a final volume of 15 μl with the following thermal cycle program: initial denaturation at 94 °C for 2 min, followed by 40 cycles of 94 °C for 15 s, 55 °C for 30 s, and 72 °C for 2 min, and final extension at 72 °C for 10 min. The PCR product is purified by the DNA cleanup spin column.
4. The amplified cDNA is sequenced with a next-generation sequencer and assembled with appropriate software (*see* **Note 9**).

3.5 RACE

1. The extreme end of the target dsRNA is reverse-transcribed with the first specific primer at 55–60 °C and poly(A)-tailed according to the instruction of the RACE kit.
2. The poly(A)-tailed cDNA is amplified with *Taq* polymerase using the oligo(dT)-anchor primer in combination with the second specific primer with the following program: initial denaturation at 94 °C for 2 min, 30 cycles of 94 °C for 30 s, 55 °C for 30 s, and 72 °C for 1 min, and final extension at 72 °C for 10 min.
3. The first PCR product is further amplified with the anchor and the third specific primers with the same thermal cycle program.
4. The second PCR product (0.2–0.4 pmol) is ligated to the vector at 16 °C for 40 min, and *E. coli* competent cell is transformed by heat shock at 42 °C for 45 s. The transformants are incubated

on the blue-white selection medium at 37 °C over night, and the insert-positive (white) clones are selected.

5. The target insert is amplified by colony PCR using U-19mer and T7 promoter primers with the following program: initial denaturation at 94 °C for 2 min, 30 cycles of 94 °C for 30 s, 50 °C for 30 s, and 72 °C for 1 min, and final extension at 72 °C for 10 min, electrophoresed on a 1.0–1.2 % agarose gel, and visualized with SYBR Safe.
6. The amplicon is extracted from the gel and purified with the spin column. Sequence reaction is performed using 10–20 ng of the purified amplicon according to the instruction of the sequence kit, and the product is sequenced with a capillary sequencer.

3.6 Sequence Analysis and Phylogeny

1. Large (>1 kbp) contigs are subjected to BLASTn search, and those showing similarities to viral sequences are analyzed further.
2. Motif searches are performed by Position-Specific Iterated BLAST (PSI BLAST) against Pfam database. Multiple sequence alignments and phylogenetic analysis with the neighbor-joining or maximum-likelihood method are implemented by MEGA5.

3.7 Detection of Mycoviruses from In Vitro Culture

1. First strand cDNA is synthesized using dsRNA prepared from the in vitro-produced spores using the virus-specific or 6N-random primer.
2. The cDNA is amplified by nested PCR using the first and second PCR primers with appropriate thermal cycle programs.
3. The final PCR product is purified, cloned, and sequenced by a capillary sequencer.

3.8 Establishment of Single Spore-Culture Lines for Virus Elimination

1. Washed river sand is put into a 6-cm pot, and seedlings of *L. japonicus* (two plants per pot) are grown for a few days in the growth chamber.
2. Spores obtained either by open pot culture or by in vitro culture are collected under a dissecting microscope.
3. A small pit (1–2 cm depth) is made in the vicinity of the roots with a plastic tip that fits to a 200-μl pipet ("yellow tip"), and a single spore is put and covered with sand.
4. The seedlings are grown for 6–7 weeks with the low-P nutrient solution in a growth chamber (first multiplication).
5. The *L. japonicus* seedlings are non-destructively transferred to a 9-cm pot with the medium (*see* **Note 10**), and then the pot is filled with sand. Marigold is sown around the seedlings (two to three seeds per pot) and further grown with the low-P

nutrient solution for 8–10 weeks in a greenhouse (second multiplication). Spores are harvested by wet sieving (*see* **Note 11**). The presence or absence of viral segments is examined by RT-PCR using a dsRNA fraction prepared from the spores.

4 Notes

1. Viruses may occasionally be lost during single spore isolation.
2. C_3 plants may be more suitable for fungal propagation in a growth chamber, i.e., under low-light conditions. Generally productivity of C_4 plants is greater than that of C_3 plants in a greenhouse (with sunlight), but not in a growth chamber.
3. Coarse sand provides aerobic conditions that are essential for hyphal growth of AM fungi.
4. The AM fungus-free root organ culture could be purchased from Glomeromycota In vitro Collection (http://www.agr.gc.ca/eng/?id=1236786816381).
5. Many companies provide sequencing and subsequent assembling service using next-generation sequencers such as Roche 454 GS Titanium and Illumina HiSeq.
6. For species that do not fit to sand culture and require a certain soil for successful colonization, a (autoclaved) soil-sand mixture at a ratio of 1:9 (v:v) may be used in the mycorrhizal compartment (within the mesh bag). The medium in the hyphal compartment from which mycelia are collected, however, should be 100 % sand for maximum RNA yield.
7. Appropriate culturing period will vary among fungal species. Some species grow rapidly, but senesce rapidly (e.g., *Rhizophagus irregularis*), for which a shorter culturing period may result in higher yield of active mycelia. Whereas some species senesce slowly (e.g., *Rhizophagus clarus* and *Acaulospora colombiana*), for which maximum yield of active mycelia may be expected 7–8 weeks after sowing.
8. The plants should be grown with minimum water (nutrient solution), which is essential to maximize mycelial yield. Too much water creates anaerobic conditions under which growth of the aerobic fungi is suboptimal.
9. The scale of sequencing depends on number of viruses in the fractions, their genome size, and amount of contaminant RNA, e.g., rRNA. Prior to deep-sequencing, we strongly recommend to construct a clone library of the randomly amplified cDNA for sequencing of at least 10–20 clones by a conventional sequencer, from which frequency of viral cDNA can be estimated.

10. The plants and medium should carefully be taken out from the pot in a non-destructive manner before transplanting. Fungal biomass is usually very small at the end of first multiplication, and thus the second multiplication is essential before examining spore production.
11. Single spore culture may usually be established from 20 to 50 % of the pots in the case of *Rhizophagus* spp.

References

1. Ghabrial SA, Suzuki N (2009) Viruses of plant pathogenic fungi. Annu Rev Phytopathol 47:353–384
2. Choi GH, Nuss DL (1992) Hypovirulence of chestnut blight fungus conferred by an infectious viral cDNA. Science 257:800–803
3. Márquez LM, Redman RS, Rodriguez RJ, Roossinck MJ (2007) A virus in a fungus in a plant: three-way symbiosis required for thermal tolerance. Science 315:513–515
4. Smith SE, Read DJ (2008) Mycorrhizal symbiosis. Academic, London
5. Simon L, Bousquet J, Levesque R, Lalonde M (1993) Origin and diversification of endomycorrhizal fungi and coincidence with vascular land plants. Nature 363:67–69
6. Hijikata N, Murase M, Tani C, Ohtomo R, Osaki M, Ezawa T (2010) Polyphosphate has a central role in the rapid and massive accumulation of phosphorus in extraradical mycelium of an arbuscular mycorrhizal fungus. New Phytol 186:285–289
7. Tani C, Ohtomo R, Osaki M, Kuga Y, Ezawa T (2009) Polyphosphate-synthesizing activity in extraradical hyphae of an arbuscular mycorrhizal fungus: ATP-dependent but proton gradient-independent synthesis. Appl Environ Microbiol 75:7044–7050
8. Bécard G, Fortin JA (1988) Early events of vesicular-arbuscular mycorrhiza formation on Ri T-DNA transformed roots. New Phytol 108:211–218
9. Ikeda Y, Shimura H, Kitahara R, Masuta C, Ezawa T (2012) A novel virus-like double-stranded RNA in an obligate biotroph arbuscular mycorrhizal fungus: a hidden player in mycorrhizal symbiosis. Mol Plant Microbe Interact 25:1005–1012
10. Tamura K, Peterson D, Peterson N, Stecher G, Nei M, Kumar S (2011) Mega5: molecular evolutionary genetics analysis using maximum likelihood, evolutionary distance, and maximum parsimony methods. Mol Biol Evol 28:2731–2739

Chapter 14

SuperSAGE as an Analytical Tool for Host and Viral Gene Expression

Hideo Matsumura, Detlev H. Krüger, Günter Kahl, and Ryohei Terauchi

Abstract

SuperSAGE is a tag-based transcript profiling method, which allows to analyze the expression of thousands of genes at a time. In SuperSAGE, 26 bp tags are extracted from cDNA using the type III restriction enzyme, EcoP15I. In SuperSAGE, the amount of transcripts was represented by tag counts. Taking advantage of uniqueness of the 26 bp tags, host and virus transcripts can be monitored in virus-infected cells. Combining next generation sequencing technology, we established High-throughput SuperSAGE (Ht-SuperSAGE), which allows the analysis of multiple samples with reduced time and cost. In this chapter, we present the protocol of Ht-SuperSAGE involving a recently available benchtop type next generation sequencer.

Key words SuperSAGE, Next generation sequencing, Transcriptome, Host-pathogen interaction, Gene expression

1 Introduction

Large-scale transcript (transcriptome) analysis is widely employed in various biological studies. Microarrays [1] have occupied the prime position as the standard transcriptome analysis tool for several decades. Recently, emerging next generation sequencing (NGS) technologies [2] were introduced, which compared to the various array platforms, represent "open architecture" technologies, so that researchers can conduct the scale of expression analysis (number of genes to be analyzed) independently of known genome or cDNA sequences. As a sequencing-based transcriptome analysis tool, serial analysis of gene expression (SAGE, [3]) has been one of several tag-based techniques in the era before NGS technologies appeared. In SAGE or its improved version SuperSAGE [4–7], a short "tag" fragment was extracted from a defined position of each transcript (cDNA). By sequencing these tags, quantitative transcript profiles could be obtained from thousands of genes as tag sequences and their counted numbers. Particularly, in SuperSAGE,

Ichiro Uyeda and Chikara Masuta (eds.), *Plant Virology Protocols*, Methods in Molecular Biology, vol. 1236,
DOI 10.1007/978-1-4939-1743-3_14, © Springer Science+Business Media New York 2015

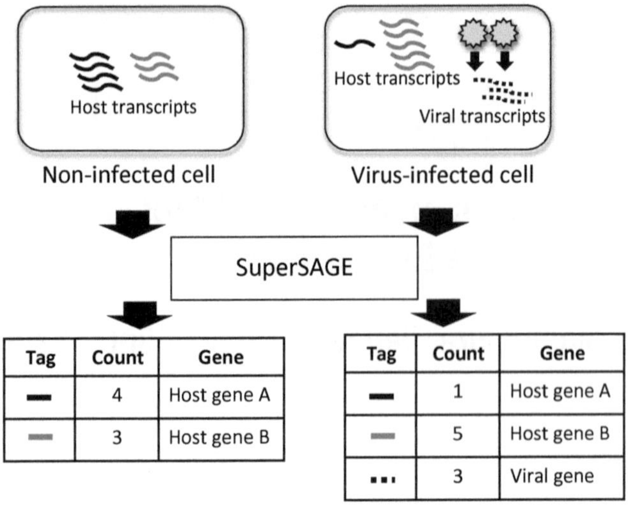

Fig. 1 Scheme of host and viral transcript analysis using SuperSAGE. By comparing tags and their counts between non-infected and viral-infected cells, both differentially expressed host genes and viral transcripts can be identified simultaneously

a 26-bp tag was extracted from each transcript using the typeIII restriction enzyme EcoP15I [4, 5]. Owing to the uniqueness of 26-bp tags, transcripts from host and pathogen could be discriminated, and gene expression of both life organisms could be simultaneously monitored in the infected cells (Fig. 1) [4, 7]. The simultaneous transcriptome analysis of host cells and parasites was also demonstrated in cytomegalovirus-infected human dendritic cells [8] and Pepper mild mottle virus-infected *Capsicum* plants [9]. As an application of SuperSAGE data, 26 bp tag sequences are directly applicable to construction of oligonucleotide microarrays for expression analysis as SuperSAGE array [10]. Also, longer cDNA fragments can be easily recovered by 3′RACE (Rapid Amplification of cDNA Ends) method using complementary primers to the SuperSAGE tag sequences even in non-model organisms [4, 10, 11].

Since SuperSAGE was highly compatible with current NGS technologies based on massively parallel sequencing, we developed the high-throughput (Ht) SuperSAGE method (also called deep-SuperSAGE), which allows the analysis of millions of tags with comparably little time and effort [12]. This improvement made expansion of SuperSAGE application as shown in several works [13–18]. As an alternative NGS application for transcript analysis, RNA-seq emerged as a tool to quantify transcript abundance by counting [19]. Additionally, it was advantageous that whole mRNA sequence information could be obtained in each transcript. However, Ht-SuperSAGE was and is superior to RNA-Seq and its variants (as for example dRNA-Seq [20] and dual RNA-Seq [21])

Table 1 Oligonucleotide sequences of IL-adapter-1 and IL-adapter-2

Adapter oligo	Oligonucleotide sequences (5′ → 3′)[a]
IL-adapter-1 sense	ACAGGTTCAGAGTTCTACAGTCCGACGATCXXXX
IL-adapter-1 antisense	NNYYYYGATCGTCGGACTGTAGAACTCTGAACCTGT-amino
IL-adapter-2-sense	CAAGCAGAAGACGGCATACGAGATAACGATGTACGCAGCAGCATG
IL-adapter-2-antisense	CTGCTGCGTACATCGTTATCTCGTATGCCGTCTTCTGCTTG -amino

[a]XXXX encodes 4-base variable index sequences and should be complementary to YYYY, *see* **Note 1**

as far as quantitativity and simplicity of analysis is concerned, because all the transcripts are represented by unique tags with equal size, and sequence assembly process is not necessary for compiling the data. Furthermore, employment of the indexing (barcoding) system allowed to sequence tags from multiple different samples in a single sequencing run, leading to an increase of analytical throughput and a reduction of analytical cost per sample [7]. Therefore, Ht-SuperSAGE recommends itself for monitoring expression kinetics of host and viral transcripts in virological studies. We here detail a protocol for Ht-SuperSAGE that was successfully employed to decipher the transcriptomes of host cells and an invading virus.

2 Materials

2.1 Adapter Preparation

1. Adapter oligonucleotides: adapter oligonucleotide synthesis and end-labeling were done by Operon Biotechnologies, Japan. Sequences of IL-adapter-1 and IL-adapter-2 oligonucleotides are shown in Table 1 (*see* **Notes 1** and **2**). These oligonucleotides were purified by an Oligonucleotide Purification Cartridge (OPC) (*see* **Note 3**).
2. LoTE buffer: 3 mM Tris–HCl, pH 7.5, 0.2 mM EDTA.
3. Polynucleotide kinase buffer (10×): 0.5 M Tris–HCl, pH 8.0, 0.1 M $MgCl_2$, 50 mM DTT.

2.2 cDNA Synthesis

1. First strand buffer (5×): 250 mM Tris–HCl, pH 8.0, 375 mM KCl, 15 mM $MgCl_2$.
2. Biotinylated adapter-oligo (dT) primer: Synthesized biotin-labeled oligonucleotides of the sequence 5′-biotin-CTGATG TAGAGGTACCGGATGCCAGCAGTTTTTTTTTTTTT TTTTT-3′, HPLC-purified, Operon Biotechnologies, Japan, (*see* **Notes 4** and **5**) were dissolved in LoTE (1 μg/μl).
3. 0.1 M DTT (dithiothreitol).
4. 10 mM dNTP: 10 mM each of dATP, dTTP, dCTP, and dGTP.

5. SuperScript III reverse transcriptase.
6. Second strand buffer: 100 mM Tris–HCl (pH 6.9), 450 mM KCl, 23 mM $MgCl_2$, 0.75 mM β-NAD^+, 50 mM $(NH_4)_2SO_4$.
7. *E. coli* DNA polymerase (10 U/μl).
8. *E. coli* DNA ligase (1.2 U/μl).
9. *E. coli* RNase H (2 U/μl).
10. Binding buffer (PB buffer) in Qiaquick PCR purification kit.
11. Qiaquick spin column in Qiaquick PCR purification kit.
12. Washing buffer (PE buffer, 5×): prepare 1× solution by adding ethanol before use.

2.3 Tag Extraction

1. NlaIII (10 U/μl): Store at −70 °C.
2. CutSmart Buffer (10×): 200 mM Tris-acetate, pH 7.9, 500 mM potassium acetate, 100 mM magnesium acetate, 1 mg/μl BSA.
3. NEBuffer 3.1 (10×): 50 mM Tris–HCl, pH 7.9, 100 mM NaCl, 10 mM $MgCl_2$, 1 mg/ml BSA.
4. Streptavidin-coated magnetic beads (Dynabeads M-270 Streptavidin, *see* **Note 6**) (10 mg/ml): Store at 4 °C.
5. Siliconized microtube (1.5 ml).
6. Binding and washing buffer (B&W buffer) (2×): 10 mM Tris–HCl, pH 7.5, 1 mM EDTA, 2 M NaCl.
7. T4 DNA ligase (2,000 U/μl): Store at −20 °C.
8. T4 DNA ligase buffer (5×): 250 mM Tris–HCl, pH 7.5, 50 mM $MgCl_2$, 5 mM ATP, 50 mM DTT, 125 μg/ml BSA.
9. EcoP15I (10,000 U/μl): Store at −20 °C.
10. 10× ATP solution (1 mM).
11. Phenol–chloroform–isoamyl alcohol (25:24:1). Store at 4 °C.
12. Ammonium acetate: 10 M solution.
13. Glycogen solution (20 mg/ml).

2.4 Indexed-Adapter Ligation and PCR

1. Phusion HF 5× Buffer.
2. dNTP solution: 10 mM each of dATP, dTTP, dCTP, and dGTP.
3. $MgCl_2$ solution (50 mM, Thermo Scientific).
4. IL-adapter-1 primer: 5′-AATGATACGGCGACCACCGAGATCTACACAGGTTCAGAGTT-3′.
   IL-adapter-2 primer: 5′-CAAGCAGAAGACGGCATACGA-3′.
   These oligonucleotides are synthesized and purified by the Oligonucleotide Purification Cartridge (OPC) and are dissolved in LoTE to a final concentration of 100 pmol/μl.

5. Phusion Hot Start II DNA polymerase (2 U/μl): Store at −20 °C.
6. Acrylamide/bisacrylamide solution (40 %, 19:1): Store at 4 °C.
7. *N*, *N*, *N*, *N*′-Tetramethyl-ethylenediamine (TEMED). Store at 4 °C.
8. Ammonium persulfate: prepare 10 % solution in sterilized water and store at 4 °C.
9. 6× loading dye: 30 %(v/v) glycerol, 0.25 % (w/v) bromophenol blue, and 0.25 % (w/v) xylene cyanol.
10. SYBR green solution: Original SYBR green stock solution was diluted 10,000 times with 1× TAE buffer. Store at 4 °C.
11. 20 bp DNA marker ladder (200 ng/μl).

2.5 Purification of PCR Product

1. ERC buffer in MinElute Reaction Cleanup kit.
2. MinElute spin column in MinElute Reaction Cleanup kit.
3. Spin-X column.

2.6 Multiplexing DNA Samples for Sequencing Analysis

1. Agilent 2100 Bioanalyzer.
2. Agilent High Sensitivity DNA kit.

2.7 Sequence Data Analysis

1. CLC Genomics Workbench version 6.5.1.

2.8 3′-RACE Analysis

1. PCR Buffer (10×): 200 mM Tris–HCl, pH 8.4, 500 mM KCl.
2. 25 mM $MgCl_2$.
3. Adapter primer: 5′-GGCCACGCGTCGACTAGTACTTTT TTTTTTTTTTTTT-3′.
4. 0.1 M DTT (dithiothreitol).
5. 10 mM dNTP: 10 mM each of dATP, dTTP, dCTP, and dGTP.
6. SuperScript II reverse transcriptase.
7. *E. coli* RNaseH (2 U/μl).
8. Abridged Universal Amplification Primer: 5′-GGCCACGCG TCGACTAGTAC-3′.
9. Taq DNA Polymerase (2 U/μl).

3 Method

NGS technologies are now widely available in many research institutes or commercial sequencing services, and their throughputs are increasing constantly. On the other hand, benchtop type NGS sequencers, like MiSeq (Illmina), IonPGM (Life Tech), or GS-Junior (Roche), are now available for sequencing small-size genomes, PCR amplicons or transcripts. These instruments also employ massively parallel sequencing, but their analytical scales are less (1–25 million reads per run) than those of conventional next generation sequencers (more than one billion reads per run). For Ht-SuperSAGE, one million tags per sample are usually sufficient for identifying differentially expressed genes and even rare transcripts, normally expected to be identified by the analysis of at least ten million tags. Therefore, Ht-SuperSAGE was a highly compatible application with these compact scale sequencers. Previously, we presented a Ht-SuperSAGE protocol for the Illumina Genome Analyzer [22], but it was not directly applicable to the Illumina benchtop type sequencer MiSeq, due to differences of immobilized oligonucleotides on its flowcell. Here, we describe an improved protocol for all the Illumina sequencers. We also show a procedure for sequence data processing, including 26-bp tag extraction from sequence reads, using the commercially available software CLC Genomics Workbench. Additionally, 3′RACE protocol is described for recovery of corresponding cDNA to the SuperSAGE tag. It can help isolating novel expressed genes independent of EST or genome sequences.

3.1 Adapter Preparation

1. Dissolve adapter oligonucleotides in LoTE buffer (100 pmol/μl). The variety of different indexed IL-adapter-1 sense depends on the number of independent samples to be multiplexed in a single run of sequencing (*see* **Note 7**).
2. Combine 10 μl of each complementary oligonucleotide solution (sense and antisense in each adapter oligonucleotide) in a tube. Add 3 μl 10× polynucleotide kinase buffer and 7 μl LoTE.
3. A total of 30 μl mixture is denatured by incubating at 95 °C for 2 min and cooled down to 20 °C for annealing complementary oligonucleotides. The annealed double-stranded DNA is designated "adapter" (like IL-adapter-1 or IL-adapter-2).

3.2 cDNA Synthesis

1. The synthesis of double-stranded cDNA followed the protocol described in the SuperScriptIII double strand cDNA synthesis kit (Invitrogen). Total RNA (2–10 μg) is dissolved in 11 μl DEPC-treated water and incubated at 70 °C for 10 min after adding 1 μl biotinylated adapter-oligo dT primer (100 pmol). Denatured RNA solution is immediately placed on ice, and 4 μl 5× First Strand buffer, 2 μl 0.1 M DTT, 1 μl 10 mM dNTP, and 1 μl SuperScriptIII reverse transcriptase are added for first strand cDNA synthesis. The reaction mixture is incubated at 45 °C for 1 h.
2. For second-strand cDNA synthesis, 30 μl 5× Second Strand buffer, 91 μl sterile water, 3 μl 10 mM dNTP, 4 μl *E. coli* DNA polymerase, 1 μl *E. coli* RNase H, and 1 μl *E. coli* DNA ligase are added to 20 μl first-strand cDNA solution, and mixed. Incubate at 16 °C for 2 h.

3. For purification of the synthesized double-stranded cDNA, 750 μl PB buffer from Qiaquick PCR purification kit (Qiagen) was added, and the mixed solution applied to a Qiaquick spin column of the same kit and centrifuged at 10,000 × *g* for 1 min. After discarding the flow-through, 750 μl washing buffer (1× PE buffer; prepared as described in Subheading 2.2) added to the column. Centrifuge at 10,000 × *g* for 1 min and discard flow-through. For completely drying, the column is centrifuged at maximum speed for 1 min. After the column is transferred to a new 1.5 ml microtube, 50 μl LoTE is added for elution. The eluate (purified cDNA) is collected by centrifugation at 10,000 × *g* for 1 min.

3.3 Tag Extraction

1. Purified double-stranded cDNA is digested with NlaIII as the anchoring enzyme (*see* **Note 8**). For digestion of cDNA with NlaIII, 20 μl CutSmart Buffer, 125 μl LoTE, 5 μl NlaIII (10 U/μl) are added to the cDNA solution, mixed, and incubated at 37 °C for 1.5 h.
2. Prepare 100 μl of a suspension of streptavidin-coated magnetic beads (Dynabeads M-270) in a siliconized 1.5 ml microtube (*see* **Note 9**). Place the tubes containing magnetic beads on a magnetic stand and remove the supernatant with a pipet. For washing the magnetic beads, 200 μl of 1× B&W solution is added and beads are suspended well by pipetting. Place the tube on a magnetic stand, and remove and discard the supernatant (*see* **Note 10**).
3. To the washed magnetic beads, 200 μl of 2× B&W solution and 200 μl of digested cDNA solution are added and suspended well by pipetting. Leave the tube for 15–20 min at room temperature with occasional mixing, so that the biotinylated cDNAs bind to streptavidin on the magnetic beads. After digested cDNAs are associated with the beads, the tube is placed on the magnetic stand, and the supernatant is discarded. Magnetic beads are washed three times with 200 μl 1× B&W and once with 200 μl LoTE.
4. IL-adapter-2 is ligated to the digested cDNAs on the beads (*see* **Note 2**). To the washed beads, 21 μl LoTE, 6 μl 5× T4 DNA ligase buffer, and 1 μl one of IL-adapter-2 solution, are added. After mixing buffer and adapter solution, the bead suspension is incubated at 50 °C for 2 min. Then the tube is cooled down at room temperature for 15 min and 2 μl T4 DNA ligase (10 U) is added. It is then incubated at 16 °C for 2 h.
5. The beads are washed four times with 1× B&W, and three times with LoTE after the ligation reaction. The beads are suspended in 75 μl LoTE.
6. For EcoP15I digestion of the fragments on the magnetic beads, 10 μl 10× NEbuffer 3, 10 μl 10× ATP solution, 1 μl

100× BSA, and 4 μl EcoP15I (10,000 U/μl) are added to the suspended magnetic beads. Incubate the tube at 37 °C for 2 h with occasional mixing (*see* **Note 11**).

7. After EcoP15I digestion, the bead suspension is placed on the magnetic stand, and the supernatant is collected into a new tube. The beads are resuspended in 100 μl 1× B&W. After separation on the magnetic stand, the supernatants are retrieved and combined to the previously collected solution (*see* **Note 12**) in each tube.
8. To the collected solution, containing adapter-tag fragments, half a volume of phenol–chloroform–isoamyl alcohol (195 μl) is added, shortly vortexed, and spun at 10,000 × *g* for a few minutes. The upper aqueous layer is transferred to a new tube (*see* **Note 13**). For ethanol precipitation, 100 μl 10 M ammonium acetate, 3 μl glycogen, and 900 μl cold ethanol are added to the collected solution (approximately 200 μl). Keep the tube at −80 °C for 1 h, and centrifuge at maximum speed for 40 min at 4 °C. The resulting pellet is washed twice with 70 % ethanol, and dried. Precipitated IL-adapter-2 ligated 26 bp-tag fragments are dissolved in 10 μl LoTE.

3.4 Indexed-Adapter Ligation and PCR

1. Prepare adapter-1 (IL-adapter-1) with defined index sequences assigned to individual samples (*see* **Note 1**). For the ligation reaction, 3 μl 5× T4 DNA ligase buffer and 0.5 μl adapter solution are added to the solution of the IL-adapter-2 ligated tags, which were released from the beads by EcoP15I digestion and purified. Incubate the tube at 50 °C for 2 min, and subsequently keep it at room temperature for 15 min. After the tubes cooled down, 1.5 μl T4 DNA ligase (7.5 U) are added, and incubated at 16 °C for 2 h (*see* **Note 14**).
2. For PCR amplification of adapter-ligated tags, a PCR reaction mixture containing 3 μl 5× Phusion HF buffer, 0.3 μl 2.5 mM dNTP, 0.1 μl 50 mM $MgCl_2$, 0.15 μl of each primer (IL-adpter-1 primer and IL-adapter-2), 10.1 μl distilled water, 1 μl ligation solution and 0.2 μl Phusion Hot Start II DNA polymerase is prepared in a tube (*see* **Note 15**).
3. PCR cycling: 98 °C for 2 min, then 10–15 cycles each at 98 °C for 30 s, and 60 °C for 30 s.
4. The size of the amplified PCR product is confirmed by polyacrylamide gel electrophoresis (PAGE, *see* **Note 16**). Prepare an 8 % PAGE gel by mixing 3.5 ml 40 % acrylamide/bisacrylamide solution, 13.5 ml distilled water, 350 μl 50× TAE buffer, 175 μl 10 % ammonium persulfate, and 15 μl TEMED. Pour the solution onto the gel plate (12 cm × 12 cm, 1 mm thickness), and insert a comb (no stacking gel).
5. Running buffer (1× TAE) is prepared and added to the upper and lower electrophoresis chambers. Then, 3 μl 6× loading dye

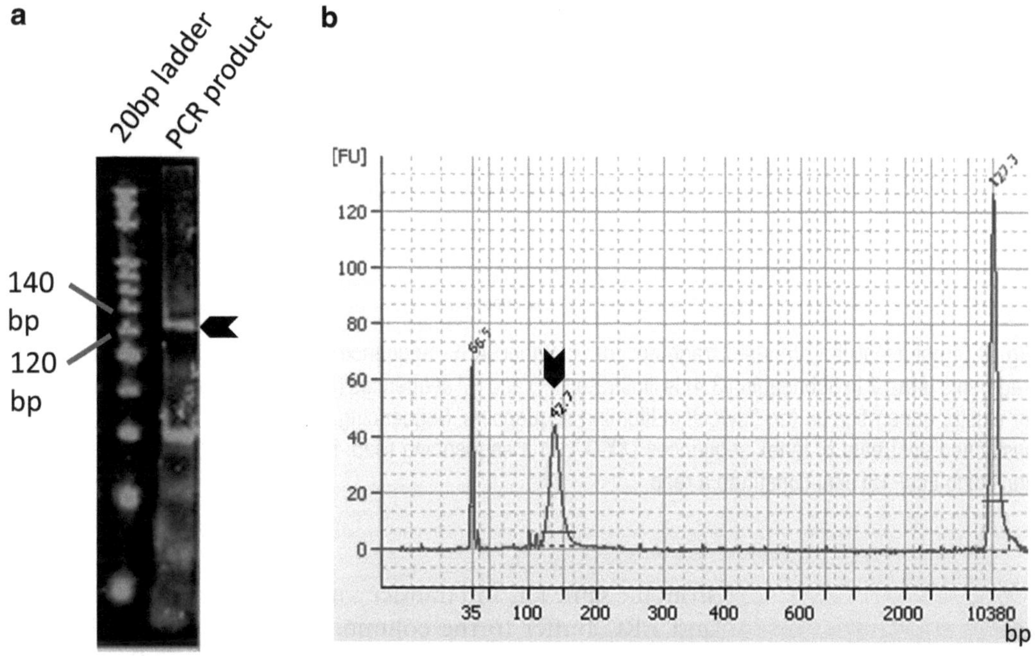

Fig. 2 Size of PCR products of adapter-ligated tag fragments. (**a**) PCR amplified products of IL-adapter-1 and IL-adapter-2 ligated tags are run on 8 % polyacrylamide gels and stained with SYBR green. A 20 bp molecular size marker is run in the *left lane* of each panel. Expected band of amplified fragments carrying the 26-bp tag is indicated by an *arrow-head* (129 bp). (**b**) Purified PCR products from the PAGE gel are run on the Agilent Bioanalyzer. The electropherogram shows peaks of detected DNA fragments. An *arrow-head* indicates the 129 bp fragment

is added to 15 μl of the PCR solution and loaded into the well. Two microliters of a 20 bp ladder is also loaded as molecular size marker. Run the gel at 75 V for 10 min, and then at 150 V for around 30 min (until the BPB dye front migrated two-third down the gel).

6. After electrophoresis, the gel is removed from the plate. Pour 1 ml SYBR green solution (diluted in 1× TAE buffer) onto the plastic wrap and place the gel on it. Further, disperse 1 ml SYBR green solution onto the gel. After a 2 min staining period, the gel is placed on a UV transilluminator. The size of the expected amplified fragment (tags sandwiched with two adapters) is 129 bp (Figs. 2a and 3).
7. After confirmation of PCR amplification of adapter-ligated tag fragments, repeat PCR reactions for the same conditions in eight tubes.

3.5 Purification of PCR Product

1. After PCR reaction, solutions from all the tubes are collected in a 1.5 ml tube and 400 μl of ERC buffer attached in MinElute Reaction Cleanup kit is added. Prepare a MinElute spin column

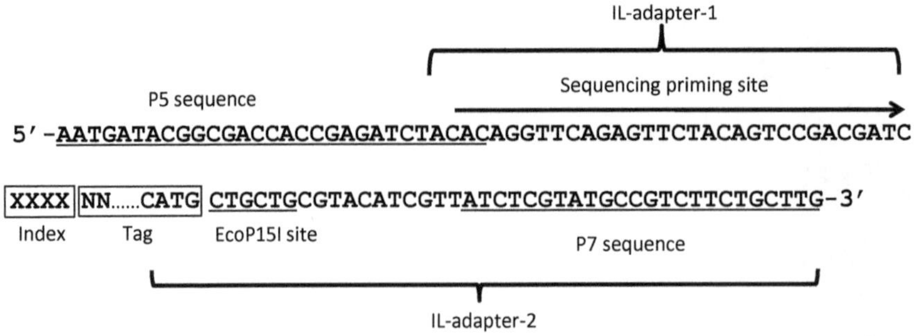

Fig. 3 Structure of the prepared fragment for Ht-SuperSAGE. Sequence structure of PCR amplified 129-bp fragments (**step 7** in Subheading 3.4) is illustrated. P5 and P7 sequences (*doubly underlined*) are priming sites for cluster generation on the flowcell of Illumina sequencers. Sequencing priming site (indicated by *arrow*) is positioned upstream of index sequences (*XXXX*). In a sequencing read, index, tag, and part of IL-adapter-2 (including EcoP15I site) constitute a read

from the same kit, and transfer a mixture between PCR solution and ERC buffer to the column. Centrifuge at 10,000 × *g* for 1 min and discard flow through. Add another 750 μl of washing buffer (PE buffer, ethanol added) to the column. Centrifuge at 10,000 × *g* for 1 min and discard flow through. For completely drying the columns, centrifuge at maximum speed for 1 min. After columns are transferred to new 1.5 ml microtubes, 15 μl LoTE is added to the column for elution. Leave the column for 1 min after adding LoTE, centrifuge at 10,000 × *g* for 1 min, and collect eluate.

2. Prepare 8 % polyacrylamide gel as described in Subheading 3.4. Add 3 μl 6× loading buffer to column-purified PCR product and load it in the well (*see* **Note 17**). After running the gel as described in Subheading 3.4, it is stained with SYBR green, and bands are visualized under UV light.
3. Only the 129 bp bands are cut out from the gel with a knife and transferred to 0.5 ml microtubes (Fig. 2a, *see* **Note 18**). Puncture holes at the top and the bottom of the tube with a needle, and place it into a 1.5 ml tube. Centrifuge the tube at maximum speed for 2–3 min. Polyacrylamide gel pieces are collected at the bottom of the tube. Add 300 μl LoTE to the gel fragments, and suspend.
4. After incubation at 37 °C for 2 h, the gel suspension is transferred to a Spin-X column, and centrifuged at maximum speed for 2 min. The eluate is once extracted by phenol–chloroform, and precipitated by adding 100 μl 10 M ammonium acetate, 3 μl glycogen, and 950 μl cold ethanol. Keep it at −80 °C for 1 h, and centrifuge at 15,000 × *g* for 40 min at 4 °C. Wash once with 70 % ethanol and dry. The resulting pellet is then dissolved in 10–15 μl LoTE.

3.6 Multiplexing DNA Samples for Sequencing Analysis

1. To confirm purity of the fragment, it is analyzed with an Agilent Bioanalyzer system (*see* **Note 19**). A DNA chip from Agilent DNA 1000 kit is prepared and filled with Gel-Dye Mix supplied with the kit. Load 1 μl purified PCR product in the well of the chip and run the chip in the Agilent 2100 Bioanalyzer.
2. After confirmation of a unique peak of the 129 bp fragments (Fig. 2b), its DNA concentration is measured with the 2100 Expert software (Agilent Technologies). Based on this quantification, an equal amount of DNA (PCR product) from each sample is combined in one single tube, and the mixture sequenced using the sequencing primer in the IL-adapter-1 (Fig. 2, *see* **Notes 1** and **20**).

3.7 Sequence Data Analysis

1. For the extraction of 26 bp-tags from the sequence reads (Fig. 3) and the estimation of tag frequency, CLC Genomics Workbench software was used.
2. First, all sequence reads were separated into independent files by the index (bar-code) using the "Multiplexing" tool in the software.
3. Demultiplexed sequence reads were converted to reverse complement sequences and adapter (IL-adapter-2) sequences were removed by the "Trimming" tool.
4. Redundant tag sequences were counted using "Extract and Count Tags" tool. The extracted tag sequence was defined to be 26 nucleotides. Consequently, a list of independent tag sequences and their counts was constructed.
5. Lists of the tags and their counts from multiple samples were compared using the "Set Up Experiment" tool for clarifying differential gene expression.

3.8 3′-RACE Analysis

1. PCR amplification primer (20–22 nucleotides) was designed such that it corresponded to part of the 26 bp SuperSAGE tag sequence (*see* **Note 21**).
2. The synthesis of single-stranded cDNA, as 3′RACE PCR template, followed the protocol described in the 3′ RACE system (Invitrogen). Total RNA (1–5 μg) was dissolved in 11 μl DEPC-treated water and incubated at 70 °C for 10 min after adding 1 μl Adapter primer (10 μM). Denatured RNA solution was immediately placed on ice, and 2 μl 10× PCR buffer, 2 μl 25 mM $MgCl_2$, 2 μl 0.1 M DTT, 1 μl 10 mM dNTP, and 1 μl SuperScriptII reverse transcriptase were added for first strand cDNA synthesis. The reaction mixture was incubated at 42 °C for 50 min.
3. Reverse transcription reaction was terminated by incubating at 70 °C for 15 min. After the tube was once placed on ice,

1 μl RNaseH (2 U/μl) was added and incubated at 37 °C for 20 min.

4. For PCR amplification of cDNA fragments, a PCR reaction mixture containing 5 μl 10× PCR buffer, 3 μl 25 mM $MgCl_2$,1 μl 10 mM dNTP, 1 μl primer for the SuperSAGE tag (10 μM), 1 μl Abridged Universal Amplification Primer (10 μM), 36.5 μl distilled water, 2 μl single-stranded cDNA, and 0.5 μl Taq DNA polymerase (2 U/μl) was prepared in a tube.
5. PCR cycling: 94 °C for 3 min, then 25–30 cycles each at 94 °C for 30 s, 55 °C for 30 s and 72 °C for 1 min.
6. PCR amplified fragments were cloned into the appropriate plasmid vector (T-vector) by conventional cloning and *E. coli* transformation. Sequences of cloned plasmids or amplified inserts were determined (*see* **Note 22**).

4 Notes

1. IL-adapter-1 contains an address site for the sequencing primer (5′-ACACAGGTTCAGAGTTCTACAGTCCGACGATC-3′), followed by an index sequence for the discrimination of different samples (Fig. 3). In these IL-adapter-1 antisense oligonucleotides, 2-bases at the 5′-end are synthesized from a mixture of deoxynucleotides (“NN”) due to ligation to various 2-base 5′-protrusion ends in EcoP15I-digested fragments.
2. IL-adapter-2 contains a recognition site for EcoP15I, and their ends should therefore be compatible with the end of fragments digested with NlaIII (5′-CATG-3′).
3. Incorrect ligation of adapters to tags was prevented by the amino-modification of the 3′-ends of antisense oligonucleotides in all the adapters.
4. Anchoring enzyme sites should not be contained in this biotinylated adapter-oligo (dT) primer.
5. An EcoP15I-recognition site (5′-CAGCAG-3′) is contained in the biotinylated adapter-oligo dT primer.
6. Dynabeads M-270 Streptavidin are paramagnetic beads with hydrophilic-surfaces, which bind nonspecific DNA fragments less strongly.
7. Different adapters are prepared in accordance with the number of samples to be analyzed. The number of tags per sample is inversely proportional to the number of multiplexed samples in a single sequencing run (e.g., one million tags per sample are expected on average, when 25 samples with different indexed adapters were multiplexed in a single sequencing run of MiSeq).

8. Here, only the protocol for NlaIII is described. However, we also have experience with three other enzymes (DpnII, BfaI, TaqI) for SuperSAGE analysis [7].
9. Employing paramagnetic beads with hydrophilic surfaces and siliconized microtubes, most of the unligated adapters can be eliminated by washing with buffers.
10. Washing of magnetic beads in other steps also follows this procedure.
11. In the IL-adapter-2 ligated cDNA fragments captured on magnetic beads, two EcoP15I sites are present. The enzyme sometimes recognizes the site adjacent to the poly-A tract, and cuts there, which leads to the release of fragments longer than adapter-tags.
12. By repeated washing of the magnetic beads, residual adapter-tag fragments can be collected.
13. Phenol–chloroform extraction in other steps also follows this procedure.
14. Tubes can be incubated for longer periods (e.g., overnight).
15. When preparing the PCR mixture, care should be taken to avoid contamination with previously amplified PCR products. Use separate pipettes and solutions, including water, from those used in the experiments after PCR. Also, use separate labware and gloves.
16. PAGE is better than agarose gel electrophoresis for a good separation of DNA fragments between 100 and 200 bp.
17. Loading too much DNA into the wells reduces the resolution of PAGE. Normally, purified PCR products from >8 reactions are loaded into two wells separately.
18. Any contamination with DNA fragments of inappropriate size (larger or smaller than 129 bp) interferes with the recovery of SuperSAGE tags from sequence data.
19. It is quite important to accurately measure the DNA amount in all samples to obtain the expected number of sequencing reads in each sample after multiplexing. Also, accurate concentration of template DNA should be determined for sequencing. For this quantification, real-time PCR is recommended recently.
20. By loading 15 pM of Ht-SuperSAGE tag fragments, >20 million high-quality reads are successfully obtained in a single run of MiSeq.
21. First 20–22 nucleotides sequence of the tag including 5′-CATG-3′ was employed as the primer (Fig. 4).
22. Corresponding cDNA to the SuperSAGE tag could be confirmed by the presence of 26 bp tag sequence in the end of amplified cDNA sequence (Fig. 4).

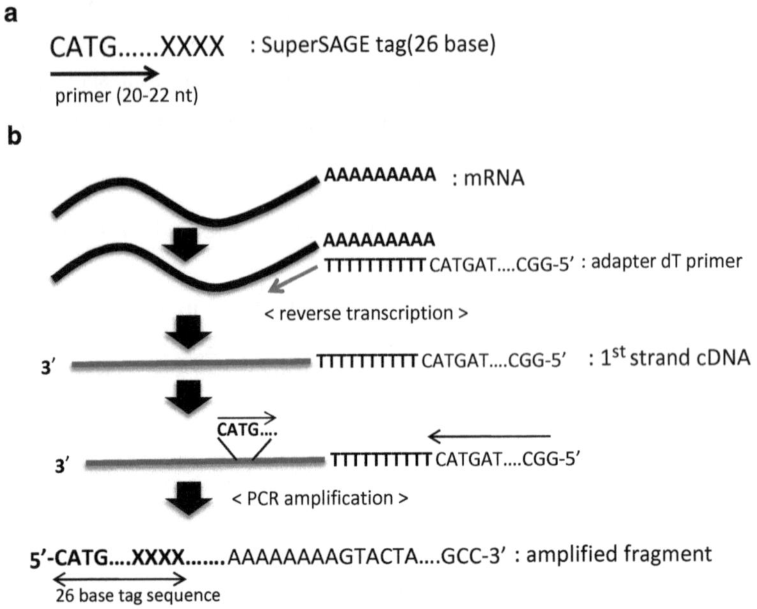

Fig. 4 Scheme of 3′RACE. (**a**) The position of the primer in the SuperSAGE tag sequence is indicated by an *arrow*. The first 20–22 nucleotides, starting from 5′-CATG, are employed as the PCR primer. (**b**) Scheme of cDNA synthesis and RACE PCR amplification is illustrated as described in Subheading 3.8. Any successfully amplified fragment should include the complete 26 base tag sequence

Acknowledgements

HM is supported by the Program for Promotion of Basic and Applied Researches for Innovations in Bio-oriented Industry (BRAIN). This work is also supported by JSPS (Japan Society for the Promotion of Science) grant no. 25450004.

References

1. Schena M, Shalon D, Davis RW, Brown PQ (1995) Quantitative monitoring of gene expression patterns with a complementary DNA microarray. Science 270:467–470
2. Metzker ML (2010) Sequencing technologies - the next generation. Nat Rev Genet 11: 31–46
3. Velculescu VE, Zhang L, Vogelstein B, Kinzler KW (1995) Serial analysis of gene expression. Science 270:484–487
4. Matsumura H, Reich S, Ito A, Saitoh H, Kamoun S, Winter P, Kahl G, Reuter M, Krüger DH, Terauchi R (2003) Gene expression analysis of host-pathogen interactions by SuperSAGE. Proc Natl Acad Sci U S A 100: 15718–15723
5. Matsumura H, Reuter M, Krüger DH, Winter P, Kahl G, Terauchi R (2007) SuperSAGE. Meth Mol Biol 387:55–70
6. Molina C, Rotter B, Horres R, Udupa SM, Besser B, Bellarmino L, Baum M, Matsumura H, Terauchi R, Kahl G, Winter P (2008) SuperSAGE: the drought stress-responsive transcriptome of chickpea roots. BMC Genomics 9:553
7. Matsumura H, Ito A, Saitoh H, Kamoun S, Winter P, Kahl G, Reuter M, Krüger D, Terauchi R (2005) SuperSAGE. Cell Microbiol 7:11–18
8. Raftery MJ, Möncke-Buchner E, Matsumura H, Giese T, Winkelmann A, Reuter M, Terauchi R, Schönrich G, Krüger DH (2009) Unravelling

the interaction of human cytomegalovirus with dendritic cells by using SuperSAGE. J Gen Virol 90:2221–2233

9. Hamada H, Matsumura H, Tomita R, Terauchi R, Suzuki K, Kobayashi K (2008) SuperSAGE revealed different classes of early resistance response genes in *Capsicum chinense* plants harboring L 3-resistance gene infected with Pepper mild mottle virus. J Gen Plant Pathol 74:313–321
10. Matsumura H, Nasir KHB, Yoshida K, Ito A, Kahl G, Krüger DH, Terauchi R (2006) SuperSAGE-array: the direct use of 26-base-pair transcript tags in oligonucleotide arrays. Nat Meth 3:469–474
11. Nasir KH, Takahashi Y, Ito A, Saitoh H, Matsumura H, Kanzaki H, Shimizu T, Ito M, Fujisawa S, Sharma PC, Ohme-Takagi M, Kamoun S, Terauchi R (2005) High-throughput in planta expression screening identifies a class II ethylene-responsive element binding factor-like protein that regulates plant cell death and non-host resistance. Plant J 43: 491–505
12. Matsumura H, Yoshida K, Luo S, Kimura E, Fujibe T, Albertyn Z, Barrero RA, Krüger DH, Kahl G, Schroth GP, Terauchi R (2010) High-throughput SuperSAGE for digital gene expression analysis of multiple samples using next generation sequencing. PLoS One 5:e1201
13. Molina C, Zaman-Allah M, Khan F, Fatnassi N, Horres R, Rotter B, Steinhauer D, Amenc L, Drevon JJ, Winter P, Kahl G (2011) The salt-responsive transcriptome of chickpea roots and nodules via deepSuperSAGE. BMC Plant Biol 11:31
14. Bonaventure G (2010) SuperSAGE analysis of the *Nicotiana attenuata* transcriptome after fatty acid-amino acid elicitation (FAC): identification of early mediators of insect responses. BMC Plant Biol 10:66
15. Sharbel TF, Voigt ML, Corral JM, Galla G, Kumlehn J, Klukas C, Schreiber F, Vogel H, Rotter B (2010) Apomictic and sexual ovules of Boechera display heterochronic global gene expression patterns. Plant Cell 22:655–671
16. Gilardoni PA, Schuck S, Jüngling R, Rotter B, Baldwin IT, Zawada AM, Rogacev KS, Rotter B, Winter P, Marell RR, Fliser D, Heine GH (2011) SuperSAGE evidence for CD14++CD16+ monocytes as a third monocyte subset. Blood 118:e50–e61
17. Kahl G, Molina C, Rotter B, Jüngling R, Frank A, Krezdorn N, Hoffmeier K, Winter P (2012) Reduced representation sequencing of plant stress transcriptomes. J Plant Biochem Biotech 21:119–127
18. Draffehn AM, Li L, Krezdorn N, Ding J, Lübeck J, Strahwald J, Muktar MS, Walkemeier B, Rotter B, Gebhardt C (2013) Comparative transcript profiling by SuperSAGE identifies novel candidate genes for controlling potato quantitative resistance to late blight not compromised by late maturity. Front Plant Sci 4:423
19. Marioni JC, Mason CE, Mane SM, Stephens M, Gilad Y (2008) RNA-seq: an assessment of technical reproducibility and comparison with gene expression arrays. Genome Res 18:1509–1517
20. Sharma CM, Hoffmann S, Darfeuille F, Reignier J, Findeiss S, Sittka A, Chabas S, Reiche K, Hackermüller J, Reinhardt R, Stadler PF, Vogel J (2010) The primary transcriptome of the major human pathogen *Helicobacter pylori*. Nature 464:250–255
21. Westermann AJ, Gorski SA, Vogel J (2012) Dual RNA-seq of pathogen and host. Nat Rev Microbiol 10:618–630
22. Matsumura H, Yoshida K, Luo S, Krüger DH, Kahl G, Schroth GP, Terauchi R (2011) High-throughput SuperSAGE. Meth Mol Biol 687: 135–148

Chapter 15

Microarray Analysis of *R*-Gene-Mediated Resistance to Viruses

Takeaki Ishihara, Yukiyo Sato, and Hideki Takahashi

Abstract

The complex process for host-plant resistance to viruses is precisely regulated by a number of genes and signaling compounds. Thus, global gene expression analysis can provide a powerful tool to grasp the complex molecular network for resistance to viruses. The procedures for comparative global gene expression profiling of virus-resistant and control plants by microarray analysis include RNA extraction, cDNA synthesis, cRNA labeling, hybridization, array scanning, and data mining steps. There are several platforms for the microarray analysis. Commercial services for the steps from cDNA synthesis to array scanning are now widely available; however, the data manipulation step is highly dependent on the experimental design and research focus. The protocols presented here are optimized for analyzing global gene expression during the *R* gene-conferred defense response using commercial oligonucleotide-based arrays. We also demonstrate a technique to screen for differentially expressed genes using Excel software and a simple Internet tool-based data mining approach for characterizing the identified genes.

Key words *Arabidopsis thaliana*, *Cucumber mosaic virus*, Gene ontology, Microarray, *R* gene, Transcriptome

1 Introduction

Plants have developed at least two strategies to defend themselves from virus attack. In virus-infected plants, the conserved molecular patterns of abundant foreign double-stranded viral RNAs are recognized as conserved molecular patterns that induce antiviral RNA silencing [1]. Because RNA silencing is often inhibited by virus-encoded suppressors, viruses are able to overcome RNA silencing and establish infection in host plants [2]. In such instances, resistance (R) proteins containing conserved nucleotide binding site (NB) and leucine-rich repeat (LRR) domains directly or indirectly recognize a broad range of viral proteins, including coat proteins, movement proteins, replicases, and silencing suppressors, thereby activating a strong defense system as a second strategy for resistance to viruses [3]. The activated R protein switches on a broad

Ichiro Uyeda and Chikara Masuta (eds.), *Plant Virology Protocols*, Methods in Molecular Biology, vol. 1236,
DOI 10.1007/978-1-4939-1743-3_15,

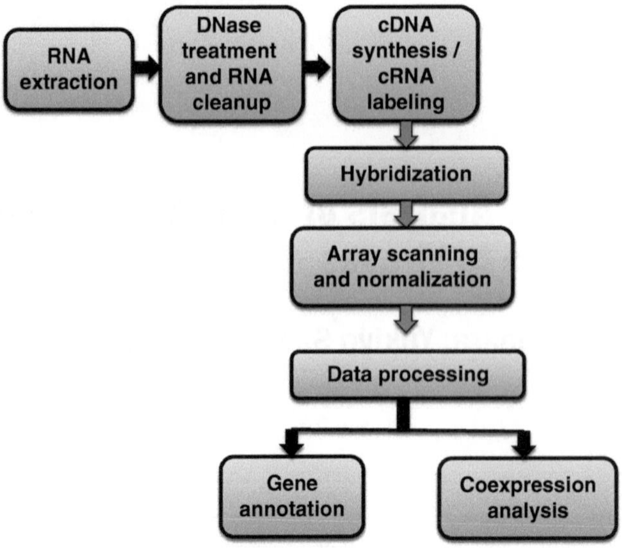

Fig. 1 Steps of the procedures for DNA microarray analysis. Commercial services for performing each step from cDNA synthesis through microarray data scanning are widely available. However, analysis of the data is highly dependent upon experimental design and research focus

array of downstream signaling pathways and induces various physiological and cytological changes to restrict virus multiplication and spread, thereby limiting virus infection to the primary infection site [3]. This complex process for *R*-mediated virus resistance to is precisely regulated by a number of genes. Global analysis of gene expression is a powerful way to dissect the complex molecular network of components essential for virus resistance in host plants carrying *R* genes, and can provide more comprehensive knowledge of host plant defenses against virus infection. In contrast, the study on the function of only single components would result in a more limited understanding of these networks.

Several protocols are now available for DNA microarray technology that allows monitoring of global gene expression during plant-virus interactions. Fig. 1 depicts the most common basic techniques required, including isolation of RNA from virus-infected plant and control samples, cDNA synthesis from the isolated mRNA by reverse transcription polymerase chain reaction (RT-PCR), cRNA labeling with the appropriate fluorescent reagent (e.g., Cyanine 3 or Cyanine 5), hybridization of the labeled cRNA pools with microarray slide, and microarray data scanning. Two major technologies for microarray fabrication have emerged. In one, the array is spotted with PCR-amplified cDNA clones, and in the other, oligonucleotides are synthesized in situ on commercially manufactured arrays using photolithography, ink-jet printing, or electrochemistry. The cDNA system can be more flexible, particularly for in-house spotted arrays, whereas in situ manufacturing seems to be more reliable, because it eliminates the complicated processes of PCR-amplified cDNA cloning and spotting on the array slide. Major commercial platforms based on in situ manufacturing include Affymetrix GeneChips, which consist of 25-mer oligonucleotides synthesized by photolithography, and Agilent microarrays, which consist of 60-mer oligonucleotides synthesized using ink-jet printing technology. For Affymetrix GeneChips, multiple independent oligonucleotides are designed to hybridize to different regions of the same RNA to minimize cross-hybridization effects. On the other hand, Agilent oligonucleotide microarrays require only one 60-mer oligonucleotide per gene, a format that provides enhancements in sensitivity. Both commercial platforms supply reliable and high-quality data. Several companies now provide commercial microarray services to perform the steps from cDNA synthesis through microarray data scanning. But subsequent data manipulation including statistical and bioinformatic analysis are highly dependent on the experimental design and research focus.

In virus–host plant pathosystems, global gene expression analysis has been conducted during compatible interactions between various viruses and host plants [4–13], because gene expression profiles are comparable to virus-infected plants with mild strain-infected or mock-inoculated controls under the same host plant genetic backgrounds. However, global gene expression profile of incompatible interactions can be limited [14–16]. In this chapter, we present a simple procedure for two-color microarray analysis using the Agilent (V2) Oligo Microarray (Agilent Technologies, Santa Clara, CA, USA) to analyze global gene expression in *RESISTANCE TO CMV(Y)* (*RCY1*)-transformed *Arabidopsis thaliana*, which accumulates high levels of the NB-LRR domain-containing RCY1 protein and thereby exhibits extreme resistance (ER) to *Cucumber mosaic virus* (CMV). The simple practical data analysis in this procedure using Microsoft Excel and Internet-based bioinformatics tools are useful for screening for genes regulating virus resistance.

2 Materials

2.1 Plants, Cultivation, and Sampling

1. *A. thaliana* ecotype Columbia-0 (Col-0, wild-type).
2. *RCY1*-transformed Col-0 line 2 (Col::pRCY1#2) [17] over-expressing the *RCY1* transgene and showing ER to a yellow strain of CMV[CMV(Y)] (*see* **Note 1**).
3. Vermiculite (Schundler Company, city, NJ, USA) and perlite (Schundler Company).
4. Fertilizer: Hyponex (HYPONeX JAPAN, Osaka, Japan).
5. Razor blades: sterile.
6. Liquid nitrogen.

2.2 RNA Preparation

1. RNeasy Plant Mini Kit (Qiagen, Hilden, Germany): including Buffer RLT, QIAshredder spin columns, RNeasy spin columns, Buffer RW1, and Buffer RPE.
2. β-mercaptoethanol, molecular biology grade.
3. Recombinant DNase I (RNase-free) (Takara, Otsu, Japan) with supplied 10× DNase I Buffer [400 mM Tris–HCl (pH 7.5), 80 mM $MgCl_2$, and 50 mM DTT] (Takara).
4. Recombinant RNase inhibitor (Takara).
5. RNase-free 1.5-ml and 2.0-ml microcentrifuge tubes.
6. Ethanol, molecular biology grade.
7. 3 M sodium acetate (pH 5.2), sterilized by autoclave.
8. RNase-free water.
9. Phenol–chloroform (phenol–chloroform = 1:1), molecular biology grade.
10. Spectrophotometer (NanoDrop, LMS Co., Tokyo, Japan).
11. Agilent 2100 Bioanalyzer (Agilent Technologies).

2.3 cDNA Synthesis and cRNA Labeling

1. Low RNA Fluorescent Linear Amplification Kit (two-color) (Agilent Technologies): T7 Promoter Primer, 5× First-Strand Reaction Buffer, 0.1 M DTT, 10 mM dNTP mix, Random Hexamers (200 ng/μl), MMLV-RT (200 U/μl), RNaseOUT (40 U/μl), 4× Transcription Buffer, 0.1 M DTT, 10 mM NTP Mix, 50 % polyethyleneglycol (PEG), Inorganic Pyrophosphatase, T7 RNA Polymerase (2,500 U/μl), 4 M lithium chloride, Cyanine 3-CTP (10 mM), and Cyanine 5-CTP (10 mM).
2. Two-color RNA Spike-In Kit (Agilent Technologies): Spike A Mix and Spike B Mix.
3. Gene Expression Hybridization Kits (Agilent Technologies): 2× Hi-RPM Hybridization Buffer, 25× Fragmentation Buffer, and 10× Gene Expression Blocking Agent.

2.4 Microarray Hybridization and Scanning

1. Two *Arabidopsis* (V2) Oligo Microarray slides (Agilent Technologies) (*see* **Note 2**).
2. Agilent Microarray Chamber Kit (Agilent Technologies): Microarray hybridization chamber assemblies (Chamber base, cover, and clamp assembly).
3. Disposable gasket slides for Agilent microarray hybridization chambers (Agilent Technologies): 1 microarray/slide format.
4. Agilent Hybridization Oven (G2545A).
5. Agilent Gene Expression Wash Buffer Kits (Agilent Technologies): Gene Expression Wash Buffer 1, Gene Expression Wash Buffer 2, and 10 % Triton X-102.
6. Acetonitrile, molecular biology grade.

7. Agilent DNA Microarray Scanner (Agilent Technologies).
8. Agilent Feature Extraction software (FE) 10.7.3.1 (Agilent Technologies).

2.5 Data Analysis

1. Microsoft Excel and a Web browser (e.g., Firefox, Internet Explorer, or Safari) installed on a personal computer with an Internet connection.

3 Methods

The *A. thaliana*–CMV system is an excellent pathosystem [3, 18, 19], for which a rich store of knowledge regarding both RNA silencing and *R*-gene-mediated virus defense has accumulated. However, when global gene expression analysis in CMV-infected *A. thaliana* is performed using DNA microarrays, differential gene expression caused by differences between the genetic backgrounds of resistant and susceptible *A. thaliana* ecotypes must be considered. Comparison of *A. thaliana* transformed with an *R* gene to its parent wild-type plant avoids this disadvantage and is more likely to be effective, as it allows for the follow up of changes in gene expression directly associated with *R*-conferred resistance to viruses in a common host plant genetic background.

The level of *RCY1* expression regulates the strength of resistance to CMV(Y) in *A. thaliana*. The transgenic Col::pRCY1#2 line exhibits ER to CMV(Y) [17]. Col::pRCY1#13 and Col::pRCY1#12 are two other independent lines transformed with HA-epitope sequence-tagged *RCY1* (*RCY1-HA*). The Col::pRCY1#13 line also exhibits ER to CMV(Y), as does Col::pRCY1#2, while the CMV(Y)-inoculated leaves of Col::pRCY1#12 [17] exhibit hypersensitive resistance (HR) accompanied by formation of local necrotic lesions. As shown in Fig. 2, the levels of *RCY1* transcript and RCY1 protein accumulated in ER-type Col::pRCY1#13 are significantly higher than those in HR-type Col::pRCY1#12 (Fig. 2a, b), because ten copies of *RCY1-HA* are integrated into the genome of Col::pRCY1#13 (Fig. 2c). In the experimental example presented here, global gene expression profiles between the Col::pRCY1#2 transformant overexpressing *RCY1* and the wild-type Col-0 were compared to investigate ER to CMV(Y) in Col::pRCY1 plants.

3.1 Plant Growth and Sampling

1. Wild-type Col-0 and Col::pRCY1#2 are grown at 25 °C under continuous illumination (8,000 lx) in a mixture of vermiculite and perlite (1:1) and are irrigated with 1,000-fold diluted Hyponex every 3 days.
2. Cut fully expanded leaves from 4-week-old plants using razor blades, weigh (tissue weight should not exceed 100 mg to avoid exceeding the RNA-binding capacity of the RNeasy spin columns), and freeze immediately in liquid nitrogen (*see* **Note 3**).

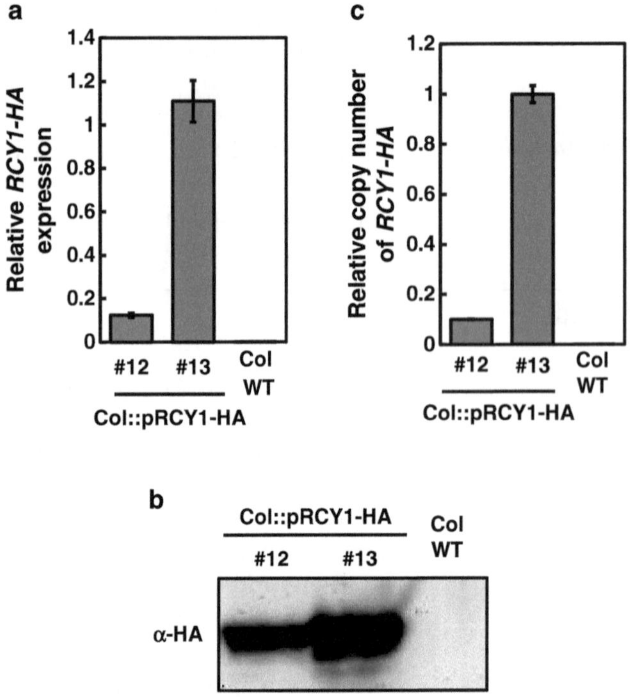

Fig. 2 Expression of the *RCY1* transgene in *A. thaliana* Col-0 transformants. *RCY1-HA*-transformed Col-0 line 13 (Col::pRCY1#13) shows ER to CMV(Y) corresponding to that of Col::pRCY1#2, which also exhibits the ER phenotype. On the other hand, in Col::pRCY1#12 exhibits HR to CMV(Y). (**a**) Levels of the *RCY1-HA* transcript accumulated in the Col::pRCY1#12 and Col::pRCY1#13 lines relative to the control, wild-type Col-0, were measured by quantitative RT-PCR according to the procedures described previously [17]. High levels of *RCY1-HA* transcript are detected in the Col::pRCY1#13 line. (**b**) Immunological detection of RCY1-HA protein in the Col::pRCY1#12 and Col::pRCY1#13 lines, and in wild-type Col-0, using an HRP-labeled anti-HA monoclonal antibody (clone 3F10; Roche, Indianapolis, IN, USA). Relatively large amounts of the RCY1-HA protein accumulate in the Col::pRCY1#13 line, which strongly expresses *RCY1-HA*. (**c**) Measurement of copy number of *RCY1-HA* transgenes in Col::pRCY1#12 and Col::pRCY1#13 lines and wild-type Col-0 *A. thaliana*. Relative amounts of the *RCY1-HA* transgene are measured by quantitative PCR using *RCY1* intron-specific primers and genomic DNA as a template. The Col::pRCY1#12 line apparently has a single copy of the *RCY1-HA* transgene, so it is likely that ten copies of *RCY1-HA* transgene are integrated into the genome of Col::pRCY1#13 line. Consistent with this result, the ER phenotype in Col::pRCY1#2 might be due to the high copy number and consequent high level of expression of the RCY1 transgene in that line

3.2 RNA Extraction

High-quality RNA is necessary for microarray analysis. The RNeasy Plant Mini Kit (Qiagen) is widely available for purification of high-quality RNA from plant tissues [20]. All centrifugation steps should be performed at 20–25 °C.

1. Grind frozen leaf samples thoroughly with a mortar and pestle. Transfer tissue powder to an RNase-free 1.5 ml microcentrifuge tube using a spatula (*see* **Note 4**).

2. Add 450 μl Buffer RLT (Qiagen) (supplied with 4.5 μl β-mercaptoethanol before use) to a maximum of 100 mg tissue powder, vortex vigorously, and incubate for 3 min at 56 °C to disrupt the tissue.
3. Transfer the lysate to a QIAshredder spin column (Qiagen) seated in a 2 ml collection tube, and centrifuge for 2 min at 13,000 rpm (approx. 10,000 × *g*). The total RNA is in the supernatant of the flow through.
4. Add 0.5 volume of ethanol to the cleared lysate and mix immediately by pipetting.
5. Transfer the sample, including any precipitate, to an RNeasy spin column (Qiagen) in a 2 ml collection tube. Centrifuge for 15 s at 10,000 rpm (approx. 8,000 × *g*). The total RNA binds to the spin column membrane. Discard the flow-through.
6. Add 700 μl Buffer RW1 (Qiagen) to the RNeasy spin column. Centrifuge for 15 s at 10,000 rpm to wash the spin column membrane. Discard the flow-through.
7. Add 500 μl of working concentration Buffer RPE (Qiagen) to the RNeasy spin column (*see* **Note 5**) and centrifuge for 15 s at 10,000 rpm to wash the spin column membrane.
8. Again add 500 μl Buffer RPE to the RNeasy spin column again and centrifuge for 2 min at 10,000 rpm to wash the spin column membrane. Place the RNeasy spin column in a new microcentrifuge tube and centrifuge for 1 min at 13,000 rpm.
9. Place the RNeasy spin column in a new microcentrifuge tube. Add 50 μl RNase-free water to the spin column membrane. Centrifuge for 1 min at 10,000 rpm to elute the RNA.

3.3 DNase Treatment and RNA Cleanup

To remove contaminating genomic DNA, treat the RNA samples with DNase I.

1. For ethanol precipitation, add 0.1 volume of 3 M sodium acetate (pH 5.2) and 2.5 volumes of ethanol to each RNA sample. Mix well and place on ice for 10 min.
2. Centrifuge for 15 min at 13,000 rpm at 4 °C. Discard the supernatant, being careful not to throw out the RNA pellet, which may be translucent. Wash the RNA pellet with 70 % ethanol and centrifuge for 5 min at 13,000 rpm. Discard the supernatant and dry the pellet.
3. Dissolve the pellet in 43 μl RNase-free water.
4. Add 5 μl 10× DNase I buffer, 1 μl RNase inhibitor, and 1 μl DNase I (final volume is 50 μl). Incubate for 30 min at 37 °C.
5. Add 50 μl RNase-free water. Add 100 μl phenol–chloroform (1:1) and vortex vigorously. Centrifuge for 5 min at 13,000 rpm at 25 °C.

6. Transfer the aqueous phase to a new microcentrifuge tube.
7. Repeat **steps 1** and **2** in this section (ethanol precipitation).
8. Dissolve pellet in 100 μl RNase-free water. Add 350 μl Buffer RLT from the RNeasy Plant Mini Kit (Qiagen), and mix well.
9. Add 250 μl ethanol and mix well by pipetting.
10. Transfer the sample to an RNeasy spin column (Qiagen) placed in a 2 ml collection tube. Centrifuge for 15 s at 10,000 rpm. Discard the flow-through.
11. Repeat **steps** 7 and **8** in Subheading 3.2.
12. Place the RNeasy spin column (Qiagen) in a new microcentrifuge tube. Add 30 μl RNase-free water to the spin column membrane. Centrifuge for 1 min at 10,000 rpm to elute the RNA.
13. Determine the quantity of each RNA sample spectrophotometrically using the NanoDrop.
14. Check the quality of each RNA sample using the Agilent 2100 Bioanalyzer (*see* **Note 6**).

3.4 cDNA Synthesis and cRNA Labeling

Commercial microarray analysis services including the cDNA synthesis, cRNA labeling, and microarray hybridization and scanning steps are widely available. Therefore, the protocol for two-color microarray analysis using the dye-swap strategy (*see* **Note** 7 and Fig. 3), provided by Hokkaido System Science (HSS) Co., Ltd., Sapporo, Japan (*see* **Note 8**), is shown as an example as follows.

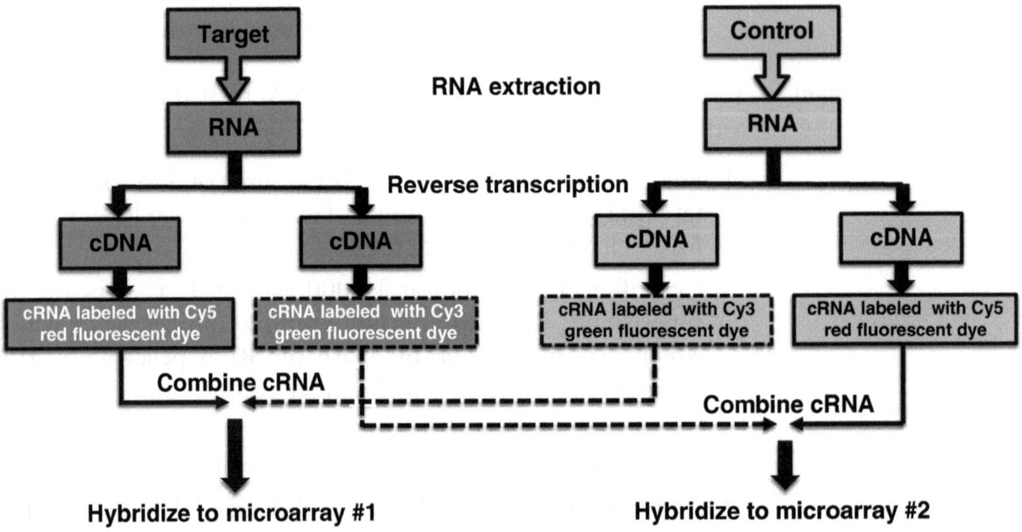

Fig. 3 Diagram of a typical two-color DNA microarray analysis performed with the dye-swapping strategy. Two DNA microarray slides (#1 and #2) are hybridized with cRNA transcribed from cDNA for Target and Control RNA samples reciprocally labeled with two different fluorescent dyes (Cyanine 3: Cy3 and Cyanine 5: Cy5 in this experiment), respectively

3.4.1 cDNA Synthesis

For cDNA synthesis, cRNA labeling, and amplification, the Low RNA Fluorescent Linear Amplification Kit (Two-color) (Agilent Technologies) is used (here in **steps 1** through **6** and below in Subheading 3.4.2) [21].

1. Add 1.2 μl T7 Promoter Primer to two sets of 500 ng RNA samples isolated from Target or Control tissue (*see* **Note 9**). Make the volume of each sample up to 9.5 μl with RNase-free water.
2. Add 2 μl Spike A from the Two-color RNA Spike-In Kit (Agilent Technologies) to the first Target RNA sample and the first Control RNA sample, respectively, for labeling with Cyanine 3 (Cy3), and add 2 μl Spike B from the Two-color RNA Spike-In Kit to the second Target RNA sample and the second Control RNA sample, respectively, for labeling with Cyanine 5 (Cy5) (*see* **Note 10**).
3. Denature RNA and primer by incubating at 65 °C for 10 min, then place on ice for 5 min.
4. Prepare four sets of cDNA master mix: 4 μl 5× First Strand Buffer (Agilent Technologies), 2 μl 0.1 M DTT, 1 μl 10 mM dNTP mix, 0.5 μl RNaseOUT, and 1 μl MMLV-RT (total 8.5 μl per reaction).
5. Add 8.5 μl cDNA master mix to each denatured RNA sample (**step 3**). Incubate for 2 h at 40 °C.
6. Incubate for 15 min at 65 °C to stop reaction, and then place on ice for 5 min.

3.4.2 cRNA Labeling

1. Prepare four sets of transcription master mix: 15.3 μl RNase-free water, 20 μl 4× Transcription Buffer (Agilent), 6 μl 0.1 M DTT, 8 μl NTP Mix, 6.4 μl 50 % PEG, 0.5 μl RNaseOUT, 0.6 μl Inorganic Pyrophosphatase, and 0.8 μl T7 RNA polymerase (total 57.6 μl per reaction), each of which is included in Low RNA Fluorescent Linear Amplification Kit.
2. For two-color microarray analysis (*see* Fig. 3), add 2.4 μl 10 mM Cyanine 3-CTP to the first Target RNA sample containing Spike A (**step 2** of Subheading 3.4.1) and 2.4 μl 10 mM Cyanine 5-CTP to the first Control RNA sample containing Spike B (**step 2** of Subheading 3.4.1), respectively. Also add 2.4 μl 10 mM Cyanine 5-CTP to the second Target RNA sample containing Spike B and 2.4 μl 10 mM Cyanine 3-CTP to the second Control RNA containing Spike A, respectively.
3. Add 57.6 μl transcription master mix to each reaction tube (**step 2**). Incubate for 2 h at 40 °C in the dark for in vitro transcription of Target or Control cRNA labeled with Cy3 or Cy5.

3.4.3 Labeled cRNA Cleanup

1. The RNeasy Mini spin column (Qiagen) is used to clean up the labeled cRNA (following **steps 9–13** of Subheading 3.3. DNase treatment and RNA cleanup). Add 20 μl nuclease-free water, 350 μl Buffer RLT (Qiagen), and 250 μl ethanol to each reaction tube.
2. Transfer 700 μl cRNA sample to the column and centrifuge for 1 min at 13,000 rpm. Discard the flow-through.
3. Add 500 μl working concentration Buffer RPE (Qiagen) (*see* **Note 5**) to the column and centrifuge for 1 min at 13,000 rpm.
4. Discard the flow-through and repeat **step 3**. Place the column in a new microcentrifuge tube.
5. Add 30 μl RNase-free water to the column membrane. Incubate for 1 min and centrifuge for 1 min at 13,000 rpm to elute the RNA.
6. Measure the quantity of labeled cRNA spectrophotometrically using the NanoDrop.

3.4.4 Preparation of cRNA Probe

1. For the dye-swapping strategy, 0.75 μg of Cy3-labeled Target cRNA and 0.75 μg of Cy5-labeled Control cRNA are combined with 50 μl 10× blocking agent (Agilent Technologies) to prepare the cRNA probe that will be applied to microarray slide #1 (*see* Fig. 3) [22]. Make the volume up to 240 μl with RNase-free water.
2. For preparation of the cRNA probe that will be applied to microarray slide #2 (*see* Fig. 3), 0.75 μg of Cy3-labeled Control cRNA and 0.75 μg of Cy5-labeled Target cRNA are combined with 50 μl 10× blocking agent (Agilent Technologies). Make the volume up to 240 μl with RNase-free water.
3. Add 10 μl 25× fragmentation buffer (Agilent Technologies) to each cRNA probe solution (**steps 1** and **2**), and incubate for 30 min at 60 °C in the dark.
4. Add 250 μl 2× Hi-RPM hybridization buffer (Agilent Technologies) to stop the fragmentation process.

3.5 Microarray Hybridization and Scanning

1. Load the gasket slide into the hybridization chamber with the gasket label facing up and dispense the entire volume (500 μl) of cRNA probe solution containing Cy5-labeled Target cRNA and Cy3-labeled Control cRNA (*see* Fig. 3 and **step 1** in Subheading 3.4.4), and mix in a drag and drop manner [22].
2. Place microarray slide #1 (Agilent *Arabidopsis* (V2) Oligo Microarray) with the active side down and assemble the chamber.
3. Load the gasket slide into the hybridization chamber with the gasket label "Agilent" facing up and dispense the entire

volume of cRNA probe solution containing Cy3-labeled Target cRNA and Cy5-labeled Control cRNA onto the gasket slide (*see* Fig. 3 and **step 2** in Subheading 3.4.4) [23], while slowly dragging the pipette across the chamber well, then place microarray slide #2 (Agilent *Arabidopsis* (V2) Oligo Microarray) with the printed side down on the probe and assemble the chamber.

4. Place the chamber cover onto the sandwiched slides and tighten the clamp onto the chamber.
5. Put assembled chambers on a rotating rack in the Agilent Hybridization Oven (G2545A).
6. Incubate for 17 h at 65 °C.
7. Wash the arrays in Agilent Gene Expression Wash Buffer 1 for 1 min at room temp.
8. Wash the arrays in Agilent Gene Expression Wash Buffer 2 for 1 min at 37 °C.
9. Wash the arrays in acetonitrile for 1 min and dry.
10. Scan the arrays using Agilent DNA Microarray Scanner at 10-μm resolution.
11. Determine the fluorescence intensity associated with each arrayed spot using Agilent Feature Extraction Software. Microarray images contain spots of poor quality, due to low or oversaturated intensity or nonuniform or non-circular shape. Such spots can be flagged in Agilent Feature Extraction Software to be removed from further analysis.
12. The data are normalized using linear and LOWESS methods [24, 25] in the Agilent Feature Extraction Software (*see* **Note 11**).

3.6 Data Analysis

3.6.1 Identification of Differentially Expressed Genes in Two-Color Microarray Analysis Using the Dye-Swap Strategy

The number of each step in this section corresponds to the annotation in bold numerals in Fig. 4.

1. Normalized data exported into a tab-delimited .txt file, are transferred into a Microsoft Excel worksheet. The example data from Col::RCY1-HA#2 and wild-type Col-0 are shown in Fig. 4. The following parameters are necessary to analyze differentially expressed genes.
   (a) Unique number (column A in Fig. 4).
   (b) Control type (positive and negative control are 1 and -1, respectively, and each gene spot is 0) (column B).
   (c) Gene name (AGI code) (column C).
   (d) Cy3 processed signal of microarray #1 (column D).
   (e) Cy5 processed signal of microarray #1 (column E).

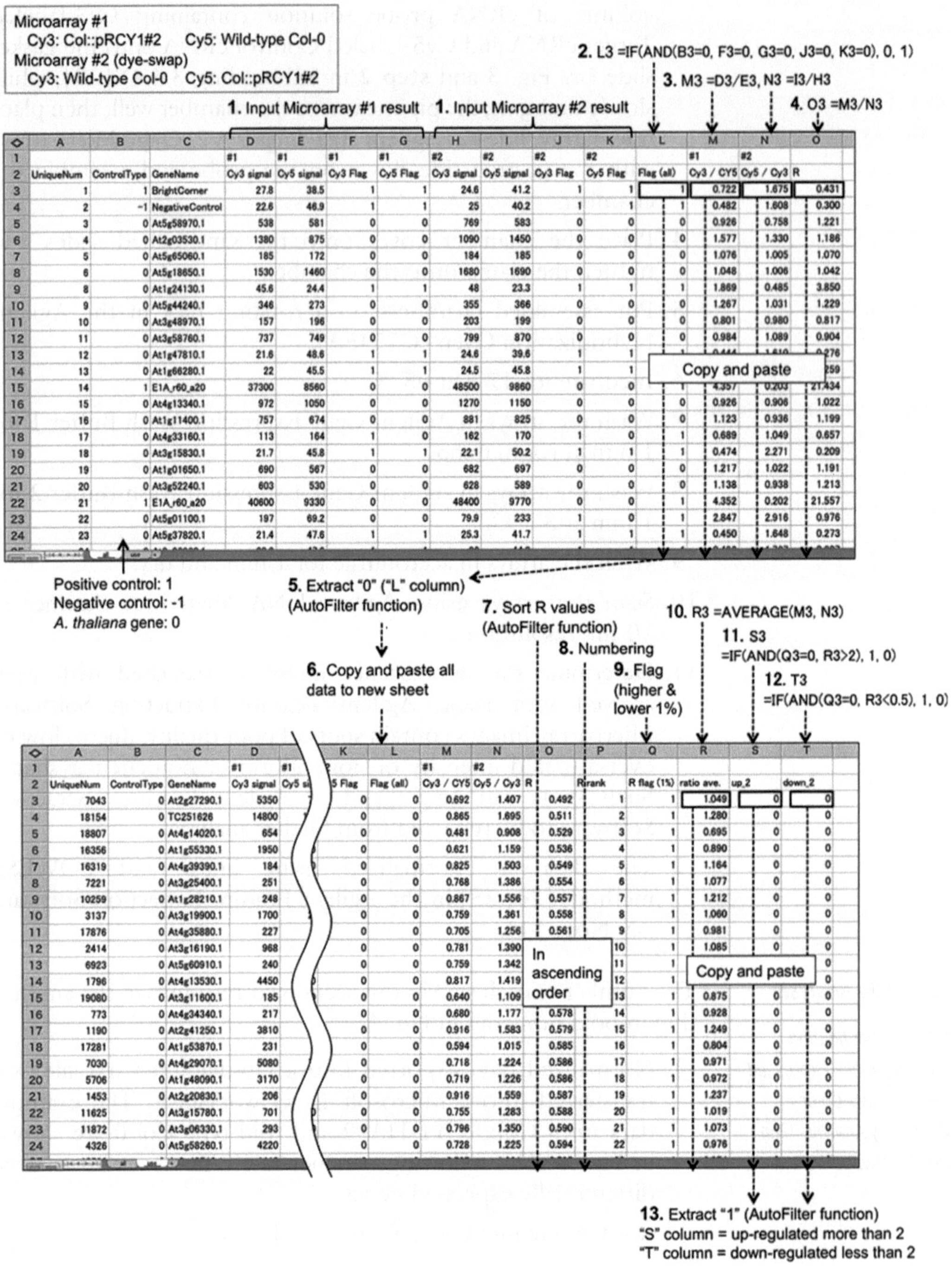

Fig. 4 Scheme for identification of differentially expressed genes. An example of the procedure for extraction of differentially expressed genes from the dye-swapping experiment data set using Microsoft Excel is shown. *Bold* numbers represent the numbers of the steps described in Subheading 3.6.1

(f) Cy3 processed signal of microarray #2 (column H).

(g) Cy5 processed signal of microarray #2 (column I).

Values for flagged spots are indicated as "1" in columns F, G, J, and K and are excluded from data analysis.

2. To remove the data for positive and negative control spots and flagged spots from the analysis, type
"=IF(AND(B3 = 0, F3 = 0, G3 = 0, J3 = 0, K3 = 0), 0, 1)"
in cell L3, copy the formula in cell L3, and paste it into the rest of the cells in the L column (*see* **Note 12**). The spots that are designated as "1" in column L in this calculation are not used for further analysis. Thus, in this manner, the spots flagged for any of the low-quality Cy3 and Cy5 signals in the microarray dye-swapping experiments were removed (*see* **Note 13**).

3. Calculate the ratio of the expression in Col::pRCY1#2 to the expression in wild-type Col-0 (i.e., the Cy3 signal/Cy5 signal in microarray #1 and the Cy5 signal/Cy3 signal in microarray #2). Type
"=D3/E3"
in cell M3 and
"=I3/H3"
in cell N3, respectively. Copy the formulas typed into cells M3 and N3, and paste into the rest of the cells in the M and N columns, respectively (*see* **Note 12**).

4. To remove spots with low reproducibility, calculate reproducibility (R) as R = (Cy3 signal/Cy5 signal on microarray #1)/(Cy5 signal/Cy3 signal on microarray #2) (*see* **Note 14**). Type
"=M3/N3"
in cell O3, copy the formula typed in cell O3, and paste it into the rest of the cells in the O column (*see* **Note 12**).

5. Extract the spots with a value of "0" in each cell in column L (**step 2**) using the AutoFilter function of Excel.

6. Copy and paste all the extracting data to a new sheet.

7. Sort the data by ascending *R*-values using the AutoFilter function of Excel.

8. Number these sequentially in column P according to rank by reproducibility (*R*).

9. Flag the highest 1 % and lowest 1 % of R-ranked samples (*see* **Note 14**, Fig. 5).

10. Calculate the averages of expression ratio. Type
"=AVERAGE(M3, N3)"
in cell R3, then copy the formula from cell R3, and paste it into the rest of the cells in column R (*see* **Note 12**).

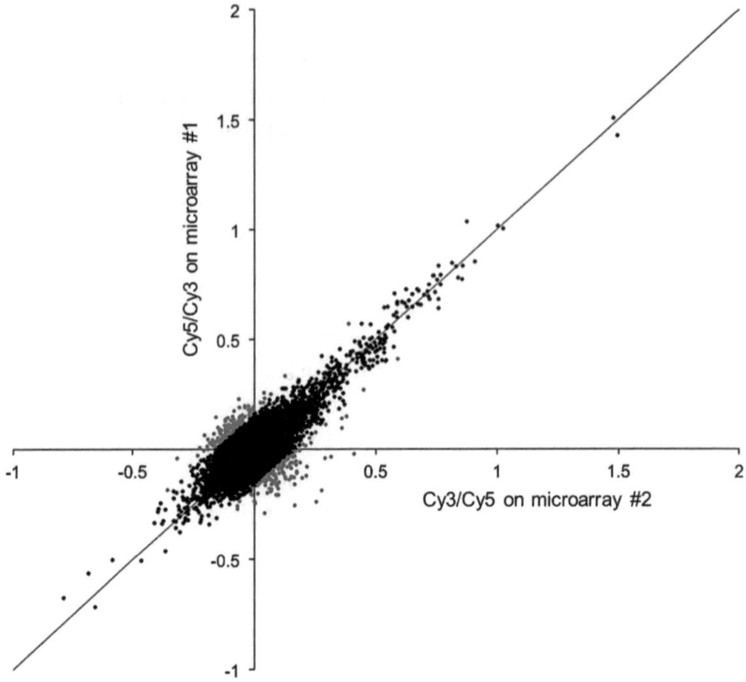

Fig. 5 Scatterplot of the logarithmic expression ratio of the Target sample the Control sample in the dye-swapping experiment for two-color microarray analysis. Two-color data (Cy3/Cy5 and Cy5/Cy3) obtained from dye-swapping experiments between Col::pRCY1#2 as the Target sample and wild-type Col-0 as the Control sample are depicted in a scatterplot. *Gray dots* represent the flagged spots at the highest 1 % or lowest 1 % of *R*-values (column O in Fig. 4). Because the spots should theoretically fall along the line log(*y*) = log(*x*), the signal of the flagged spots exhibits low reproducibility in this dye-swapping experiment and should be excluded from analysis. *Black dots* represent the spots retained for further analysis. In this example, 12,988 spots were extracted in **step 5** of Subheading 3.6.1. Therefore, in this example, the top 130 and bottom 130 spots according to *R*-value rank (i.e., spots numbered from 1 to 130 and from 12,859 to 12,988 according to their *R*-value rank, respectively) were flagged (in column Q in Fig. 4)

11. To separate the twofold upregulated spots from the non-flagged spots in **step 9**, type
"=IF(AND(Q3 = 0, R3 > 2), 1, 0)"
in cell S3, then copy the formula from cell S3, and paste it into the rest of the cells in column S (*see* **Notes 12** and **15**).
12. To separate 1/2-fold downregulated spots from the non-flagged spots in **step 9**, type
"=IF(AND(Q3 = 0, R3 < 0.5), 1, 0)"
in cell T3, then copy the formula from cell T3, and paste it into the rest of the cells in column T (*see* **Notes 12** and **15**).

13. Extract the spots with 2-fold upregulated or 1/2-fold downregulated expression signal using the AutoFilter function of Excel (*see* **Note 16**).

3.6.2 Gene Annotation

Gene annotation information including the Gene Ontology of *A. thaliana* can be obtained from TAIR (http://www.arabidopsis.org). Because Gene Ontology terms are precisely defined and have controlled vocabularies, a large number of genes can be grouped according to them and the sizes of these groups are then easy to compare. However, to decide that enrichment of gene term is significantly different from what would be expected by chance, it must be compared with the relative abundance of the term for all genes on the genome. To address this problem, we used DAVID (Database for Annotation, Visualization, and Integrated Discovery) [26, 27] to provide enrichment analysis for each term compared among all genes on the genome. Further, terms with similar meanings are grouped into single clusters, and overall enrichment of those clusters is calculated and ranked. DAVID can be used for analyzing various species, including *A. thaliana* in this example, and helps to extract biological meaning from large lists of differentially expressed genes.

1. Access the DAVID Web site (http://david.abcc.ncifcrf.gov/home.jsp), and click the "Functional Annotation" button at the top left side of the page under "Shortcut to DAVID Tools".
2. Copy the AGI codes of upregulated genes without the decimal point or following digits (i.e., At5g43470, rather than At5g43470.1), and paste into the "Paste a list" entry field of "Step 1: Enter Gene List".
3. Select "TAIR_ID" at "Step 2: Select Identifier" to analyze *A. thaliana* genes.
4. At "Step 3: List Type", select "Gene List" as the List Type, and click the "Submit List" button.
5. Click "Functional Annotation Table" to display the results as a table in the Web browser (*see* Fig. 6).
6. Click "Download File" in the upper right-hand corner of the screen to download the list of the annotation clusters containing similar Gene Ontology terms or keywords for the query genes in statistically significant order (*see* Table 1 and **Note 17**).
7. Repeat **steps 1–6** for downregulated genes.

3.6.3 Analysis Using Gene Coexpression Database

To date, analyses of vast amounts of transcriptome data for genes expressed under various conditions have been performed. Analyses of these data can identify gene pairs with similar expression patterns; such coexpressed genes might be members of gene clusters or biological pathways that work in concert to perform particular functions (e.g., a transcription factor that is coexpressed with

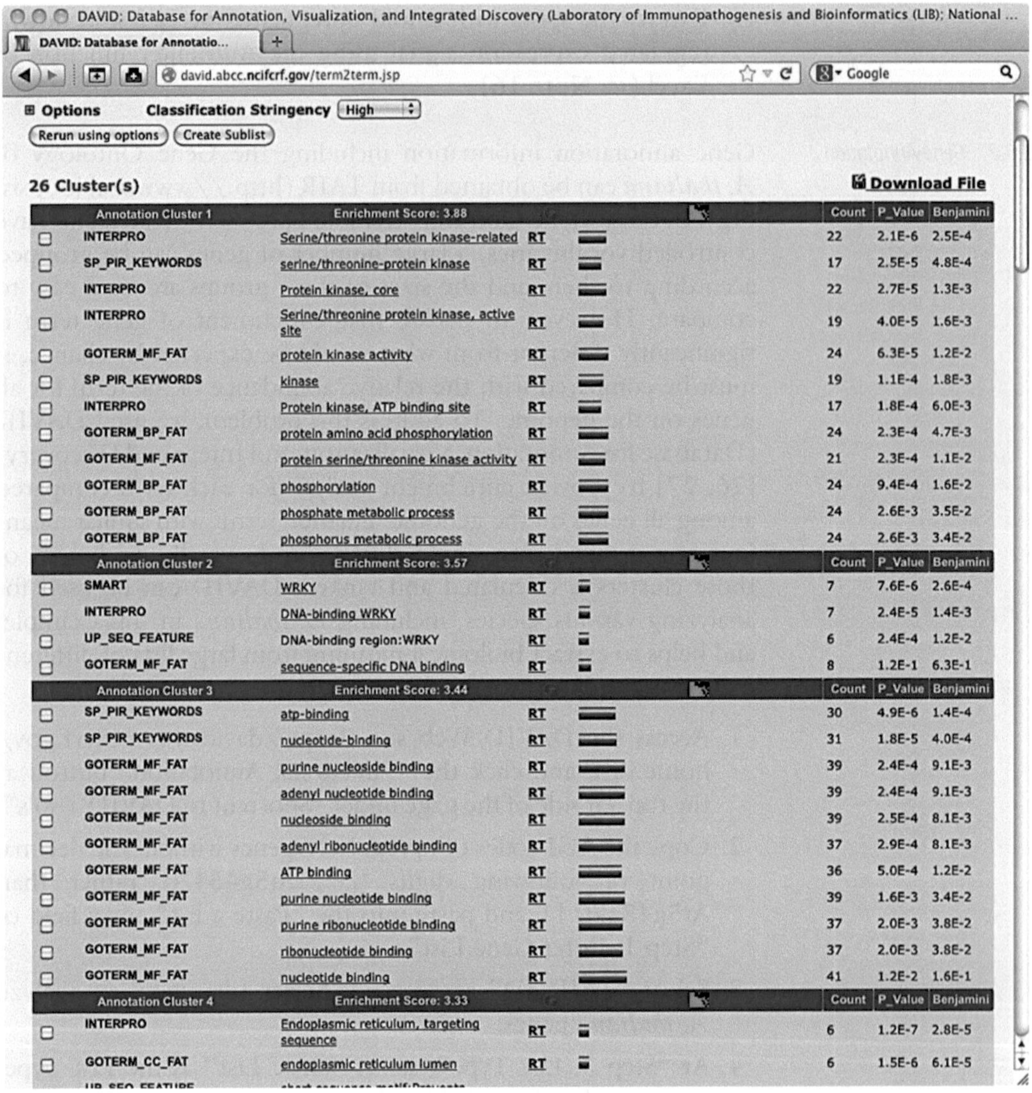

Fig. 6 Screenshot of results from DAVID. Functional annotations of upregulated genes were analyzed using DAVID. Genes are grouped by annotation and are numbered in statistically significant order. In the present example, the top 4 of 26 clusters are shown. Cluster 1 contains mainly serine/threonine protein kinase genes, cluster 2 contains WRKY transcription factor genes, cluster 3 contains genes encoding ATP-binding proteins, and cluster 4 contains genes encoding endoplasmic reticulum-targeting proteins

another gene might be expected to regulate the expression of that gene). Analysis of coexpression data should assist with formation of hypotheses regarding the functions of uncharacterized genes of interest and aid in the search for new genes that are functionally related to previously characterized genes. To this end, ATTED-II, a database of coexpressed genes for *A. thaliana* and some other plant species, has been constructed [28].

1. Access the ATTED-II Web site (http://atted.jp), and click "Search". To obtain coexpression information for each single gene, use the "GeneTable" form (Fig. 7) (*see* **Note 18**).

Table 1
Summary of the functional annotations of upregulated genes obtained from DAVID

Annotation cluster[a]	Enrichment score[b]	Representative term(s)[c]
1	3.883	Serine/threonine protein kinase-related, Protein kinase activity
2	3.567	DNA-binding WRKY
3	3.443	ATP binding, Nucleotide binding
4	3.329	Endoplasmic reticulum lumen
5	2.993	Cell death, Programmed cell death
6	2.120	EGF-type aspartate/asparagine hydroxylation conserved site
7	2.075	Transmembrane
8	1.922	Thioredoxin-like subdomain, protein disulfide isomerase activity
9	1.847	Anti-sense to fibroblast growth factor protein GFG, NUDIX hydrolase domain
10	1.530	Positive regulation of immune response
11	1.286	Glutathione transferase activity
12	1.285	Cytochrome P450
13	1.284	Heme binding, Oxygen binding, Iron ion binding
14	1.276	Chitinase activity, Polysaccharide catabolic process
15	0.890	Cation binding
16	0.860	Positive regulation of macromolecule biosynthetic process
17	0.790	Ethylene mediated signaling pathway
18	0.654	Calcium-binding EF-hand
19	0.637	Ion channel activity
20	0.554	Transcription factor activity
21	0.518	Cell surface receptor linked signal transduction
22	0.510	Repeat: LRR 1, Repeat: LRR 2, Repeat: LRR 3, Repeat: LRR 4

[a]Terms having similar biological meaning are grouped
[b]Statistical enrichment of each term was calculated, and the enrichment score for the group was calculated based on the scores of each term within the cluster. The higher scores mean that the term is more enriched
[c]Only representative term(s) are shown, even if multiple terms are included in the cluster

2. Input the AGI code number of query gene into the entry field "Input AGI codes" in the "GeneTable" form and click the "submit" button (e.g., At5g01540 as in Fig. 7). In the resulting table, information regarding the query gene such as gene function and predicted subcellular localizations of its gene product are displayed. The table contains links to the "Coexpressed gene list" and "Locus page" for each result (Fig. 7).

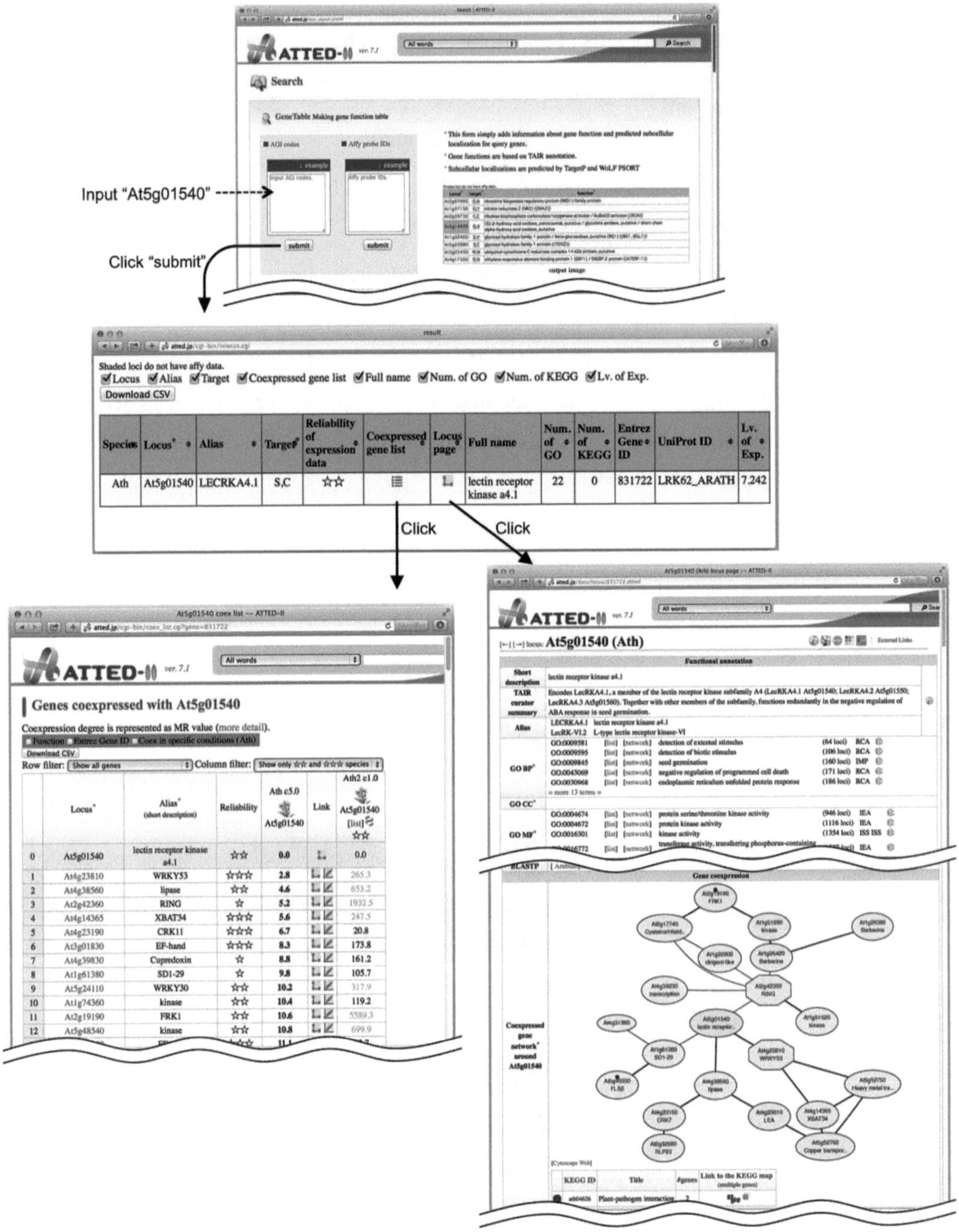

Fig. 7 Screenshots of the ATTED-II Web site. Search result obtained using an upregulated gene (AGI code At5g01540) is shown. Coexpression information for At5g01540, which encodes a serine/threonine protein kinase-related protein that is the most highly upregulated (expression ratio = 4.6) in Cluster 1 (Table 1), is shown in the "GeneTable" form. Coexpressed genes and precise information for At5g01540 are shown at the *lower left* as "Coexpressed gene list" and the *lower right* as the "Locus page", respectively. The schematic of the network of genes coexpressed with the query gene is illustrated at *lower right*. The gene encoding the *WRKY53* transcription factor (At4g23810) is strongly coexpressed with At5g01540

3. Click the link to the "Coexpressed gene list" and browse the list of genes strongly coexpressed with the query gene (Fig. 7).
4. Click the link to the "Locus page" in **step 2** and go to "Locus page" to view more information about the query gene (Fig. 7). On the "Locus page", information regarding functional annotation, including Gene Ontology, is supplied. A schematic for a network of genes coexpressed with the query gene is presented on this page, which is useful for understanding the relationships among coexpressed genes (Fig. 7).

4 Notes

1. Col::pRCY1#2 exhibits ER to CMV(Y) [17]. Col::pRCY1#13 also shows ER to CMV(Y), as does Col::pRCY1#2; however, CMV(Y)-inoculated Col::pRCY1#12 leaves exhibit HR accompanied by necrotic lesion formation [17]. The ER phenotype might be due to high copy numbers of the *RCY1* transgene and consequent high level of RCY1 protein accumulation in the ER lines (*see* Fig. 2).
2. The 60mer oligonucleotide sequences arrayed on the Agilent *Arabidopsis* (V2) Oligo Microarray can be viewed in eArray (https://earray.chem.agilent.com/earray/) using Internet Explorer or Mozilla browsers (but not Safari).
3. Excess leaf sample causes low RNA yield and quality.
4. Use sufficient liquid nitrogen and take care not to dissolve the sample before adding the buffer.
5. Add 4 volumes of 96–100 % ethanol to the Buffer RPE concentrate to obtain a working solution.
6. The Agilent 2100 Bioanalyzer is a microfluidics-based platform for quality control of RNA. To assess RNA sample integrity, the 18S and 28S rRNA peaks in the RNA sample should be checked with the Bioanalyzer according to the manufacturer's instructions: (http://gcf.pbrc.edu/docs/Agilent/Agilent%20 Manual.pdf).
7. Two-color microarray analysis using the dye-swap strategy involves two DNA microarray slides that are hybridized with cRNA transcribed from cDNA to Target and Control RNA samples reciprocally labeled with two different fluorescent dyes (e.g., Cyanine 3: Cy3 or Cyanine 5: Cy5), respectively (Fig. 3).
8. In the microarray analysis service provided by HSS (http://www.hssnet.co.jp/e/2/2_4_3_1.html), RNA samples are subjected to quality control upon receipt using an Agilent Technologies Bioanalyzer.

9. In the example of global gene expression analysis of the Col::pRCY1#2 transformant (*see* **Note 1** and Fig. 2) presented here, the Target RNA sample is isolated from the Col::pRCY1#2 transformant and the Control RNA sample is isolated from wild-type Col-0 *A. thaliana*.
10. A known amount of an RNA Spike-In, an RNA transcript that is used to calibrate measurements in a DNA microarray, is mixed with the experimental sample. The measured degree of hybridization of the Spike-In with a specific probe on the array is used to normalize the signal intensity of the sample RNA.
11. Each fluorescent dye using for microarray analysis has slightly different biochemical properties that may affect subsequent data manipulation. To eliminate data artifacts, linear and LOWESS [24, 25] methods are widely used.
12. On a PC, data can be copied and pasted by double-clicking on the bottom-right corner of the selected cell to be copied. The formula in that cell is pasted into the cells in the column, until the row number in the neighboring column.
13. The goals and focus of the research project will determine how to treat any flagged features on the microarray. For example, only spots that are flagged as low quality in both the Cy3 and Cy5 signals of the microarray might be removed, thus increasing the number of genes analyzed.
14. A scatter plot of the logarithm of the (Cy3 signal/Cy5 signal) on microarray #1 against the logarithm of the (Cy5 signal/Cy3 signal) on microarray #2 is shown in Fig. 5. Considering that the same RNA samples were used in the dye-swap experiment, plotting of all spots along the line $\log(y) = \log(x)$ is ideal, and any differences in the $\log(y) = -\log(x)$ direction are due to differences in the dye properties. The number of spots excluded from analysis should vary depending on the quality of the microarray data or the research focus.
15. In this example, the ratio of Target sample (Col::pRCY1#2) to Control sample (wild-type Col-0) is 2. However, this ratio will vary depending on the quality of the samples or the research focus.
16. In this example, 189 and 21 spots were extracted as upregulated and downregulated genes, respectively.
17. The enrichment score of each annotation cluster is the geometric mean of all the enrichment *P*-values of each annotation term in the group. A higher score indicates that annotation term members in the annotation cluster are more enriched in a given study, because minus log transformation is applied to the average *P*-values. For example, an enrichment score of 1.3 is equivalent to *P*-value 0.05. Although more attention is typically given to

annotation clusters with scores above 1.3, clusters with lower enrichment scores could still potentially be of interest, and should be explored if possible [26].

18. There are two other advanced search forms that accept multiple gene queries: "EdgeAnnotation" and "CoExSearch". Using "EdgeAnnotation", coexpression information among multiple query genes themselves can be retrieved. This analysis can be used to search for coexpressed gene groups from the query gene list. Member genes of such groups may be involved in same biological pathway. When such strongly coexpressed genes are used as query genes, the "CoExSearch" form returns a unified coexpressed gene list, allowing additional genes involved in the same biological pathway to be found. To use these search functions, as with the "GeneTable" form, input AGI codes into the entry field of each form, and click the "submit" button.

Acknowledgement

We would like to thank Dr. Ken-Taro Sekine (Iwate Biotechnology Research Center, Kitakami, Japan) for kindly sharing microarray data.

References

1. Ding SW (2010) RNA-based antiviral immunity. Nat Rev Immunol 10:632–644
2. Díaz-Pendón JA, Ding S-W (2008) Direct and indirect roles of viral suppressors of RNA silencing in pathogenesis. Annu Rev Phytopathol 46:303–326
3. Cournoyer P, Dinesh-Kumar SP (2011) NB-LRR immune receptors in plant virus defense. In: Caranta C, Aranda MA, Tepfer M, Lopez-Moya JJ (eds) Recent advances in plant virology. Caister Academic Press, Norfolk, UK, pp 149–176
4. Golem S, Culver JN (2003) *Tobacco mosaic virus* induced alterations in the gene expression profile of *Arabidopsis thaliana*. Mol Plant Microbe Interact 16:681–688
5. Whitham SA, Quan S, Chang H-S, Cooper B, Estes B, Zhu T, Wang X, Hou Y-M (2003) Diverse RNA viruses elicit the expression of common sets of genes in susceptible *Arabidopsis thaliana* plants. Plant J 33:271–283
6. Huang Z, Yeakley JM, Garcia EW, Holdridge JD, Fan J-B Whitham SA (2005) Salicylic acid-dependent expression of host genes in compatible Arabidopsis-virus interactions. Plant Physiol 137:1147–1159
7. Yang C, Guo R, Jie F, Nettleton D, Peng J, Carr T, Yeakley JM, Fan J-B, Whitham SA (2007) Spatial analysis of *Arabidopsis thaliana* gene expression in response to *Turnip mosaic virus* infection. Mol Plant Microbe Interact 20:358–370
8. Catoni M, Miozzi L, Fiorilli V, Lanfranco L, Accotto GP (2009) Comparative analysis of expression profiles in shoots and roots of tomato systemically infected by *Tomato spotted wilt virus* reveals organ-specific transcriptional responses. Mol Plant Microbe Interact 22:1504–1513
9. Góngora-Castillo E, Ibarra-Laclette E, Trejo-Saavedra DL, Rivera-Bustamante RF (2012) Transcriptome analysis of symptomatic and recovered leaves of geminivirus-infected pepper (*Capsicum annuum*). Virol J 9:295
10. Lu J, Du Z-X, Kong J, Chen L-N, Qiu Y-H, Li G-F, Meng X-H, Zhu S-F (2012) Transcriptome analysis of *Nicotiana tabacum* infected by *Cucumber mosaic virus* during systemic symptom development. PLoS One 7:e43447
11. Postnikova OA, Nemchinov LG (2012) Comparative analysis of microarray data in Arabidopsis transcriptome during compatible interactions with plant viruses. Virol J 9:101

12. Allie F, Rey MEC (2013) Transcriptional alterations in model host, *Nicotiana benthamiana*, in response to infection by *South African cassava mosaic virus*. Eur J Plant Pathol 137: 765–785
13. Mochizuki T, Ogata Y, Hirata Y, Ohki ST (2014) Quantitative transcriptional changes associated with chlorosis severity in mosaic leaves of tobacco plants infected with *Cucumber mosaic virus*. Mol Plant Pathol 15:242
14. Ishihara T, Sakurai N, Sekine K-T, Hase S, Ikegami M, Shibata D, Takahashi H (2004) Comparative analysis of expressed sequence tags in resistant and susceptible ecotypes of *Arabidopsis thaliana* infected with cucumber mosaic virus. Plant Cell Physiol 45:470–480
15. Marathe R, Guan Z, Anandalakshmi R, Zhao H, Dinesh-Kumar SP (2004) Study of *Arabidopsis thaliana* resistome in response to cucumber mosaic virus infection using whole genome microarray. Plant Mol Biol 55:501–520
16. Zheng W, Ma L, Zhao J, Li Z, Sun F, Lu X (2013) Comparative transcriptome analysis of two rice varieties in response to rice stripe virus and small brown planthoppers during early interaction. PLoS One 8:e82126
17. Sekine K-T, Kawakami S, Hase S, Kubota M, Ichinose Y, Shah J, Kang H-G, Klessig DF, Takahashi H (2008) High level expression of a virus resistance gene, *RCY1*, confers extreme resistance to *Cucumber mosaic virus* in *Arabidopsis thaliana*. Mol Plant Microbe Interact 21: 1398–1407
18. Ouibrahim L, Caranta C (2013) Exploitation of natural genetic diversity to study plant-virus interaction: what can we learn from *Arabidopsis thaliana*? Mol Plant Pathol 14:844–854
19. Westwood JH, Carr JP (2012) *Cucumber mosaic virus - Arabidopsis* interaction: interplay of virulence strategies and plant responses. In: Sessa G (ed) Molecular plant immunity. Wiley-Blackwell, Oxford, UK
20. QIAGEN (2001) RNeasy mini handbook, 3rd edn. QIAGEN, Valencia, CA, pp 75–81
21. Agilent Fluorescent Linear Amplification Kit Protocol (2002) version 3.0, Agilent Technologies
22. Two-Color Microarray-Based Gene Expression Analysis with Tecan HS Pro Protocol (2008) version 5.7, Agilent Technologies
23. Agilent G2534A microarray hybridization chamber user guide (2009) version 1.0, Agilent Technologies
24. Cleveland WS (1979) Robust locally weighted regression and smoothing scatterplots. J Am Stat Assoc 74:829–836
25. Cleveland WS, Devlin S (1988) Locally weighted regression: an approach to regression analysis by local fitting. J Am Stat Assoc 83: 596–610
26. Huang DW, Sherman BT, Lempicki RA (2009) Systematic and integrative analysis of large gene lists using DAVID bioinformatics resources. Nat Protoc 4:44–57
27. Huang DW, Sherman BT, Lempicki RA (2009) Bioinformatics enrichment tools: paths toward the comprehensive functional analysis of large gene lists. Nucleic Acids Res 37: 1–13
28. Obayashi T, Okamura Y, Ito S, Tadaka S, Aoki Y, Shirota M, Kinoshita K (2014) ATTED-II in 2014: evaluation of gene coexpression in agriculturally important plants. Plant Cell Physiol 55:e6

Chapter 16

Construction of Infectious cDNA Clones Derived from the Potyviruses *Clover Yellow Vein Virus* and *Bean Yellow Mosaic Virus*

Kenji S. Nakahara, Kei Nishino, and Ichiro Uyeda

Abstract

Infectious cDNA clones are now indispensible tools for the genetic analysis of viral factors involved in viral virulence and host resistance. In addition, infectious cDNA-derived virus vectors that express foreign genes in infected plants enable the production of useful proteins at low cost and can confer novel crop traits. We constructed infectious cDNA clones derived from two potyviruses, *Clover yellow vein virus* and *Bean yellow mosaic virus*, which infect legume plants and cause disease. Here, we present our procedure for constructing these potyvirus infectious clones.

Key words Infectious cDNA clone, Potyvirus

1 Introduction

Cloned *Cauliflower mosaic virus* (CaMV) genomic DNA was shown to infect turnip plants in 1981 [1]. However, most plant viruses have RNA genomes. In 1984, *Brome mosaic virus* became the first plant RNA virus for which a full-length cDNA clone was shown to be infectious [2]. In the subsequent decade, infectious cDNA clones were produced from diverse RNA viruses, including *Tobacco vein mottle virus*, a potyvirus [3].

Potyviruses in the family *Potyviridae* infect a broad range of plant species and cause economic damage to diverse crops. Potyviruses are composed of a single stranded positive sense RNA approximately 10 kb in length that has a protein linked to the 5′ end and is covered by 2,000 copies of coat protein (CP). The potyvirus genome encodes a large polypeptide that is cleaved by virus-encoded proteases into at least ten functional proteins, including P1, HC-Pro, P3, 6K1, CI, 6K2, VPg, NIaPro, NIb, and CP.

Ichiro Uyeda and Chikara Masuta (eds.), *Plant Virology Protocols*, Methods in Molecular Biology, vol. 1236, DOI 10.1007/978-1-4939-1743-3_16, © Springer Science+Business Media New York 2015

An additional protein, P3N-PIPO has recently been shown to be expressed as a result of a ribosomal frameshift or transcriptional slippage [4].

We constructed a highly infectious cDNA derived from *Clover yellow vein virus* (ClYVV) in a plasmid vector in which the full-length cDNA corresponding to the ClYVV genomic RNA was positioned downstream of a CaMV 35S promoter with a 10-nucleotide poly-A tail added to the 3′ end of the cDNA [5]. We then developed a ClYVV vector derived from the ClYVV infectious clone to express a foreign gene in infected plants [6, 7]. ClYVV causes severe disease in legume plants including pea and broad bean. Viral and host factors involved in ClYVV virulence and host resistance have been identified using the ClYVV vector [8–13]. An infectious cDNA clone derived from another potyvirus, *Bean yellow mosaic virus* (BYMV)-CS, has been constructed recently in a similar manner. BYMV-CS was isolated from a red clover and also infects legume plants [14]. Here, we present the construction of the BYMV infectious clone to exemplify our procedure for producing a potyvirus infectious clone.

2 Materials

2.1 Construction of the BYMV Infectious cDNA Clone

1. TRIzol reagent (Life Technology).
2. Cloned AMV reverse transcriptase (Life Technology).
3. KOD-plus (TOYOBO, Otsu, Shiga, Japan).
4. pBluescript II SK– (Agilent Technologies, CA, USA).
5. T4 DNA ligase.
6. Restriction enzymes.
7. *E. coli* DH5α.
8. Synthetic DNAs for polymerase chain reaction (PCR) and cloning.

   P1: 5′-TTTGGAGAGGAAAATATAAAAACCAGAC-3′

   P2: 5′-TCTATTCCTTATGTACCGAAC-3′

   P3: 5′-GGCCACCTCGAGCCCGGGCACAAAACGCGATC ACTT-3′

   P4: 5′-CCTCTTGTCTCAATCTCTGTC-3′

   P5: 5′-GGCCACCTCGAGCCCGGGGGTGGGAYAATCAA CTGCAGA-3′

   P6: 5′-GGAAGAATTCGCGGCCGCAGATCACGATTGACA TCTCCT-GCTGTG-3′

   P7: 5′-ACACCCTTATGGTGATCATGG-3′

   P8: 5′-TCTGTCGACTTTTTTTTTTCTCGCTCTACAAA GATCAGGC-3′

P9: 5′-GGGATCAGGTACCCTAGAGATCCGTCAACA-3′

P10: 5′-GTTTTTATATTTTcctctccaaatgaaa-3′

LinkerF: 5′-GATCAGGTACCCTGATCTAGAATGACTAG TCTGTGCATGCATTGCTGCAGCCTGGTCGACA-3′

LinkerR: 5′-GATCTGTCGACCAGGCTGCAGCAATGCAG CACAGACTAGTCATTCTAGATCAGGGTACCT-3′

2.2 Inoculation of Plants with the Infectious cDNA Clone

1. A PDS-1000/He Particle Delivery System (Bio-Rad).
2. Rupture disks (900 psi).
3. Macrocarriers.

3 Methods

Following is the procedure to construct a BYMV-CS infectious cDNA clone in the plasmid vector pBluescript II SK– containing the 35S promoter of CaMV followed by the full-length cDNA of the viral genome with an added poly-A tail (*see* **Note 1**).

3.1 Cloning of BYMV Genomic cDNAs

1. Total RNA was extracted from BYMV-CS-infected broad bean leaves. One hundred milligrams of BYMV-infected broad bean leaves was ground with liquid nitrogen and dissolved in 1 ml of TRIzol reagent. After centrifugation at 19,000 × *g* for 5 min, the supernatant was transferred to a new microcentrifuge tube containing 200 μl of chloroform. After vortexing followed by centrifugation at 19,000 × *g* for 10 min, the precipitated RNA was dissolved in 400 μl of distilled water. The RNA was further purified by phenol–chloroform extraction and precipitation with ethanol.
2. Complementary DNAs (cDNAs) were reverse-transcribed from the extracted RNA. The extracted RNA (0.5 μg) was heat-denatured by incubation at 65 °C for 5 min and cDNAs corresponding to the RNA were synthesized in a 20-μl reaction solution containing 50 pmol of random 9-mers, 1 mM of dNTPs, 5 mM DTT, 40 U RNase inhibitor, and 15 U of cloned AMV reverse transcriptase at 45 °C for 1 h. Enzymes in the reaction solution were then denatured by incubation at 90 °C for 5 min.
3. Three cDNA fragments were amplified by PCR to construct a full-length BYMV genomic cDNA using a thermostable high-fidelity DNA polymerase, KOD-plus. The fragments were amplified using 15 pmol of primer pairs P3 and P4, P5 and P6, and P7 and P8 (Fig. 1) in 1× KOD-plus PCR buffer containing 0.2 mM dNTPs, 2.5 mM $MgSO_4$, 1 U of KOD-plus DNA polymerase, and cDNA corresponding to 25 ng of the extracted RNA (*see* **Note 2**).

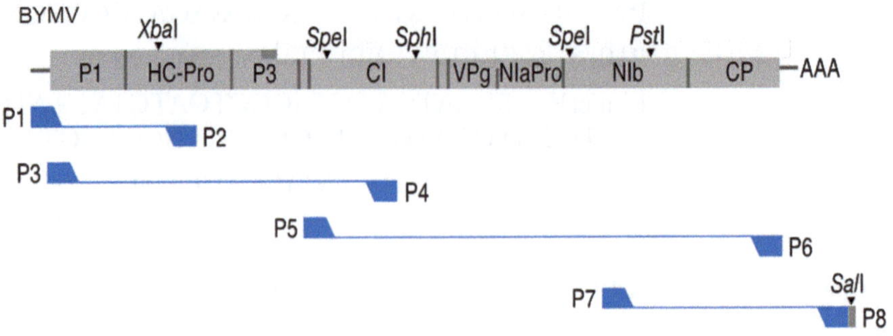

Fig. 1 Cloned cDNAs derived from BYMV-CS for construction of an infectious cDNA clone

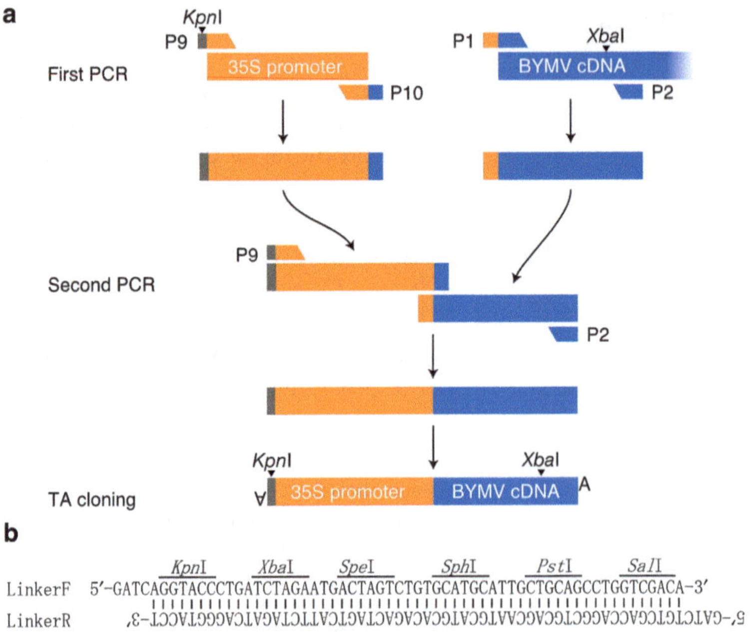

Fig. 2 Schematic representation of the combination of the CaMV 35S promoter and BYMV cDNA (**a**) and the polylinker inserted into the modified pBluescript II SK– vector to construct the BYMV infectious clone (**b**)

4. Amplified cDNA fragments were purified by phenol–chloroform extraction and precipitation with ethanol for cloning (*see* **Note 3**). After fractionation of the cDNAs by agarose gel electrophoresis, cDNA fragments of the expected sizes were cloned and sequenced (*see* **Note 4**). Clones having the expected nucleotide sequences were used in the subsequent steps.

3.2 Joining of the 35S Promoter to the 5′ Terminus of the BYMV cDNA

1. An initial PCR reaction containing KOD-plus, the 35S promoter, and the P9 and P10 primers (Fig. 2a) attached a *Kpn*I site and the 5′ terminal sequence of the BYMV genomic cDNA to the 5′ and 3′ ends, respectively, of the 35S promoter (*see* **Note 5**).

A second PCR reaction using the P1 and P2 primers (Figs. 1 and 2a) attached the 3′ terminal sequence of the 35S promoter to the 5′ end of the BYMV genomic cDNA (*see* **Note 6**).

2. After fractionation by agarose gel electrophoresis, both PCR products were used as templates for PCR using the P9 and P2 primers to combine the 35S promoter and the 5′ end of the BYMV genomic cDNA (Fig. 2a).
3. The combined fragment was cloned and sequenced. Clones having the expected sequence were selected for subsequent steps.

3.3 Modification of pBluescript II SK– (pBS) Vector with a Polylinker for Cloning the BYMV cDNA (pBS-Polylinker)

1. The multicloning site of pBS was removed, except for the *Bam*HI site, by polymerase chain reaction (PCR) using a high fidelity DNA polymerase such as KOD-plus or Phusion (Finnzymes, Espoo, Finland) and the 5′-ATATATGGATCC CGTACCCAATTCGCCCTATAGTGAGTCGTA-3′ and 5′-TT AAGGGATCCACCCAGCTTTTGTTCCCTTTAGTGAG GGTTA-3′ primers.
2. After PCR, the restriction enzyme *Dpn*I was added to the PCR mix, followed by incubation at 37 °C for 1 h to digest the template pBS vector. The PCR product, a linear pBS vector lacking the multicloning site but retaining the original *Bam*HI site (pBS-BamHI), was fractionated by agarose gel electrophoresis.
3. pBS-BamHI was digested with *Bam*HI and incubated with T4 DNA ligase. *E. coli* DH5α competent cells were transformed with the T4 DNA ligase-treated pBS-BamHI to clone and amplify the vector.
4. Two synthetic oligonucleotides (Fig. 2b, LinkerF and LinkerR) were annealed by incubating them in 25 mM Tris–HCl (pH 8.0) and 50 mM KCl at 90 °C for 10 min followed by slow cooling to room temperature.
5. After digestion with *Bam*HI, pBS-BamHI was ligated to the annealed oligonucleotides. DH5α was transformed with the ligated DNA to clone and amplify pBS-polylinker. The plasmid clones were sequenced to select plasmids possessing the expected polylinker sequence (Fig. 2b).

3.4 Construction of a Full-Length cDNA Derived from BYMV-CS Genomic RNA

1. The P3-P4 primer pair partial BYMV cDNA clone was digested with *Xba*I and *Spe*I and the digested cDNA fragment was fractionated by agarose gel electrophoresis (Fig. 1). After linearization by digestion with *Xba*I and *Spe*I, pBS-polylinker was ligated to the digested BYMV cDNA (Fig. 2b). The ligated product, pBS-polylinker containing *Xba*I–*Spe*I BYMV cDNA, was cloned (*see* **Note 7**).

2. The P5-P6 primer pair partial BYMV cDNA clone was digested with *Sph*I and *Pst*I to prepare the *Sph*I–*Pst*I fragment. The fractionated *Sph*I–*Pst*I fragment was cloned into pBS-polylinker.
3. The P7-P8 primer pair partial BYMV cDNA clone was digested with *Pst*I and *Sal*I. The fractionated *Pst*I-*Sal*I fragment was cloned into pBS-polylinker.
4. The combined 35S promoter and P1-P2 primer pair fragment clone (Fig. 2a) was digested with *Kpn*I and *Xba*I. The fractionated *Kpn*I-*Xba*I fragment was cloned into pBS-polylinker.
5. Finally, the pBS-polylinker containing the BYMV cDNAs and the P5-P6 primer pair partial BYMV cDNA was digested with *Spe*I. The *Spe*I–*Spe*I BYMV cDNA was ligated and cloned into pBS-polylinker (*see* **Note 8**). The pBS-polylinker construct contained the full-length BYMV genomic cDNA positioned downstream of the 35S promoter with a 10 nucleotide poly-A tail inserted at 3′ end of the BYMV cDNA between the polylinker *Kpn*I and *Sal*I sites (*see* **Note 9**).

3.5 Maintenance and Modification of the Infectious cDNA Clone

1. Infectious cDNA clones derived from *Pea seed borne mosaic virus* [15], *Potato virus Y* [16, 17], and *Watermelon mosaic virus* [18] were reported to be toxic to or unstable in *E. coli*. An infectious cDNA clone derived from a potyvirus may also be toxic and unstable in *E. coli*. This possibility should be considered when cloning and maintaining infectious cDNA plasmids in *E. coli* (*see* **Note 10**).
2. Infectious cDNA constructs derived from potyviruses can be used to express a foreign gene in plants. Foreign genes were inserted in frame between the P1 and HC-Pro or NIb and CP protein-coding sequences and expressed as a polyprotein, which was subsequently cleaved into several polypeptides by viral proteases (*see* **Note 11**).

3.6 Inoculation of Plants with the Infectious cDNA Clone

1. Detached pea and broad bean leaves were inoculated with the infectious plasmid using the following biolistics procedure (*see* **Note 12**).
2. Aliquots (0.18 mg) of tungsten particles (Bio-Rad) were mixed with 0.5 μg of the infectious plasmid in distilled water and calcium nitrate (1.25 M, pH 10.5) and incubated at room temperature for 5–10 min.
3. Ethanol (250 μl) was added to the plasmid and tungsten mixture and mixed followed by centrifugation at 500 rpm at room temperature for 3 min and removal of the supernatant.
4. The precipitate was suspended in 60 μl of ethanol. Bombardment was carried out with a PDS-1000/He Particle Delivery System using leaves from 1-week-old plants. After bombardment, the leaves were incubated in a petri dish with water-damped filter paper at 24 °C in darkness.

4 Notes

1. Determination of the precise nucleotide sequence of a potyviral genomic RNA is important to obtain a highly infectious cDNA clone. Accurate 5′ and 3′ terminal sequences (the number of sequential adenines in the 5′ end of many potyviruses, in particular), are especially important. We believe the best method to obtain the correct 5′ terminal sequence of a potyviral genome is the use of genomic RNA extracted from purified virions for 5′ RACE (rapid amplification of cDNA end) methods.
2. PCR primers were designed based on the BYMV-CS genome sequence (accession no. AB373203 in DDBJ/EMBL/GenBank).
3. Because of KOD-plus lacks terminal deoxynucleotidyl transferase activity, the amplified fragments have blunt ends. To clone molecules using the TA cloning technique, the fragments were incubated in PCR reaction solution containing Taq DNA polymerase, which has terminal deoxynucleotidyl transferase activity, at 72 °C for 30 min.
4. Prepare cloned cDNA fragments that are sequenced and make sure they have correct sequence.
5. Viral sequence is inserted downstream of 35S promoter sequence in a way that transcription starts exactly +1 of the viral sequence.
6. Two types of 5′ terminal sequence have been reported for BYMV. That is, they have either A4 or A6. We constructed full-length clones that have either A4 or A6 at 5′ terminus and tested its infectivity. Both were infectious. A recovered virus with A4 showed authentic symptoms on broad bean and pea, while that with A6 showed no symptoms on broad bean and systemic movement is much delayed.
7. First step is to create a plasmid that have only 5′ and 3′ with oligo A portions of the viral sequence.
8. Clones containing the *Spe*I–*Spe*I fragment in the correct orientation must be selected.
9. Infectivity of the infectious full-length cDNA clones varied depending on the length of oligo A tract at its 3′ terminus. cDNA clones devoid of oligo A tract is still infectious, although its infectivity is lower than that with oligo A. Transcriptional termination signal is not inserted following poly A tail.
10. We do not culture *E. Coli* transformants with the BYMV and ClYVV infectious clones or their derivatives for more than 14 h.

11. A foreign gene with the viral protease cleavage site, NIaPro, at the 3′ end can be inserted between P1 and HC-Pro or between NIb and CP [6, 7]. However, insertion of a foreign gene into these positions may significantly reduce the infectivity of the infectious clone. Which position is optimal for obtaining expression of an inserted foreign gene is determined empirically.
12. Although we usually inoculate the BYMV and ClYVV infectious clones into broad bean and pea leaves using a biolistics procedure [8, 19, 20], the clones can be inoculated using conventional mechanical inoculation with carborundum powder (600 mesh) [5].

References

1. Gronenborn B, Gardner RC, Schaefer S, Shepherd RJ (1981) Propagation of foreign DNA in plants using cauliflower mosaic virus as vector. Nature 294:773–776
2. Ahlquist P, French R, Janda M, Loesch-Fries LS (1984) Multicomponent RNA plant virus infection derived from cloned viral cDNA. Proc Natl Acad Sci U S A 81:7066–7070
3. Domier LL, Franklin KM, Hunt AG, Rhoads RE, Shaw JG (1989) Infectious in vitro transcripts from cloned cDNA of a potyvirus, tobacco vein mottling virus. Proc Natl Acad Sci U S A 86:3509–3513
4. Chung BY, Miller WA, Atkins JF, Firth AE (2008) An overlapping essential gene in the Potyviridae. Proc Natl Acad Sci U S A 105: 5897–5902
5. Takahashi Y, Takahashi T, Uyeda I (1997) A cDNA clone to clover yellow vein potyvirus genome is highly infectious. Virus Genes 14: 235–243
6. Masuta C, Yamana T, Tacahashi Y, Uyeda I, Sato M, Ueda S et al (2000) Development of clover yellow vein virus as an efficient, stable gene-expression system for legume species. Plant J 23:539–546
7. Wang ZD, Ueda S, Uyeda I, Yagihashi H, Sekiguchi H, Tacahashi Y et al (2003) Positional effect of gene insertion on genetic stability of a clover yellow vein virus-based expression vector. J Gen Plant Pathol 69: 327–334
8. Sato M, Masuta C, Uyeda I (2003) Natural resistance to Clover yellow vein virus in beans controlled by a single recessive locus. Mol Plant Microbe Interact 16:994–1002
9. Yambao ML, Yagihashi H, Sekiguchi H, Sekiguchi T, Sasaki T, Sato M et al (2008) Point mutations in helper component protease of clover yellow vein virus are associated with the attenuation of RNA-silencing suppression activity and symptom expression in broad bean. Arch Virol 153:105–115
10. Andrade M, Abe Y, Nakahara KS, Uyeda I (2009) The *cyv-2* resistance to *Clover yellow vein virus* in pea is controlled by the eukaryotic initiation factor 4E. J Gen Plant Pathol 75:241–249
11. Nakahara KS, Shimada R, Choi SH, Yamamoto H, Shao J, Uyeda I (2010) Involvement of the P1 cistron in overcoming eIF4E-mediated recessive resistance against *Clover yellow vein virus* in pea. Mol Plant Microbe Interact 23:1460–1469
12. Nakahara KS, Kitazawa H, Atsumi G, Choi SH, Suzuki Y, Uyeda I (2011) Screening and analysis of genes expressed upon infection of broad bean with Clover yellow vein virus causing lethal necrosis. Virol J 8:355
13. Choi SH, Hagiwara-Komoda Y, Nakahara KS, Atsumi G, Shimada R, Hisa Y et al (2013) Quantitative and qualitative involvement of P3N-PIPO in overcoming recessive resistance against Clover yellow vein virus in pea carrying the cyv1 gene. J Virol 87:7326–7337
14. Uyeda I, Takahashi T, Shikata E (1991) Relatedness of the nucleotide sequence of the 3′-terminal region of clover yellow vein potyvirus RNA to bean yellow mosaic potyvirus RNA. Intervirology 32:234–245
15. Johansen IE (1996) Intron insertion facilitates amplification of cloned virus cDNA in Escherichia coli while biological activity is reestablished after transcription in vivo. Proc Natl Acad Sci U S A 93:12400–12405
16. Jakab G, Droz E, Brigneti G, Baulcombe D, Malnoe P (1997) Infectious in vivo and in vitro

transcripts from a full-length cDNA clone of PVY-N605, a Swiss necrotic isolate of potato virus Y. J Gen Virol 78(Pt 12):3141–3145

17. Chikh Ali M, Said Omar A, Natsuaki T (2011) An infectious full-length cDNA clone of potato virus Y(NTN-NW), a recently reported strain of PVY that causes potato tuber necrotic ringspot disease. Arch Virol 156:2039–2043
18. Desbiez C, Chandeysson C, Lecoq H, Moury B (2012) A simple, rapid and efficient way to obtain infectious clones of potyviruses. J Virol Methods 183:94–97
19. Tacahashi Y, Uyeda I (1999) Restoration of the 3′ end of potyvirus RNA derived from Poly(A)-deficient infectious cDNA clones. Virology 265:147–152
20. Gal-On A, Meiri E, Elman C, Gray DJ, Gaba V (1997) Simple hand-held devices for the efficient infection of plants with viral-encoding constructs by particle bombardment. J Virol Methods 64:103–110

Chapter 17

Virus-Induced Gene Silencing of *N* Gene in Tobacco by *Apple Latent Spherical Virus* Vectors

Chunjiang Li and Nobuyuki Yoshikawa

Abstract

Virus infections induce an RNA-mediated defense that targets viral RNAs in a nucleotide sequence-specific manner in plants, commonly referred to as virus-induced gene silencing (VIGS). When the virus carries sequences of plant genes, it triggers virus-induced gene silencing (VIGS) and results in the degradation of mRNA of endogenous homologous gene. VIGS has been shown to have great potential as a reverse-genetics tool for studying of gene functions in plants, and it has several advantages over other functional genomics approaches. Here, we describe VIGS of *N* gene in tobacco cv. Xanthi nc by ALSV vectors containing fragments of *N* gene from *Nicotiana glutinosa*.

Key words *Apple latent spherical virus* (ALSV), Virus-induced gene silencing (VIGS), *N* gene in tobacco, *Tomato mosaic virus*

1 Introduction

Virus-induced RNA silencing (VIGS) is rapidly emerging as a powerful technology that exploits an RNA-mediated antiviral defense mechanism, and VIGS vectors have been widely applied to the investigation of the gene function by silencing of the plant endogenous genes without plant transformation [1–3]. Although the VIGS technology is simple and could rapidly produce experimental results, it also has some limitations. One is that most VIGS vectors can be only applied in limited host plant species, especially in the model plant *N. benthamiana* [4]. The other limitation is that some of the viruses used in VIGS induce symptoms that would confuse the phenotype caused by silencing the target gene [5]. Of course, the instability of VIGS-based vector carrying some foreign gene fragments from plants should also be considered, especially when it is used for silencing woody plants.

Thus far, several kinds of plant virus VIGS vectors have been used for investigating a wide variety of plant genes, and they are especially powerful for dissecting signaling components involved in

Ichiro Uyeda and Chikara Masuta (eds.), *Plant Virology Protocols*, Methods in Molecular Biology, vol. 1236,
DOI 10.1007/978-1-4939-1743-3_17, © Springer Science+Business Media New York 2015

disease resistance [6–12]. Until recently, many *R* genes have been identified in plants [13]. The tobacco *N* gene (an *R* gene derived from *N. glutinosa*) encodes a Toll-interleukin-1 receptor/nucleotide binding site/leucine-rich repeat (TIR-NSB-LRR) class protein and confers resistance to *Tobacco mosaic virus* (TMV) and *Tomato mosaic virus* (ToMV) [14]. *Tobacco rattle virus* (TRV) vector bearing the *N* gene fragment has been successfully used for silencing the *N* gene (the minimum *N* sequence required to confer resistance to TMV)-containing transgenic *N. benthamiana* plants which show resistance to TMV and restrict virus spread by the induction of hypersensitive response (HR). The Rar1-, EDS1-, and NPR1/NIM1-like genes required for N-mediated resistance were also investigated by using TRV vectors and the N gene-containing transgenic *N. benthamiana* plants [15]. However, there is still no VIGS vector suitable for investigating the genes associated with the N-mediated signaling pathway in original tobacco plants.

The *Apple latent spherical virus* (ALSV), a Cheravirus that has isometric virus particles c. 25 nm in diameter, contains two ssRNA species (RNA 1 and RNA 2) and three capsid proteins (Vp25, Vp20, and Vp24) [16, 17]. The ALSV infectious cDNA clones have been constructed, and the ALSV-RNA2 cDNA clone has been modified into viral vectors for the expression of foreign genes in plants [17, 18]. By now, ALSV vectors have also been shown to be a powerful tool for reliable and effective VIGS among a broad range of plants, including tobacco, tomato, *Arabidopsis*, cucurbits, legumes, and *Rosaceae* fruit trees [19–23].

Here, we describe the construction of ALSV vectors containing fragments of the *N* gene from *Nicotiana glutinosa*, as well as the induction of VIGS of the *N* gene in tobacco by inoculation of the ALSV vectors. In tobacco plants pre-inoculated with the ALSV vectors, ToMV moved systemically from inoculated leaves into upper leaves, indicating that expression of the *N* gene in tobacco was suppressed by VIGS.

2 Materials

All solutions are prepared using RNase-free deionized water in RNA experiments.

2.1 Propagation of Plants

1. Seeds of *Chenopodium quinoa* are planted in pots and grown in a greenhouse (18–25 °C, long-days condition) until reaching seven to eight true-leaf stages.
2. Seeds of *N. tabacum* nc. Xanthi nc and *N. glutinosa* are planted in moistened soil at 25 °C in a growth chamber until rooting.

2.2 Construction of Infectious cDNA Clones Containing Target Sequence

1. Infectious cDNA clone of ALSV-RNA2 (pEALSR2L5R5) (Fig. 1a).
2. *N. glutinosa*.

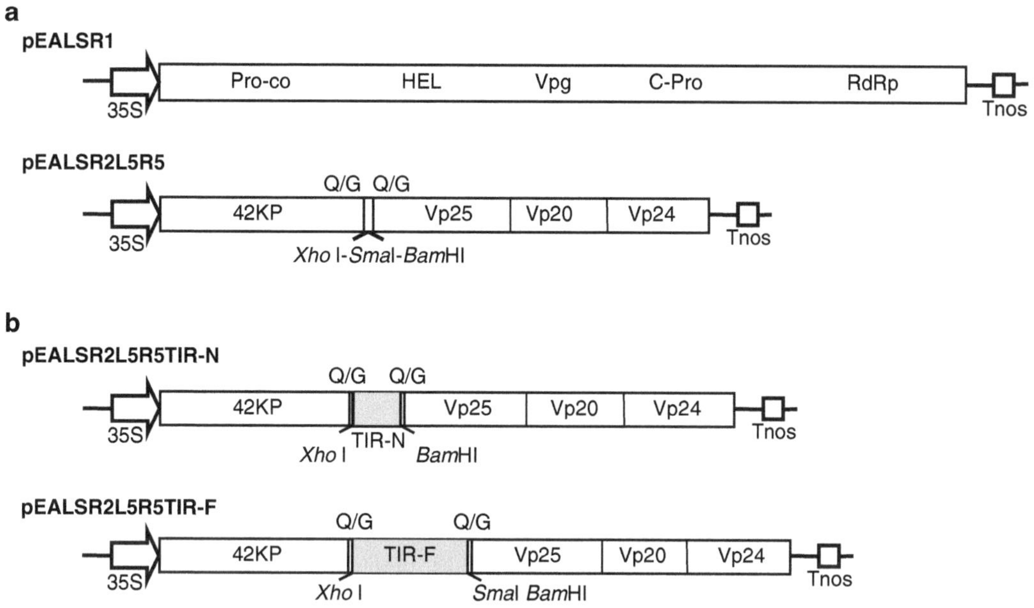

Fig. 1 (**a**) Schematic representation of the ALSV-RNA1 infectious cDNA clone (pEALSR1) and ALSV-RNA2 based vector (pEALSR2L5R5), which is modified from ALSV-RNA2 infectious cDNA clone pEALSR2 (ref. 17) for cloning foreign genes. (**b**) ALSV-RNA2 vectors containing TIR-N and TIR-F. The TIR-N and TIR-F fragments are inserted into the *Xho*I-*Bam*HI and *Xho*I-*Sma*I cloning sites of pEALSR2L5R5, respectively, and are shown by *grey boxes*. 35S, enhanced CaMV35S promoter; Tnos, nopaline synthase terminator; PRO-Co, protease cofactor; HEL, NTP-binding helicase; C-PRO, cysteine protease, POL, RNA polymerase; 42KP, 42K movement protein; Vp25, Vp20, and Vp24, capsid proteins

3. RNA extraction buffer: 0.1 M Tris–HCl (pH 8.0), 10 mM EDTA (pH 8.0), 0.1 M LiCl, and 1 % SDS.
4. 4 M LiCl.
5. Chloroform.
6. RNase free DNase I (Takara).
7. Oligo$(dT)_{12}$ primer, dNTPs, 5× RT buffer and ReverTra Ace (Toyobo) for first strand cDNA synthesis (Reverse Transcription, RT).
8. GeneAmp PCR System (Perkin Elmer).
9. *Taq* DNA polymerase, *Taq* DNA polymerase buffer, dNTPs for Polymerase chain reaction (PCR).
10. A primer pairs for amplification of the 479 bp fragment (corresponding to nt 52–530) coding for Tool-interleukin-1 receptor (TIR) region of *N* gene from *N. glutinosa* [Ngene-TIR(+): 5′-ATGGCATCTTCTTCTTCTTC-3′ and Ngene-TIR(−): 5′-TTGTCACGATTATCACAGGA-3′].
11. Two primer pairs, respectively for amplification of the 201 bp fragment (corresponding to nt 52–252) of the TIR region

[Ngene-TIR-Xho(+): 5′-CCGCTCGAGATGGCATCTTCTT CTTCTTC-3′, containing *Xho*I site (underlined) and Ngene-TIR-Bam(−): 5′-CGGGATCCTTGAGACTCTTCTATAGC TT-3′, containing *Bam* HI site (underlined)] and the 477 bp coding sequences (corresponding to nt 52–528) of the TIR region [Ngene-TIR-Xho(+) and Ngene-TIR-Sma(−): 5′-TCC CCCGGGGTCACGATTATCACAGGAGC-3′, containing *Sma*I site (underlined)].

12. DNA Ligation Kit (Takara, Kyoto, Japan).
13. pGEM®-T Easy vector (Promega, USA).
14. Competent cells: *Escherichia coli* (*E. coli*) DH5α or JM109.
15. SOC liquid medium.
16. Luria Broth (LB) liquid media and LB agar plate.
17. Ampicillin (50 mg/ml).

2.3 Purification of the Infectious cDNA Clones

1. LB agar plate, LB liquid media.
2. Ampicillin.
3. QIAGEN Plasmid Midi Kit (QIAGEN, Düsseldorf, Germany).
4. 2-propanol.
5. 70 % ethanol.
6. TE buffer: 10 mM Tris–HCl (pH 8.0), 1 mM EDTA (pH 8.0).

2.4 Inoculation to C. quinoa Plants

1. Purified ALSV-RNA1 infectious cDNA clone (pEALSR1) (Fig. 1a), pEALSR2L5R5-derived cDNA clones containing TIR sequences (Fig. 1b).
2. *C. quinoa* (7–8 leaf stages).
3. 600-mesh carborundum (NAKARAI TESQU INC., Kyoto, Japan).
4. Latex finger cots or Latex gloves.

2.5 Induction of VIGS in N. Tabacum cv. Xanthi nc by Inoculation of TIRN-ALSV and TIRF-ALSV

1. Inoculated leaves and upper leaves of *C. quinoa* plants systemically infected with TIRN-ALSV or TIRF-ALSV.
2. ALSV inoculation buffer: 0.1 M Tris–HCl (pH 7.8), 0.1 M NaCl, 0.005 M $MgCl_2$. Sterilized by autoclave.
3. Infected *C. quinoa* leaves with SU-ALSV (ref. 20).
4. Mortar and pestle.

2.6 Inoculation of Tomato Mosaic Virus (ToMV)

1. Tobacco plants infected with wtALSV, TIRN-ALSV, or TIRF-ALSV.
2. Purified ToMV (100 ng/ml).

2.7 Detection of the N Gene Fragments Inserted into ALSV Vector from Tobacco Plants Infected with TIRN-ALSV or TIRF-ALSV

1. RNA extraction buffer and etc.: *see* Subheading 2.2, **items 3–9**.
2. A primer pair for detection of the TIRN- and TIRF fragment of *N* gene: R2ALS1363(+) (5′-GCGAGGCACTCCTTA-3′) and R2ALS1511(−) (5′-GCAAGGTGGTCGTGA-3′). This primer pair is designed for the amplification of the foreign gene fragment, which is inserted into the cloning site of *XhoI-SmaI-BamHI* of ALSV-RNA2-derived vector.

2.8 Detection of ToMV from Inoculated Leaves of Tobacco Plants by Enzyme Linked Immunosorbent Assay (ELISA)

1. Microplates (Greiner bio-one, Germany).
2. 0.05 M carbonate buffer (coating buffer): 1.59 g Na_2CO_3, 2.93 g $NaHCO_3$ (pH 9.6). Sterilized by autoclave.
3. TTBS buffer: 0.02 M Tris–HCl (pH 7.5), 0.05 M NaCl, 2% Tween-20.
4. Anti-TMV serum.
5. Anti-rabbit IgG-AP linked antibody (Cell signaling).
6. *p*-nitrophenyl phosphate (Wako, Japan).
7. 10 % diethanolamine buffer.
8. Microplate reader (Bio-Rad, USA).
9. Ant-TMV serum.
10. Anti-rabbit IgG-AP linked antibody (Cell signaling).
11. 0.2 M Tris–HCl (pH 7.2).
12. Fast Red TR salt and Naphthol AS-MIX phosphate (Sigma).

3 Methods

3.1 Plants for Propagation of ALSV Vectors

1. *C. quinoa* is grown in a greenhouse under natural conditions at a minimum temperature of 20 °C with supplementary light (16 h of daylight) in the winter season until reaching the 7–8 true-leaf stage.
2. *N. tabacum* and *N. glutinosa* plants are grown in a growth chamber under 25 °C conditions and with 16 h of day length.

3.2 Construction of Apple Latent Spherical Virus Vectors Containing N Gene Fragments

3.2.1 Cloning of the TIR Region of N Gene from N. Glutinosa

1. *N. glutinosa* leaves (0.1 g) are homogenized with an RNA extraction buffer. After centrifugation (16,120×*g*, 4 °C, 5 min), the supernatants are re-extracted with chloroform and then mixed with an equal volume of 4 M LiCl. The pellets are collected by centrifugation, washed with 70 % ethanol, and dissolved in 50 μl of TE buffer. The nucleic acid samples are then treated with RNase free DNase I (Takara, Japan), and the resulting RNAs are finally dissolved in RNase-free deionized water or TE buffer at a concentration of 1 μg/μl.

2. The first strand cDNA is synthesized using 1 μl of RNA, 1 μl of Oligo(dT)12 primer, 4 μl 2.5 mM dNTP mixture (TaKaRa), 2 μl of 5× RT buffer (Toyobo), and 1 μl of ReverTra Ace (Toyobo) and 1 μl of RNase-free water. The reverse transcription (RT) reaction is conducted on GeneAmp PCR System 2400 (Perkin Elmer) as follows: 42 °C, 50 min; 99 °C 5 min; 4 °C, 5 min.
3. The sequence coding for TIR region (479 bp, nt 52–530) of tobacco *N* gene is amplified using a primer pair, Ngene-TIR(+) and Ngene-TIR(−) designed from *N* gene (Accession no. U15605), first strand cDNAs, and TaKaRa *Ex Taq*. The PCR reactions are conducted as follows: initial denaturation at 94 °C for 5 min, followed by 35 cycles of denaturation at 94 °C for 30 s, annealing at 55 °C for 30 s, and extension at 72 °C for 1 min.
4. The amplified TIR DNA fragment is cloned into pGEM®-T Easy vector, and the sequence of TIR is confirmed by sequencing.

3.2.2 Construction of ALSV Vectors Containing the N Gene Fragments

1. To construct ALSV vectors containing the *N* gene fragments (*see* **Note 1**), an ALSV-RNA2 vector (pEALSR2L5R5) is used (Fig. 1a).
2. The 201 bp fragment designated as TIRN (nt 52–252) and the 477 bp fragment named TIRF (nt 52–528) are amplified from the TIR DNA cloned in the pGEM®-T vector (Subheading 3.2.1) by using the primer pairs described in materials.
3. The TIRN PCR products are double-digested with *Xho*I and *Bam*HI and the TIRF products are double-digested with *Xho*I and *Sma*I. These DNA fragments are ligated with pEALSR2L5R5 restricted with the same enzymes. Competent *E. coli* DH5 α or JM 109 cells are used for ligation products of transformation.
4. Transformants are selected on LB agar plates containing 50 μg/ml ampicillin.
5. The plasmid DNAs are extracted from the transformed *E. coli* cells cultured in 2 ml of LB liquid medium at 37 °C by the alkaline lysis method.
6. The inserts are identified by digestion of plasmid DNAs with *Xho*I and *Bam*HI or *Sma*I, and electrophoresed on a 1 or 2 % agarose gel. The plasmids containing inserts with the desired length are then sequenced.
7. The resulting pEALSR2L5R5-based vector containing the TIRN fragment (201 bp) and the TIRF fragment (477 bp) is designated as pEALSR2L5R5TIR-N and pEALSR2TIR-F, respectively (Fig. 1b).

3.3 Purification of Infectious cDNA Clones

1. Plasmids, pEALSR2L5R5TIR-N, pEALSRL5R5TIR-F, and pEALSR1 (Fig. 1) are propagated in *E. coli* DH5α or JM109 cells.
2. *E. coli* cells transformed with the above plasmid are prepared on a freshly streaked LB agar medium containing 50 μg/ml of ampicillin.
3. A single colony of transformed *E. coli* is inoculated into a 100 ml of LB liquid medium (containing 50 μg/ml of ampicillin) in flask. Shake it vigorously at 37 °C overnight.
4. Plasmid DNAs are purified by the QIAGEN Plasmid Midi Kits (QIAGEN, Japan) according to the manufacturer's protocols, dissolved in TE buffer at a concentration of 2 μg/μl and stored at −20 °C or −80 °C until use for inoculation.

3.4 Inoculation of the Infectious cDNA Clones to C. quinoa

1. Purified pEALSR2L5R5TIR-N or pEALSR2L5R5TIR-F is mixed with pEALSR1 at a final concentration of 1 μg/μl each.
2. The surface of *C. quinoa* leaves are dusted with a thin coating of 600 mesh carborundum (NAKARAI TESQU INC. Kyoto, Japan) just before inoculation with plasmids.
3. The mixture of pEALSR1 and pEALSR2L5R5TIR-N or pEALSR2L5R5TIR-F (10 μl) is dropped onto the tip of a finger wrapped with latex finger cots (or latex gloves) and mechanically inoculated onto a leaf of *C. quinoa* plant (7–8 true-leaf stage). The inoculation of infectious cDNA clones is conducted onto four leaves (forth to seventh true leaves) per plant.
4. A surface of inoculated leaves is flushed with a sufficient amount of water to remove carborundum.
5. The symptoms consisting of chlorotic spots and mosaic appeared on the upper leaves 2–3 weeks post-inoculation (Fig. 2). The pALSR2L5R5TIR-N and pALSR2L5R5TIR-F-derived viruses are designated as TIRN-ALSV and TIRF-ALSV, respectively.
6. Leaves with symptoms are homogenized with 2–3 volumes (w/v) of inoculation buffer in a mortar and pestle. The extract is mechanically inoculated onto leaves of *C. quinoa* (7–8 leaf-stage).
7. Inoculated leaves and upper leaves with symptoms (7–10 dpi) are collected and stored at −80 °C until use (*see* **Note 2**).

3.5 Induction of VIGS by Inoculation of TIRN-ALSV and TIRF-ALSV Onto Tobacco cv. Xanthi nc

1. Leaves of *C. quinoa* (stored at −80 °C) infected with TIRN-ALSV, TIRF-ALS and wtALSV are ground in 2–3 volumes of inoculation buffer (i.e., 1:2 or 1:3, w/v) in a mortar and pestle.
2. A cotton swab is soaked with leaf sap and rubbed firmly but gently on the surface of the leaves. These viruses are mechanically inoculated onto three leaves (the first to third true leaves

Fig. 2 Chlorotic and mosaic symptoms on leaves of *C. quinoa* plants infected with ALSV vector

of four-leaf stage) of tobacco (cv. Xanthi nc). The surface of inoculated leaves is flushed with a sufficient amount of water to remove carborundum.

3. Inoculated plants are grown in a growth chamber (*see* **Note 3**).

3.6 Inoculation of ToMV

1. At 30 days after inoculation of wtALSV, TIRN-ALSV, or TIRF-ALSV, the ninth to eleventh upper leaves of tobacco plants are inoculated with 50 μl of purified ToMV (100 ng/ml).

3.7 Detection of ToMV in Inoculated Leaves of Tobacco Plants Pre-infected with ALSV Vectors by ELISA

1. At 20 days after inoculation with ToMV, ten areas with necrotic lesion and ten green regions on tobacco leaves infected with wtALSV, TIRN-ALSV, or TIRF-ALSV are randomly sampled with cork borer. As a negative control for ELISA, three leaf samples from healthy tobacco are used.
2. Samples are homogenized with 500 μl of 0.05 M carbonate buffer (pH 9.6) in a mortar and pestle. After centrifuging for 5 min at 16120×*g* (at 4 °C), 150 μl of supernatant is added into the wells (two wells per sample) of plate and incubated at 37 °C for 1 h.
3. After washing with TTBS 10 times, anti-TMV serum (150 μl) absorbed with crude sap with healthy tobacco is added into the well and incubated overnight at 4 °C.
4. After washing with TTBS, an AP-conjugated anti-ALSV rabbit IgG diluted at 1:1,000 in TTBS (150 μl) is added into the well and the plate was incubated at 37 °C for 1 h.
5. The wells are washed 10 times with TTBS and the color is developed by incubation with a solution of *p*-nitrophenyl phosphate substrate (0.67 mg/ml, Wako, Japan) in 10 % diethanolamine buffer.
6. Finally, the absorbance is detected at 405 nm with a microplate reader (Bio-rad, USA).

3.8 Silencing of the *N* Gene in Tobacco with ALSV Vectors

3.8.1 Local Lesion on Inoculated Leaves of Tobacco Plants Pre-inoculated with ALSV Vectors

To monitor the progress of VIGS in each leaf of tobacco, the SU-ALSV (ref. 20) was also inoculated, and the result showed that the upper leaves of tobacco started turning yellow after 7–10 days (Fig. 3a). Tobacco plants inoculated with TIRN-ALSV and TIRF-ALSV did not show any symptoms (Fig. 3a). After 30 days, upper leaves (the ninth to eleventh leaves) of tobacco plants infected with wtALSV (wtALSV-tobacco), TIRN-ALSV (TIRN-ALSV-tobacco), and TIRF-ALSV (TIRF-ALSV-tobacco) were then inoculated with purified ToMV (100 ng/ml). In wtALSV-tobacco inoculated with ToMV, local necrotic lesions appeared on the inoculated leaves at 2–3 dpi (Fig. 3b). ToMV had localized near local lesions and never moved to upper uninoculated leaves. On the other hand, in TIRN-ALSV-tobacco and TIRF-ALSV-tobacco inoculated with ToMV, necrotic lesions started to develop at 3–4 dpi, and the lesion numbers were about half of that in wtALSV-tobacco (Table 1 and Fig. 3b). ELISA of ToMV-inoculated leaves showed that ToMV was distributed on green tissues among necrotic lesions in TIRN-ALSV-tobacco and TIRF-ALSV-tobacco leaves. In contrast, ToMV was detected only on tissues containing local lesions in wtALSV-tobacco plants.

3.8.2 Systemic Infection of ToMV in Tobacco Pre-inoculated with ALSV Vectors

TIRN-ALSV-tobacco and TIRF-ALSV-tobacco inoculated with ToMV developed necrosis on the stem and petioles of newly developed upper leaves and stunted severely (Fig. 3c, right), indicating that ToMV moved to upper uninoculated leaves from the inoculated leaves. In contrast, upper leaves of wtALSV-tobacco inoculated with ToMV developed normally (Fig. 3c, left).

Among 11 TIRN-tobacco inoculated with ToMV, 2 plants were systemically infected with ToMV (2/11, 18 %). On the other hand, all TIRF-ALSV-tobacco were systemically infected with ToMV (5/5, 100 %). Western blot analysis showed that ToMV is only found on the necrotic area of upper leaves, and the green area did not contain ToMV, indicating that the expression of *N* gene could be silenced by ALSV vector, but the effect of silencing is not complete.

4 Notes

1. Because ALSV-RNA2 is expressed as polyprotein, the foreign gene fragments must be inserted in-frame at the *Xho*I, *Sma*I, and/or *Bam*HI sites, and they must not contain a stop codon in the foreign gene sequence.
2. Leaves of *C. quinoa* grown in a greenhouse do not show obvious chlorotic spot symptoms, when the temperature of greenhouse exceeds 30 °C.
3. Inoculated plants should determine whether the virus lost the target sequence by RT-PCR. The stability of the foreign gene in the ALSV vector depends on the size and sequence of the inserted gene (ref. 20).

Fig. 3 (**a**) Tobacco plants infected with SU-ALSV (*left*) and TIRF-ALSV (*right*). (**b**) Appearance of necrotic lesions on ToMV-inoculated leaves of healthy tobacco (H), and tobacco pre-infected with wtALSV, and TIRF-ALSV. Necrotic local legions appeared on inoculated leaf of healthy plants and plants pre-infected with wtALSV 3–4 days after ToMV inoculation and of plants pre-infected with TIRF-ALSV 4–5 days after ToMV inoculation. (**c**) Severe stunting of tobacco plants pre-infected with TIRF-ALSV (*right three plants*) compared with normal development of plants pre-infected with wtALSV (*left three plants*) subsequent to 24 days after ToMV inoculation

Table 1
Number of local lesions induced by ToMV in leaves of tobacco (cv. Xanthi nc) preinoculated with wtALSV, TIRN-ALSV, and TIRF-ALSV

Virus vector	Number of inoculated leaves	Number of lesions[a] ± S.D
wtALSV	18	14.97 ± 4.88
TIRN-ALSV	15	5.97 ± 3.67
TIRF-ALSV	15	7.41 ± 3.86

[a]Average number of local lesions ± standard deviation per 1 cm^2 of inoculated leaves

References

1. Lu R et al (2003) Virus-induced gene silencing in plants. Methods 30:296–303
2. Robertson D (2004) VIGS vector for gene silencing: many targets, many tools. Annu Rev Plant Biol 55:495–519
3. Purkayastha A, Dasgupta I (2009) Virus-induced gene silencing: a versatile tool for discovery of gene functions in plants. Plant Physiol Biochem 47:967–976
4. Goodin MM et al (2008) Nicotiana benthamiana: its history and future as a model for plant-pathogen interactions. Mol Plant-Microbe Interact 21:1015–1026
5. Burch-Smith TM et al (2004) Applications and advantages of virus-induced gene silencing for gene function studies in plants. Plant J 39: 734–746
6. Hofgen R et al (1994) A visible marker for antisense mRNA expression in plants: inhibition of chlorophyll synthesis with a glutamate-1-semialdehyde aminotransferase antisense gene. Proc Natl Acad Sci U S A 91:1726–1730
7. Romeis T et al (2001) Calcium-dependent protein kinases play an essential role in a plant defence response. EMBO J 20:5556–5567
8. Hiriart JB et al (2002) Suppression of a key gene involved in chlorophyll biosynthesis by means of virus-inducing gene silencing. Plant Mol Biol 50:213–224
9. Liu Y, Schiff M, Dinesh-Kumar SP (2002) Virus-induced gene silencing in tomato. Plant J 31:777–786
10. Cai X-Z et al (2006) Development of a virus-induced gene-silencing system for functional analysis of the RPS2-dependent resistance signalling pathways in Arabidopsis. Plant Mol Biol 62:223–232
11. Deng X et al (2012) Virus-induced gene silencing for Asteraceae - a reverse genetics approach for functional genomics in Gerbera hybrid. Plant Biotechnol J 10:970–978
12. Manmathan H et al (2013) Virus-induced gene silencing of *Arabidopsis thaliana* gene homologues in wheat identifies genes conferring improved drought tolerance. J Exp Bot 64:1381–1392
13. Kang B, Yeam I, Jahn MM (2005) Genetics of plant virus resistance. Annu Rev Phytopathol 43:581–621
14. Whitham S et al (1994) The product of the tobacco mosaic virus resistance gene N: similarity to Toll and the interleukin-1 receptor. Cell 78:1101–1115
15. Liu Y et al (2002) Tobacco Rar1, EDS1 and NPR1/NIM1 like genes are required for N-mediated resistance to tobacco mosaic virus. Plant J 30:415–429
16. Li C et al (2000) Nucleotide sequence and genome organization of Apple latent spherical virus: a new virus classified into the family Comoviridae. J Gen Virol 81:541–547
17. Li C et al (2004) Stable expression of foreign proteins in herbaceous and apple plants using Apple latent spherical virus RNA2 vectors. Arch Virol 149:1541–1558
18. Takahashi T, Yoshikawa N (2008) Analysis of cell-to-cell and long-distance movement of Apple latent spherical virus in infected plants using green, cyan, and yellow fluorescent proteins. In: Forester G, Johansen E, Hong Y, Nagy P (eds) Plant virology protocols, vol 451, Methods in molecular biology. Humana Press, Totowa, NJ, pp 545–554
19. Yaegashi H et al (2007) Characterization of virus-induced gene silencing in tobacco plants infected with apple latent spherical virus. Arch Virol 152:1839–1849
20. Igarashi A et al (2009) Apple latent spherical virus vectors for reliable and effective virusinduced gene silencing among a broad range of plants infecting tobacco, tomato, Arabidopsis thaliana, cucurbits, and legumes. Virology 386:407–416

21. Yamagishi N, Yoshikawa N (2009) Virus-induced gene silencing in soybean seeds and the emergence of soybean plants with Apple latent spherical virus vectors. Plant Mol Biol 71:15–24
22. Yamagishi N et al (2011) Promotion of flowering and reduction of a generation time in apple seedlings by ectopical expression of the Arabidopsis thaliana FT gene using the Apple latent spherical virus vector. Plant Mol Biol 75:193–204
23. Yamagishi N, Kishigami R, Yoshikawa N (2014) Reduced generation time of apple seedlings to within a year by means of a plant virus vector: a new plant-breeding technique with no transmission of genetic modification to the next generation. Plant Biotechnol J 12:60

Chapter 18

Simplified Methods for the Construction of RNA and DNA Virus Infectious Clones

Tatsuya Nagata and Alice Kazuko Inoue-Nagata

Abstract

Infectious virus clones are one of the most powerful tools in plant pathology, molecular biology, and biotechnology. The construction of infectious clones of RNA and DNA viruses, however, usually requires laborious cloning and subcloning steps. In addition, instability of the RNA virus genome is frequently reported after its introduction into the vector and transference to *Escherichia coli*. These difficulties hamper the cloning procedures, making it tedious and cumbersome. This chapter describes two protocols for a simple construction of infectious viruses, an RNA virus, the tobamovirus *Pepper mild mottle virus*, and a DNA virus, a bipartite begomovirus. For this purpose, the strategy of overlap-extension PCR was used for the construction of infectious tobamovirus clone and of rolling circle amplification (RCA) for the construction of a dimeric form of the begomovirus clone.

Key words *Tobamovirus*, *Begomovirus*, Infectious clone, PMMoV

1 Introduction

Pepper mild mottle virus (PMMoV) is a member of the genus *Tobamovirus* that has a single stranded RNA genome in the positive sense polarity. The virus genome has 6,356 nt, four ORFs and no poly(A) extension. Infectious cDNA clones of plant RNA viruses are very useful tools for the basic studies on viral replication, virus–host interactions and, more recently, for biotechnological purposes [1–4]. Typically, an infectious clone under the T7 RNA polymerase promoter, which normally requires an in vitro transcription step before inoculation, can be constructed by a simple RT-PCR procedure since the minimum size of the T7 promoter is small (17 nt). Hence the promoter can be incorporated into the forward primer [5–10]. Despite the convenience of this cloning procedure and the small size of the resulting plasmids, this technique requires careful handling of the RNA transcripts which are prone to easy degradation. A more convenient strategy relies in

Ichiro Uyeda and Chikara Masuta (eds.), *Plant Virology Protocols*, Methods in Molecular Biology, vol. 1236,
DOI 10.1007/978-1-4939-1743-3_18, © Springer Science+Business Media New York 2015

the insertion of the genomic cDNA into a binary vector downstream of a plant-recognizable promoter [11, 12]. However, this strategy is, in many cases, troublesome due to the plasmid instability in the transformed *Escherichia coli*. Larger plasmids are unstable for various reasons [13]. Therefore, the cloning steps are often challenging due to the lower transformation efficiency and to the occurrence of deletions in essential parts of the viral genome within the *E. coli* host [1, 14, 15]. Recently, the insertion of multiple plant-gene introns has been used to improve the insert stability of plant viral vectors in *E. coli* and plant cells [3, 16, 17]. However, this strategy requires a more laborious cloning strategy.

In this protocol, we propose an approach using a minimum number of cloning steps to successfully construct an agro-infectious cDNA clone containing the whole PMMoV genome in a binary vector. For this purpose, the RT-PCR procedure was optimized, the more suitable *E. coli* strain Stbl4 was used, and the incubation temperature of *E. coli* culture was lowered.

A completely distinct approach is suggested for producing infectious clones from a circular single stranded DNA virus. A geminivirus has a single stranded circular DNA as the genome, and is unique amongst viruses as the virion is geminated. The family *Geminiviridae* is formed by seven genera, and except for one genus, the members possess a monopartite genome. They replicate by a mechanism of rolling circle, and a typical origin of replication (*ori*) is present in all members present within the sequence TAATATTAC (with some variations) [18, 19]. The genus *Begomovirus* contains monopartite (genome of 2.6–2.8 kb) and bipartite (ca. 5.6 kb) species. For the bipartite viruses, both DNA components (known as DNA-A and DNA-B) are necessary for infectivity [20]. As a conventional approach for a begomovirus genome cloning, the replicative form, i.e., double stranded DNA, is first isolated by differential centrifugation and ethanol precipitation procedures. These molecules are digested by restriction enzymes and cloned into plasmid vectors for sequencing. For an infectious clone, after restriction enzyme mapping of the genome and localization of the *ori* region, a new construct is prepared which contains the complete genome and a duplicated *ori* region [21], usually referred as 1.5 genomic unit copy clone. This duplicated *ori* enables the release of the entire genome into the cell and recovering of the intact virus genome. In 2004, a simple but powerful method was described for cloning a begomovirus genome based on the rolling circle amplification (RCA) method using the Illustra TempliPhi Genome Amplification kit, patented by GE Healthcare Life Sciences [22]. It uses the bacteriophage derived *phi*-29 DNA polymerase. This polymerase has an exceptional strand displacement property, generating high molecular weight molecules formed mainly by double stranded DNA. This method can easily produce a large amount of DNA, in an isothermal reaction. It facilitated the cloning procedures of the circular viral DNA genome, especially useful for those viruses in a low titer in plants. Here, a simple method is described for generation of infectious begomovirus clones using RCA for genome enrichment and partial digestion for isolation of dimeric molecules with posterior insertion in a binary vector for agroinoculation in plants [23].

2 Materials

2.1 RNA Genome Amplification by PCR and Cloning in a Binary Vector

1. Purified total RNA extracted from a *Nicotiana benthamiana* plant infected with PMMoV.
2. Reverse primer *Eco*RI-MluI PMMoV Rev (5′-TGA GAG AAT TCA CGC GTG GGC CGC TAC CCG C-3′, *Eco*RI and *Mlu*I sequences are underlined) and forward primer BamHI-PMMoV For (5′-TGA GAG GAT CC*T AAT ACG ACT CAC TAT A*GT AAA TTT TTC ACA ATT TAA CAA C-3′, BamHI site is underlined and the T7 RNA polymerase promoter is in italics) (Fig. 1a).
3. SuperScript III transcriptase (Invitrogen).
4. RNaseOUT RNase inhibitor (Invitrogen).

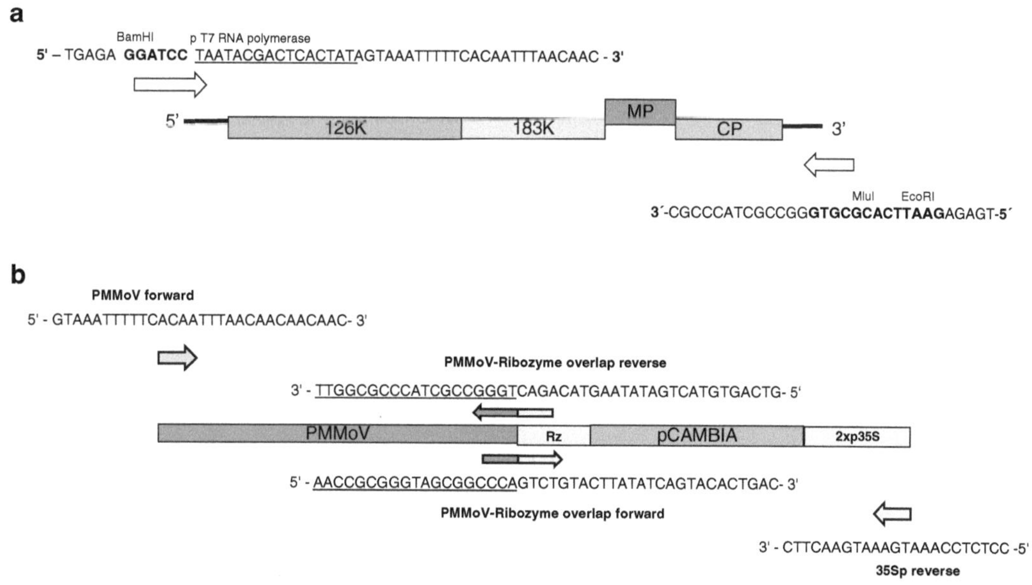

Fig. 1 Schematic view of the construction of an infectious clone of PMMoV based on PCR and overlap-extension PCR. (**a**) Primers are designed for amplification and cloning the complete genome in a TA-vector. The T7 promoter is *underlined* and restriction enzymes indicated. (**b**) Primers used for overlap-extension PCR producing a DNA fragment with the viral genome and the vector. The overlap region (sense and antisense orientation) consisting of the 3′ upmost end of PMMoV is *underlined*

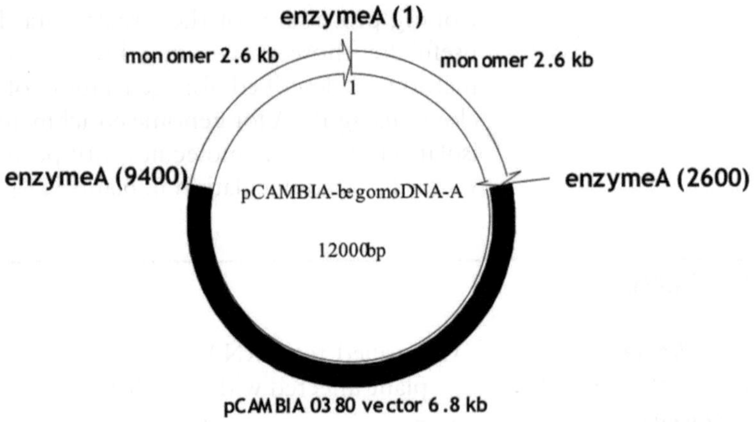

Fig. 2 Schematic view of the infectious begomovirus DNA-A clone in pCAMBIA 0380. The vector is linearized by enzyme A and ligated with the dimer, produced after RCA of viral DNA and partially digested with enzyme A

5. RNase H (Invitrogen).
6. LongAmp *Taq* DNA polymerase (New England Biolabs—NEB).
7. Cristal violet.
8. Illustra GFX PCR DNA and Gel Band Purification kit (GE Healthcare Life Sciences).
9. pCR4-TOPO TA cloning kit (Invitrogen).
10. *E. coli* strain DH5α and Stbl4 (Invitrogen).
11. Primer set for overlap-extension PCR: PMMoV forward primer (5′-GTA AAT TTT TCA CAA TTT AAC AAC AAC AAC-3′) (Fig. 1b); PMMoV-Ribozyme overlap reverse primer (5′-GTC AGT GTA CTG ATA TAA GTA CAG ACT GGG CCG CTA CCC GCG GTT-3′, the 3′ upmost terminal end of PMMoV is underlined) (Fig. 2b). The pTRV-MCS backbone (pCAMBIA 0390 derivative) was amplified using the PMMoV-Ribozyme overlap forward primer (5′-AAC CGC GGG TAG CGG CCC AGT CTG TAC TTA TAT CAG TAC ACT GAC-3′, the 3′ upmost end of PMMoV genome is underlined) and the 35Sp reverse primer (5′-CCT CTC CAA ATG AAA TGA ACT TC-3′) (Fig. 1b).
12. Phusion High-Fidelity DNA polymerase (NEB).
13. pTRV2-MCS, containing the duplicated CaMV 35S promoter and the *Subterranean clover mottle virus* satellite RNA ribozyme (AF406991).
14. QIAEX II Gel Extraction kit (Qiagen).
15. T4 Polynucleotide Kinase (PNK) (NEB).
16. MF-Millipore membrane mixed cellulose esters, 0.025 μm (Millipore).

2.2 DNA Genome Amplification by RCA and Cloning in a Binary Vector

1. Total DNA extracted from a plant infected with a bipartite begomovirus.
2. Illustra TempliPhi Amplification Kit (GE Healthcare Life Sciences).
3. Binary vector pCAMBIA 0380 (Cambia Labs) or any vector of choice.
4. Restriction enzymes present in the multiple cloning site of the vector.
5. Specific probe for DNA-A; specific probe for DNA-B (for example, nonradioactive, digoxigenin labeled probes).
6. T4 DNA ligase.
7. Electrocompetent *E. coli* cells strain DH5α.
8. DNA-A and DNA-B specific primers for PCR detection.
9. Plasmid extraction kit.

2.3 Agrobacterium Transfection and Infiltration of Infectious Clones

1. *Agrobacterium tumefaciens* (GV3101 strain).
2. MES (Sigma-Aldrich).
3. Acetosyringone (3′, 5′-Dimethoxy-4′-hydroxyacetophenone) (Sigma-Aldrich).
4. Murashige-Skoog medium powder (M5524, Sigma-Aldrich).
5. *Nicotiana benthamiana* plants (wild-type).
6. Syringe (1–2.5 ml) without needle.

3 Methods

For the construction of infectious clones of RNA viruses, basically two strategies are used: cloning of the genomic DNA in a small plasmid vector harboring a bacteriophage promoter (such as T7 RNA polymerase promoter; *see* **Note 1**) or in a binary vector with a plant recognizable promoter. The selection of the best restriction enzyme for cloning may be difficult depending on the limited options of restriction enzyme sites in the plasmid vector and the virus genome sequence. The procedure described here for the construction of infectious cDNA in a binary vector does not rely on the restriction enzyme digestion steps, but in a two-step cloning method, the first based on the TA-cloning step, followed by sub-cloning in a binary vector using the overlap-extension PCR (Fig. 1), then a step of ligation. Restriction enzymes and the minimum T7 promoter were added in the primers used in the first cloning step to allow the synthesis of in vitro transcripts-based infectious clones to perform a preliminary infectivity testing of the cDNA. These regions can be eliminated from the primers if the in vitro transcripts will not be considered. The dimer cloning strategy used for begomoviruses requires the generation of two genomic units as the presence of the duplicated *ori* region increases the infection efficiency (Fig. 2). These dimeric units can be generated by partial digestion of the multimeric genome concatemers yielded by the *phi*-29 DNA polymerase reaction. This method can be applied for any circular DNA virus, especially for geminiviruses. Making a 1.5 genome copy construct by conventional subcloning procedures is not complicated, but after the advent of the RCA procedure, the generation of constructs with genomic dimers became simpler and more convenient. The most laborious step is the definition of the partial digestion condition to achieve the highest amount of dimeric molecules.

3.1 RNA Genome Amplification by RT-PCR and Cloning into pCR4-TOPO Vector

1. Purified total RNA extracted from 100 mg of PMMoV infected plant is stored at −80 °C in 70–80 % ethanol solution (*see* **Note 2**). Immediately before using it, pellet the RNA at 13,000 × *g* for 10 min. Dry the pellet placing the open tube onto the ice for 10 min, then resuspend it with 20 μl of RNase-free water.
2. In a cold temperature condition (either on ice or inside the cold room), add 1 μl of 10 mM dNTPs and 1 μl of 50 μM *EcoRI-MluI* PMMoV Rev reverse primer to 10 μl of resuspended RNA (half volume of **step 1**); incubate the solution mix at 75 °C in a 0.2 ml tube for 5 min, then quickly chill the solution on ice for >2 min. Reverse transcription is done by adding 4 μl of 5× first strand buffer, 2 μl of DTT (100 mM), 1 μl of RNaseOUT (40 U/μl) and 1 μl of SuperScript III (200 U/μl) (*see* **Note 3**). Incubate the solution at 47 °C for 60 min, then heat-inactivate the enzyme at 85 °C for 10 min. For the fragmentation of the RNA strand, add 1 μl of RNase H (2 U/μl) to the solution and incubate at 37 °C for 30 min.
3. For PCR (*see* **Note 4**), use LongAmp *Taq* DNA polymerase (NEB). In a PCR tube, add 2 μl of the cDNA solution, 5 μl of 5× LongAmp buffer, 3 μl of 2.5 mM dNTPs, 1 μl each forward (BamHI-PMMoV For) and reverse (*Eco*RI-*Mlu*I PMMoV Rev) primers (10 μM), 1 μl of LongAmp *Taq* DNA polymerase (2.5 U; *see* **Note 5**), and DNase-free water up to 25 μl. PCR cycling condition: 80 °C for 1 min (manual Hot start, transfer samples from the ice to the hot thermal-cycler block during this period); 94 °C for 2 min (pre-denaturing step), 35 cycles of 94 °C for 30 s/55 °C for 1 min/65 °C for 7 min, then 65 °C for 20 min as post-PCR incubation.
4. Amplified PCR fragments are gel-purified. Perform agarose gel (0.8 %) electrophoresis with 0.5× TBE buffer. Visualize DNA bands by crystal violet staining (40 μg/ml for 10–15 min; *see* **Note 6**). Excise the DNA band under normal light and purify the DNA using Illustra GFX PCR DNA and Gel Band Purification kit (GE Healthcare Life Science; *see* **Note 7**).

5. Purified genomic cDNA is inserted in pCR4-TOPO vector (Invitrogen; *see* **Note 8**). Add 50 ng of purified cDNA into 10 ng vector solution (1 μl). Add 1 μl of the diluted salt solution of the kit (300 mM NaCl, 15 mM $MgCl_2$) and DNase-free water up to 6 μl. Incubate for 30 min at room temperature and transform *E. coli* cells (DH5α) by electroporation.
6. Mini-scale plasmid extraction is done by a standard protocol and select colonies with plasmids containing a larger insert confirmed by *Eco*RI digestion and electrophoresis profile analysis, and then confirm the correct insertion by sequencing.

3.2 Overlap-Extension PCR of the Binary Vector and the cDNA of the RNA Virus Genome

1. Two separate PCRs are done for PMMoV genome and the backbone [(pTRV2-MCS, including 2× 35S promoter of *Cauliflower mosaic virus* and *Subterranean clover mottle virus* satellite RNA ribozyme (AF406991)] using each primer pair, PMMoV forward/PMMoV-Ribozyme overlap reverse, PMMoV-Ribozyme overlap forward/35Sp reverse (Fig. 1b). Use Phusion DNA polymerase to generate blunt-ended PCR products (*see* **Note 9**). Both reactions are done with 5 μl of Phusion HF buffer 5× (NEB), 2.5 μl de dNTP 2.5 mM, 1 μl of each primer, 1 μl of cDNA, and 0.25 μl of Phusion DNA polymerase (2 U/μl) (NEB). PCR cycling condition: 80 °C for 1 min (manual Hot start); 96 °C for 1 min (pre-denaturing step), 25 cycles of 96 °C for 30 s/53 °C for 10 s/72 °C for 4 min, then 72 °C for 10 min as post-PCR incubation. Amplified PCR fragments are gel-purified as described above (Subheading 3.1, **step 4**).
2. The first step of overlap-extension PCR is done without adding primers, but the two large DNA fragments: the complete PMMoV genome cDNA and the pTRV-MSC backbone DNA. As the first step add the two DNA fragments at 1:1 molecular ratio (PMMoV, 84 ng and pTRV-MSC backbone = pCAMBIA-35Sp-ribo, 100 ng) into a 0.2 ml PCR microtube. Then add 5 μl of Phusion HF buffer 5× (NEB), 2.5 μl of dNTP (2.5 mM), 0.25 μl of Phusion DNA polymerase (2 U/μl) and fill up to 25 μl with DNase-free water. PCR cycling condition: 80 °C for 1 min (manual Hot start); 96 °C for 1 min (pre-denaturing step), 12 cycles of 96 °C for 30 s/50 °C for 60 s/72 °C for 7 min, then 72 °C for 10 min as post-PCR incubation resulting in pCAMBIA-PMMoV. As the second reaction, add 9 μl of Phusion HF buffer 5× (NEB), 5 μl of dNTP (2.5 mM), 1 μl of primer PMMoV forward (10 μM), 1 μl of primer 35Sp reverse (10 μM), and 0.5 μl de Phusion DNA polymerase (2 U/μl) (NEB). Use the same cycling condition described in the last step, except for the annealing temperature of 53 °C, in a total of 25 cycles.
3. The resulting DNA, pCAMBIA-PMMoV, is gel-purified using QIAEX II Gel Extraction kit (Qiagen) produced for large DNA

isolation (*see* **Note 7**). Purified DNA fragments are then phosphorylated with T4 Polynucleotide Kinase (PNK) (NEB). For 200 ng of purified DNA fragments, add 0.5 μl of 10× T4 DNA ligase buffer (NEB), which is compatible for the kinase reaction, 0.4 μl of T4 PNK and fill up with water up to 5 μl. In a thermo-block, incubate at 37 °C for 30 min. After this incubation, add 0.5 μl T4 DNA ligase buffer (NEB), 0.6 μl of T4 DNA ligase and fill up to 10 μl with water. This solution is ligated at 16 °C overnight in a thermo-block. The ligation solution is dialyzed against water using a cellulose ester membrane (Millipore) for 15 min. Part of this ligation solution (1–1.5 μl) is added into the *E. coli* Stbl4 (Invitrogen; *see* **Note 10**) competent cell suspension and electroporated. Then, add 1 ml of SOC medium without antibiotics and incubate the bacterial suspension at 30 °C for 2 h with agitation at 200 rpm. The bacterial suspension (200 μl) is plated onto LB medium-agar containing 50 μg/ml of kanamycin and incubated at 30 °C up to 48 h.

4. For screening the correct clones, mini-scale plasmid extractions are done by a standard protocol, the plasmids are digested by the single cutter enzyme, the plasmid size is confirmed by electrophoresis and the insert sequenced. Efficiency of correct transformant recovery can reach 30 %.

3.3 DNA Genome Amplification by RCA and Cloning in a Binary Vector

1. The first step is the amplification of circular viral DNA by RCA. Add 5 μl of TempliPhi Sample buffer (GE Healthcare Life Sciences) to a microtube, and add 0.2–0.5 μl of the DNA sample (total DNA extract of an infected plant, *see* **Notes 11** and **12**). There is no need of denaturing the DNA, as the genome of a begomovirus is single stranded. By denaturing, the replicative forms will serve as template for amplification and it may result in an improved yield of amplified viral DNA. If desired, heat the mixture to 95 °C for 3 min and allow it to cool to room temperature. In a separate tube, combine 5 μl of TempliPhi Reaction Buffer and 0.2 μl TempliPhi Enzyme mix. Mix these two portions (Sample buffer with DNA + Enzyme buffer with TempliPhi enzyme) in one tube and incubate at 30 °C for 4–18 h. Heat-inactivate the enzyme incubating the solution mixture at 65 °C for 10 min, then cool the solution to 4 °C. The resulting solution is viscous, and a high molecular weight DNA band can be seen in an agarose gel electrophoresis (0.7 %). Scale up the reaction according to the volume of RCA products needed for the next steps.
2. Once the viral DNA is amplified by RCA, it is necessary to select the single-cutter enzyme which will be used for cloning (*see* **Note 13**). The selection is best done by Southern blot hybridization. Use 1 μl of the RCA product for a digestion reaction of 10 μl, using 1–4 U of restriction enzyme per reaction.

Test all enzymes present in the multiple cloning site of the plasmid vector of choice (*see* **Note 13**). Perform agarose gel (0.7 %) electrophoresis with the digested RCA products, adding the size marker and an undigested control. Make two equal gels and transfer them to nylon or nitrocellulose membranes. Hybridize each membrane using a DNA-A probe and a DNA-B probe, separately (*see* **Note 14**). Select the single-cutter enzyme that yields a DNA genome fragment of the monomeric size of 2.6 kb. By comparison of the hybridization using both probes, it is possible to select the same enzyme to clone the DNA-A and DNA-B fragments, if not use two different restriction enzymes to clone each genomic segment.

3. Once the enzyme is selected, the dimeric units are produced by partial digestion for cloning. First, the partial digestion condition is evaluated by a series of enzyme dilution. Prepare five tubes with 10 μl of restriction enzyme digestion solutions using 1 μl of RCA products. In the first tube, add 1 U of restriction enzyme, and dilute 2× in the next four tubes, i.e., 0.5, 0.25, 0.125, and 0.0625 U. Incubate the tubes at the appropriate temperature for optimal enzyme function for 30 min. After the incubation period, heat-inactivate the enzyme and perform electrophoresis of 0.7 % agarose gel with undigested RCA solution in one lane as control. Select the best reaction condition to achieve a high amount of the dimeric molecules (ca. 5.2 kb). You will find an increasing amount of monomers (2.6 kb) by increasing the enzyme concentration. If necessary, other trials with different enzyme concentrations or reaction times will be needed to find the optimal condition for maximum recovery of partially digested RCA products, in the form of dimers. Once the enzyme is selected, digest at least 20 μg of RCA using the condition established earlier but in a proportionally scale-upped volume. Collect the desired dimeric DNA fragments from the agarose gel by using a sharp and clean razor blade and elute it from the agarose gel. If different restriction enzymes were selected for both fragments, repeat the procedure with the second enzyme.
4. Prepare the binary vector for ligation with the insert. Select a binary vector (for example, pCAMBIA 0380; *see* **Note 15**) with the desired selection marker. As a first step, linearize at least 1 μg of the plasmid with the selected restriction enzyme. After the complete digestion, dephosphorylate the vector using calf or shrimp alkaline phosphatase by standard procedures. After phenol and phenol–chloroform extraction and ethanol precipitation, run the plasmid in an agarose gel to separate from undigested plasmids (*see* **Note 16**). Run the eluted fragment in an agarose gel to confirm quality and the amount of purified DNA.

5. The ligation is done using standard procedures. Use at least 3:1 (insert–vector) ratio for efficient recovery of the desired clones (using >50 ng of the vector in a 10 μl. reaction). Transform *E. coli* cells by electroporation.
6. The screening of the transformants may be done by colony PCR using primers specific for the DNA-A or the DNA-B segment. Select the positive colonies and perform the plasmid extraction. Digest the plasmids with the restriction enzyme used for cloning and a second enzyme known to digest the plasmid in one site and that does not cut the insert. Select the clones that produce: (1) when digested with the enzyme used for cloning, two fragments, 2.6 kb (monomeric unit) and 6.8 kb (vector size); (2) when digested with the enzyme that linearize the construction, one fragment of 12 kb (dimeric genome segment + vector size).
7. Confirmation of the correct cloning. Sequence the insert in both directions to confirm the correct insertion of the dimeric molecule.

3.4 Agrobacterium Transfection and Infiltration of Infectious Clones

1. *A. tumefaciens* GV3101 is transformed with the binary vector containing the insert by electroporation. After confirmation of the presence of the correct plasmids by colony PCR, inoculate the bacterial colony to LB medium and incubate at 28 °C with agitation (180–200 rpm) for 1–2 days with the proper antibiotics. For the induction of *vir* genes, incubate 1 ml of culture with 3 ml of complete induction buffer [60 mM K_2HPO_4, 33 mM KH_2PO_4, 7.5 mM $(NH_4)_2SO_4$, 0.17 mM Sodium citrate, 1 mM $MgSO_4$, 0.2 % (w/v) Sucrose, 0.5 % (w/v) Glycerol, 50 μM acetosyringone, 10 mM MES without antibiotics] overnight (*see* **Note 17**). Centrifuge the bacterial suspension for 10 min at 1,200×*g* at 4 °C. Resuspend pellet with 2 ml of Murashige-Skoog (MS) complete medium (30 g Sucrose, 4 g MS powder/l of water containing 10 mM MES, pH 5.8, and 150 μM acetosyringone). Measure OD_{600} and dilute suspension with the same MS medium to $OD_{600} = 0.5$. Agroinfiltrate the suspension on the abaxial part of the leaves for inoculation. Alternatively, use a needle to puncture the petiole axillary buds and deposit 100 μl of bacterial suspension. When inoculating the bipartite begomovirus infectious clones, mix in equal amounts the DNA-A and DNA-B bacterial suspensions prior to agroinoculation.

4 Notes

1. Description of primers useful for amplification of the complete PMMoV genome containing the minimum T7 RNA polymerase promoter, the *Bam*HI site in the forward primer, and

the *Mlu*I and *Eco*RI sites in the reverse primer. These restriction sites were introduced for subcloning the insert to pUC19 in *Bam*HI/*Eco*RI sites and for run-off of in vitro transcription using T7 RNA polymerase. This construct was produced to test the infectivity of the transcripts, synthesized using RiboMax Large Scale RNA production System (Promega) with Ribo m7G Cap Analog (Promega).

2. Purified total RNA can be prepared using any column-based RNA extraction kit adapted for plant material or Plant RNA reagent (Invitrogen) with a standard protocol.
3. The use of SuperScript III is important to raise the transcription reaction temperature (47–50 °C) and for increasing the yield of full-length cDNA, as it reduces the secondary and tertiary RNA template conformation.
4. The one-step PCR amplification of the complete genome of RNA viruses with a poly(A) tail may be troublesome. In this case, the genomic cDNA may be cloned separately into two portions (5′ end and 3′ end) with an overlapping region in the middle of the two portions. These two parts can be joined by subcloning using a single cutter restriction enzyme and subsequent ligation or by overlap-extension PCR, and then, this fused complete genomic cDNA can be cloned in a smaller-sized plasmid before subcloning to a binary vector.
5. LongAmp *Taq* DNA polymerase is an enzyme mix containing a proof-reader DNA polymerase and *Taq* DNA polymerase and generates PCR products with and without A-overhangs. However, the resulting DNA fragments can be inserted into TA cloning vector without any A-overhang pretreatment due to the higher amount of DNA fragments with the A-overhangs.
6. Crystal violet for DNA staining is used to avoid UV irradiation to visualize DNA fragments, which causes unwanted DNA cleavage.
7. For DNA purification from the band generated by agarose-gel electrophoresis, the column-based purification kit is convenient. However, for longer DNA fragments (larger than 7 kbp), it is better to use silica slurry-based purification kit as QIAEX II Gel Extraction kit (Qiagen) to avoid DNA shearing, which is invisible on the agarose gel, but disturbs the cloning of long inserts. For the same reason, gel-purified large DNA fragments are better stored in the refrigerator at 4–6 °C than in freezers to avoid DNA shearing by freezing–thawing steps.
8. The pCR-4 TOPO TA cloning kit is more permissive for long DNA fragment insertion than pGEM-T cloning kit (Promega). However, the success of insertion can be influenced not only by the size of inserts but also by the conformation of the insert (and plasmid) DNA.

9. Phusion DNA polymerase is more robust and accurate in fidelity among the thermostable DNA polymerases. On the other hand, LongAmp *Taq* DNA polymerase is more suitable for amplifying larger DNA fragments by PCR, such as those over 10 kbp.
10. The binary vector containing cDNA of RNA virus genome is very unstable for *E. coli* transformation. Therefore, the use of suitable *E. coli* strains is very important. In our case, the desired clones were obtained when Stbl4 strain was used, and only with the incubation temperature of 30 °C. Trials of transformation failed when using *E. coli* Stbl4 with the incubation of 37 °C or with other *E. coli* strains as OmniMax, DH10B, and DH5α at neither 30 °C nor 37 °C.
11. DNA extraction can be done using one of many plant DNA extraction kits or simple protocols such as the one described by Doyle and Doyle [24].
12. Avoid using too much template DNA in the RCA reaction. Only a small amount is required for amplification, such as 1 ng. The presence of inhibitors from plants can hamper an efficient amplification, and therefore use less than 1 μl of DNA extract for the reaction. If necessary, dilute the sample before adding to the reaction solution.
13. The choice of the restriction enzyme is critical. The cloning procedure relies on the separation of the DNA in monomeric units of the genome. An enzyme that digests the genome in a single point is essential. It is possible that the enzyme digests the DNA in two sites in close proximity and if the genome is not complete it will not be infectious.
14. Hybridization. For a bipartite begomovirus, there will be two DNA components (segments), DNA-A and DNA-B (See an example of Southern blot hybridization using both probes in ref. 22). They share a common region of ca. 200 bases, but the remaining genome is completely different. It is possible to select the restriction enzymes for cloning both components by using the specific probe for only one component (for example, the DNA-A). The selection of the enzyme that digests the DNA-B probe in one site can be selected by comparison of the stained gel and the DNA-A hybridization. You will select the enzyme that produces a fragment of 2.6 kb, visible in the gel, but not hybridizing to the DNA-A probe. Hence, this fragment is of DNA-B.
15. Any binary vector can be used for cloning. Relatively small vectors, such as pCAMBIA 0380 (6.8 kb), are easier to manipulate.
16. Alternatively, you may amplify the vector by RCA, and prepare the plasmid with the desired ends without the requirement of agarose gel separation. In RCA products, the amount of circular plasmid is low and may be neglected.

17. Agro-infiltration is an efficient procedure due to the induction step with acetosyringone. This step is not essential for virus recovery and inoculation with the pelleted bacteria resuspended in water can be used. In this case, the incubation (latent) time is usually longer. The simplest method is to apply the agrobacterium colony with the aid of a toothpick by puncturing the leaf veins.

References

1. Boyer JC, Haenni AN (1994) Infectious transcripts and cDNA clones of RNA viruses. Virology 198:415–426
2. Yusibov V, Shivprasad S, Turpen TH, Dawson W, Koprowski H (1999) Plant viral vectors based on tobamoviruses. Curr Top Microbiol Immunol 240:81–84
3. Marillonnet S, Thoeringer C, Kandzia R, Klimyuk V, Gleba Y (2005) Systemic *Agrobacterium tumefaciens*-mediated transfection of viral replicons for efficient transient expression in plants. Nat Biotechnol 23: 718–723
4. Lindbo JA (2007) TRBO: a high-efficiency Tobacco mosaic virus RNA-based overexpression vector. Plant Physiol 145:1232–1240
5. Staczek J, Bendahmane M, Gilleland LB, Beachy RN, Gilleland HE (2000) Immunization with a chimeric *Tobacco mosaic virus* containing an epitope of outer membrane protein F of *Pseudomonas aeruginosa* provides protection against challenge with *P. aeruginosa*. Vaccine 18:2266–2274
6. Yoon JY, Min BE, Choi SH, Ryu KH (2001) Completion of nucleotide sequence and generation of infectious transcripts to cucurbits from full-length cDNA clone of *Kyuri green mottle mosaic virus*. Arch Virol 146: 2085–2096
7. Wu L, Jiang L, Zhou Z, Fan J, Zhang Q, Zhu H, Han Q, Xu Z (2003) Expression of foot-and-mouth disease virus epitopes in tobacco by a *Tobacco mosaic virus* based vector. Vaccine 21:4390–4398
8. Fujiyama K, Saejung W, Yanagihara I, Nakado J, Misaki R, Honda T, Watanabe Y, Seki T (2006) *In planta* production of immunogenic poliovirus peptide using *Tobacco mosaic virus-based* vector system. J Biosci Bioeng 101: 398–402
9. Jiang L, Li Q, Li M, Zhou Z, Wu L, Fan J, Zhang Q, Zhu H, Xu Z (2006) A modified TMV-based vector facilitates the expression of longer foreign epitopes in tobacco. Vaccine 24:109–115
10. Chandran D, Shahana PV, Rani GS, Sugumar P, Shankar CR, Srinivasan VL (2010) Display of neutralizing epitopes of *Canine parvovirus* and a T-cell epitope of the fusion protein of *Canine distemper virus* on chimeric tymovirus-like particles and its use as a vaccine candidate both against Canine parvo and Canine distemper. Vaccine 28:132–139
11. Turpen TH, Turpen AM, Weinzettl N, Kumagai MH, Dawson WO (1993) Transfection of whole plants from wounds inoculated with *Agrobacterium tumefaciens* containing cDNA of *Tobacco mosaic virus*. J Virol Methods 42:227–240
12. Kapila J, Rycke RD, Montagu MV, Angenon G (1997) An Agrobacterium-mediated transient gene expression system for intact leaves. Plant Sci 122:101–108
13. Al-Allaf FA, Tolmachov OE, Zambetti LP, Tchetchelnitski V, Mehmet H (2013) Remarkable stability of an instability prone lentiviral vector plasmid in *Escherichia coli* STBL3. Biotechnology 3:61–70
14. Lai MMC (2000) The making of infectious viral RNA: no size limit in sight. Proc Natl Acad Sci U S A 97:5025–5027
15. Satyanarayana T, Gowda S, Ayllón MA, Dawson WA (2003) Frameshift mutations in infectious cDNA clones of *Citrus tristeza virus*: a strategy to minimize the toxicity of viral sequences to *Escherichia coli*. Virology 313:481–491
16. Gils M, Kandzia R, Marillonet S, Klimyuk V, Gleba Y (2005) High-yield production of authentic human growth hormone using a plant virus-based expression system. Plant Biotechnol J 3:613–620
17. Sainsbury F, Cañizares CM, Lomonossoff GP (2010) The plant virus-based biotechnology workhorse. Annu Rev Phytopathol 48: 437–455
18. Lazarowitz SG, Wu LC, Rogers SG, Elmer JS (1992) Sequence-specific interaction with the viral AL1 protein identifies a geminivirus DNA replication origin. Plant Cell 4:799–809
19. Heyraud-Nitschke F, Schumacher S, Laufs J, Schaefer S, Schell J, Gronenborn B (1995) Determination of the origin cleavage and joining domain of geminivirus Rep proteins. Nucleic Acids Res 23:910–916

20. Stanley J (1983) Infectivity of the cloned geminivirus genome requires sequences from both DNAs. Nature 305:643–645
21. Buragohain AK, Sung YK, Coffin RS, Coutts RHA (1994) The infectivity of dimeric potato yellow mosaic geminivirus clones in different hosts. J Gen Virol 75:2857–2861
22. Inoue-Nagata AK, Albuquerque LC, Rocha WB, Nagata T (2004) A simple method for cloning the complete begomovirus genome using the bacteriophage phi-29 DNA polymerase. J Virol Meth 116:209–211
23. Ferreira PTO, Lemos TO, Nagata T, Inoue-Nagata AK (2008) One-step cloning approach for construction of agroinfectious begomovirus clones. J Virol Meth 147:351–354
24. Doyle JJ, Doyle JL (1987) A rapid DNA isolation procedure for small quantities of fresh leaf tissue. Phytochem Bull 19:11–15

Chapter 19

Efficient Double-Stranded RNA Production Methods for Utilization in Plant Virus Control

Andreas E. Voloudakis, Maria C. Holeva, L. Peter Sarin, Dennis H. Bamford, Marisol Vargas, Minna M. Poranen, and Francisco Tenllado

Abstract

Double-stranded RNA (dsRNA) is an inducer molecule of the RNA silencing (RNA interference, RNAi) pathway that is present in all higher eukaryotes and controls gene expression at the posttranscriptional level. This mechanism allows the cell to recognize aberrant genetic material in a highly sequence specific manner. This ultimately leads to degradation of the homologous target sequence, rendering the plant cell resistant to subcellular pathogens. Consequently, dsRNA-mediated resistance has been exploited in transgenic plants to convey resistance against viruses. In addition, it has been shown that enzymatically synthesized specific dsRNA molecules can be applied directly onto plant tissue to induce resistance against the cognate virus. This strongly implies that dsRNA molecules are applicable as efficacious agents in crop protection, which will fuel the demand for cost-effective dsRNA production methods. In this chapter, the different methods for dsRNA production—both in vitro and in vivo—are described in detail.

Key words dsRNA, *Escherichia coli*, *Pseudomonas syringae*, Phi6, RNA silencing, RNAi

1 Introduction

The first attempt to utilize the concept of pathogen derived resistance, proposed by Sanford and Johnston [1], resulted in the pioneering work on coat-protein mediated resistance against Tobacco mosaic virus by the Beachy lab [2]. Since then multiple strategies have been developed to engineer resistance into transgenic plants. This includes expression of the target protein (protein-based resistance) or the corresponding RNA molecule (RNA-mediated resistance).

RNA silencing is an endogenous cellular mechanism found in all higher eukaryotes that regulates gene expression. RNA silencing operates by two mechanisms: (a) by suppressing transcription through epigenetic modifications dependent on RNA-directed DNA methylation (Transcriptional Gene Silencing, TGS) or (b) by

Ichiro Uyeda and Chikara Masuta (eds.), *Plant Virology Protocols*, Methods in Molecular Biology, vol. 1236, DOI 10.1007/978-1-4939-1743-3_19, © Springer Science+Business Media New York 2015

inducing a highly sequence specific mRNA degradation or translation inhibition at the posttranscriptional level (named Posttranscriptional Gene Silencing, PTGS in plants).

RNA silencing, also known as RNA interference (RNAi), is a mechanism that protects the cell against "invading" nucleic acids, such as viruses and transposons. RNA silencing plays a key role in plant defense against pathogens [3, 4] and has therefore been exploited in plant biotechnology to induce resistance against plant diseases [5].

The RNA silencing mechanism has been studied extensively and the main players have been identified [6, 7]. DsRNA or stem loop RNA structures are the effector molecules that initiate the RNAi mechanism. Upon recognition, dsRNA-specific ribonucleases (RNases) termed Dicer or Dicer-like (DCL) proteins cleave the dsRNA molecules into 21–27 base pair (bp) long small interfering RNAs (siRNAs). The siRNAs are subsequently incorporated into the RNA-induced silencing complex (RISC), which is an effector complex where Argonaute proteins play a key role. Upon unwinding of the siRNA, one of the strands pairs with its sequence counterpart in the mRNA target, leading to mRNA degradation via the enzymatic action of the RISC. The mechanism also includes an amplification loop as well as a secondary siRNA production process, both of which reinforce the RNAi response.

Enzymatic production of dsRNA molecules for RNAi applications utilizes the catalytic activity of polymerases derived from bacteriophages [8, 9]. In this chapter, five dsRNA production methods are described, three of which are performed in vitro using purified enzymes (Methods 1 to 3) and two in vivo in genetically modified bacteria (Methods 4 and 5). All methods utilize the DNA-dependent RNA polymerase (DdRP) from bacteriophage T7 for transcription of target-specific sequences placed downstream of the promoter sequence. Complementary ssRNA molecules produced by T7 DdRP are hybridized in vitro (Methods 1 and 2) or anneal in vivo (Method 4) to form dsRNA. Furthermore, the T7 DdRP-mediated transcription can be combined with: (a) in vitro RNA replication catalyzed by the RNA-dependent RNA polymerase (RdRP) of *Pseudomonas* phage phi6 (Method 3; *see* Fig. 3) [8, 10], or (b) in vivo amplification by phage phi6 polymerase complexes in a phi6 carrier state cell line (Method 5; *see* Fig. 5) [8, 11].

Escherichia coli or *Pseudomonas syringae* bacteria harboring the gene for the expression of T7 DdRP (*gene 1* of phage T7), either in the DE3 prophage or in a plasmid construct, are utilized as hosts for in vivo dsRNA production (Methods 4 and 5). Target specific DNA placed under the T7 promoter is transformed into the host cells, where it is transcribed by the T7 DdRP. In the *E. coli* system the plasmids are stably replicated in the host cell and upon induction of the T7 DdRP, the target sequences are continuously transcribed in both directions to produce complementary ssRNA molecules that anneal to yield dsRNA (Method 4, Fig. 4) [12]. In these *E. coli* cell the *rnc14* gene, which encodes the dsRNA-specific RNase III, has been inactivated to inhibit degradation of the dsRNA.

The production of dsRNA in *P. syringae* is based on the formation of a stable carrier state cell line in which RNA is amplified by the polymerase complex of phi6 [11]. The genetic information encoding for the phi6 polymerase complex (cDNA of the phi6 genome segment L) is introduced into the *Pseudomonas* host together with target specific DNA flanked by phage-specific packaging and replication signals (*see* Fig. 5) [8]. Both constructs are cloned into plasmids downstream of the T7 promoter. Since *E. coli* plasmids do not replicate in *Pseudomonas*, transformation of these constructs into *P. syringae* leads only to transient transcription of the T7 DdRP regulated sequences. Amplification takes place at the RNA level inside the polymerase complexes of phi6 and is well protected from host RNases (*see* Fig. 5). These bacterial dsRNA production systems (Methods 4 and 5) can be scaled up, making the methods cost-effective.

Previous studies have shown that the direct application of in vitro produced dsRNA, derived from viral sequences, onto plant tissue at the time of virus inoculation, confers resistance against the homologous virus [13–17]. In addition, crude extracts of bacterially produced dsRNAs effectively protect plants against virus infections when sprayed onto plants [14, 15, 18, 19]. It is proposed that the exogenously supplied dsRNA mimics the viral dsRNA intermediate, which is involved in viral replication, and thereby triggers RNAi [20]. This may lead to the production of siRNAs that are incorporated into the nuclease complex (RISC) responsible for degradation of the cognate target RNA.

2 Materials

2.1 In Vitro dsRNA Production (See Note 1)

2.1.1 Methods 1 and 2: One- and Two-Step PCR, Transcription, and Hybridization

1. Target specific DNA, e.g., plasmid DNA containing the target specific DNA or cDNA sequence.
2. M-MuLV Reverse transcriptase (New England BioLabs, Inc.).
3. Primers for target DNA amplification. The forward and reverse target-specific primers (Primer Forward/Primer Reverse [PRF/PRR]) are designed for each target sequence and have restriction sites incorporated in their 5′ ends to facilitate subcloning into the transcription vector. These restriction sites should be absent from the target sequence and present in the polylinker of the transcription vector. The target sequence can be the whole or part of a gene (*see* **Notes 2** and **3**).
4. Sequencing primers to confirm plasmid constructs:

   M13 Forward (-20): 5′ GTAAAACGACGGCCAG 3′.

   M13 Reverse: 5′ CAGGAAACAGCTATGAC 3′.

5. T7 primer: 5′ TAATACGACTCACTATAGG 3′ (New England BioLabs, Inc.).
6. Primers for target DNA amplification carrying a Linker sequence (GGGGATCC) at the 5′ end and target specific sequence (N^{17-22}) at the 3′ end:

   Linker-PRF: 5′ GGGGATCCN^{17-22} 3′.

   Linker-PRR: 5′ GGGGATCC N^{17-22} 3′.
7. Primer T7-Linker: 5′ N^x*TAATACGACTCACTATA*GGG GATCC 3′ (a restriction site, N^x, may be incorporated at the 5′ end, the promoter sequence is in italic font, the Linker sequence is underlined).
8. $Vent_R$® DNA Polymerase (New England BioLabs, Inc.).
9. Phusion High-Fidelity DNA Polymerase (Thermo Scientific).
10. dNTPs (Promega).
11. T4 Ligase (New England BioLabs, Inc.).
12. PCR cloning vector: pCR™ IITopo (Invitrogen, CA).
13. In vitro transcription vector: LITMUS 28i (New England BioLabs, Inc.), containing an extensive set of restriction sites in polylinker, T7 promoters on both sides of the polylinker.
14. *Escherichia coli* Mach 1 Topo cells (Invitrogen).
15. *Escherichia coli* DH5α cells.
16. QIAquick Gel Extraction Kit (Qiagen).
17. QIAquick PCR Purification Kit (Qiagen).
18. QIAprep Spin Miniprep Kit (Qiagen).
19. In vitro transcription T7 RiboMAX™ Express RNAi System (Promega).
20. DNase and RNase, provided with the T7 RiboMAX™ Express RNAi System (Promega).
21. 20× SSC solution: Dissolve 175.3 g of NaCl and 88.2 g of Sodium Citrate in 800 ml water. Adjust the pH to 7.0. Adjust the volume to 1 l with ultrapure water. Dispense into aliquots. Sterilize by autoclaving.
22. Phenol (equilibrated with 8-hydroxyquinoline):Chloroform–isoamyl alcohol (25:24:1 v/v): Store the mixture in a clean brown glass bottle in a fume hood at 25 °C.
23. 100 % ethanol.
24. 70 % ethanol (stored at 4 °C).
25. 3 M sodium acetate pH 5.2 (store at 4 °C).
26. RNase-free water: provided with the RiboMAX™ Express RNAi System kit (Promega) or prepared as follows: add 0.1 % diethylpyrocarbonate (DEPC) in distilled water, mix and let sit at room temperature overnight and autoclave.

27. Agarose, Molecular Grade (Bioline Reagents Ltd.) Tris–borate EDTA (TBE) buffer (50 mM Tris base, 50 mM H_3BO_3, 2.5 mM EDTA, pH 8.3), 10× TD loading dye (50 % glycerol, 40 mM EDTA pH 8.0, 0.1 % bromophenol blue (w/v), 0.1 % xylene cyanol) (w/v), and ethidium bromide (EtBr) or SYBR Gold (Invitrogen) for agarose gel electrophoresis.
28. 100 bp DNA Ladder (New England BioLabs, Inc.).

2.1.2 Method 3: One-Step PCR, Transcription, and RNA Replication

1. Target specific DNA, e.g., genomic viral DNA or plasmid DNA containing the target specific DNA or cDNA sequence.
2. Primers for target DNA amplification. The reverse and forward primers are designed separately for each target sequence and need to contain the promoter sequences for T7 and phi6 RNA polymerases, respectively, at their 5′ ends (*see* **Note 4**). The 3′ end of the primers should contain 17–22 nucleotides of target gene-specific sequence.

   Forward primer (T7 promoter): 5′ *TAATACGACTCACT ATA<u>G</u>GG*N^{17-22} 3′.

   Reverse primer (phi6 promoter): 5′ *<u>G</u>GAAAAAAA*N^{17-22} 3′.

   The promoter sequences are in italic font and the target-specific sequence marked with N^{17-22}. The first nucleotides to be incorporated into the final dsRNA product are underlined.
3. Replicator RNAi Kit (Thermo Scientific). The Replicator RNAi kit contains components for target-specific DNA amplification (Phusion High-Fidelity DNA Polymerase, 10 mM dNTP mix, 5× Phusion HF buffer) and dsRNA production (T7 RNA Polymerase, phi6 RNA Replicase, Pyrophosphatase, 10× dsRNA Synthesis Buffer, 5× NTP mix, 50 mM $MnCl_2$).
4. RNase-free water produced by reverse osmosis, filtration and deionization using Milli-Q Plus Ultra Pure Water System (Millipore Corporation) and subsequent sterilization by autoclaving.
5. Wizard SV Gel and PCR Clean-Up System (Promega), QIAquick PCR Purification Kit (Qiagen) or QIAquick Gel Extraction Kit (Qiagen) for the purification of the PCR product.
6. Agarose, Molecular Grade (Bioline Reagents Ltd) Tris-borate EDTA (TBE) buffer (50 mM Tris base, 50 mM H_3BO_3, 2.5 mM EDTA, pH 8.3), 10× TD loading dye (50 % glycerol, 40 mM EDTA pH 8.0, 0.1 % bromophenol blue (w/v), 0.1 % xylene cyanol) (w/v), and EtBr or SYBR Gold (Invitrogen) for agarose gel electrophoresis.
7. 8 M LiCl (included in the Replicator RNAi Kit of Thermo Scientific), 70 and 96 % (v/v) ethanol and 10 M ammonium acetate for purification and precipitation of RNA.

2.2 In Vivo dsRNA Production (see Note 1)

2.2.1 Method 4: Production of dsRNA in Escherichia coli Using T7 RNA Polymerase

1. AMV Reverse Transcriptase (Promega).
2. Forward and reverse gene-specific primers (PRF/PRR) designed to have restriction sites incorporated in their 5′ ends to facilitate subcloning in the transcription vector. These restriction sites should be absent in the target gene sequence and present in the polylinker of the transcription vector. The target sequence can be the whole or part of a gene.

   PRF: 5′ restriction site-forward target specific sequences 3′.

   PRR: 5′ restriction site-reverse target specific sequences-3′.
3. Phusion High-Fidelity DNA Polymerase (Thermo Scientific).
4. dNTPs (Promega).
5. L4440 is a transcription vector that has two convergent T7 promoters flanking the multiple cloning site.
6. HT115 (DE3) (Genotype *F-, mcrA, mcrB, IN (rrnD-rrnE)1, rnc14::Tn10* (DE3 lysogen: *lavUV5 promoter-T7 polymerase*) is an RNase III-deficient *E. coli* strain, which expresses T7 RNA polymerase from an isopropyl β-D-1-thiogalactopyranoside (IPTG)-inducible promoter. HT115 (DE3) was made in a W3110 background, a derivative from strain K-12.
7. Luria–Bertani (LB) broth (1 l): add 10 g of Tryptone, 5 g of Yeast Extract, 5 g of NaCl, and deionized water up to 1 l (adjust pH 7 with NaOH). Dissolve components and sterilize by autoclaving at 121 °C for 15 min.
8. Ampicillin (Amp), Tetracycline (Tet).
9. IPTG.

2.2.2 Method 5: Production of dsRNA in Pseudomonas syringae Using phi6 Polymerase Complexes

1. *Pseudomonas syringae* LM2691 strain [11] harbors plasmid pLM1086 that constitutively expresses T7 RNA polymerase (*see* **Note 5**).
2. Plasmid pLM991 [11] contains the cDNA copy of phi6 genome segment L under T7 promoter in vector pT7T319U. *NtpII* gene (kanamycin resistance marker) has been placed in the non-translated region of the L-segment.
3. Plasmid pS_{insert} in which the target specific DNA is placed between the RNA packaging and replication signals specific for the phi6 genome segment S. The whole construct is under T7 promoter (*see* Fig. 5). An example of such construct is pPS9 [8].
4. LB medium. *See* above (Subheading 2.2.1).
5. LB agar plates supplemented with Kanamycin: prepare LB medium and add agar to a final concentration of 1.5 %. Heat the mixture to boiling to dissolve agar and sterilize by autoclaving. After autoclaving, allow the medium to cool to 50 °C before adding kanamycin to a final concentration of 25 μg/ml.

Pour the medium into petri dishes. Let the agar harden. Store at 4 °C.

6. Kanamycin stock solution: 10 mg/ml in sterile deionized water, stored at –20 °C.
7. SOC medium [add 2.0 g Bacto®-tryptone, 0.5 g Bacto®-yeast extract, 1 ml 1 M NaCl, and 0.25 ml 1 M KCl to 97 ml distilled water. Stir to dissolve. Autoclave and cool to room temperature. Add 2 M $MgCl_2$ stock (filter sterilized) and 2 M glucose (filter sterilized), each to a final concentration of 20 mM. Bring to 100 ml with sterile, distilled water. The final pH should be 7.0].
8. 10 % (v/v) glycerol for the preparation of competent cells.
9. TRIsureTM (Bioline Inc.) or equivalent acidic phenol agent and chloroform for dsRNA isolation.
10. 8 M LiCl and 70 % (v/v) ethanol for dsRNA precipitation.
11. Agarose, Molecular Grade (Bioline Reagents Ltd, UK) Tris–borate EDTA (TBE) buffer (50 mM Tris base, 50 mM H_3BO_3, 2.5 mM EDTA, pH 8.3), 10× TD loading dye (50 % glycerol, 40 mM EDTA pH 8.0, 0.1 % bromophenol blue (w/v), 0.1 % xylene cyanol) (w/v), and EtBr or SYBR Gold (Invitrogen) for agarose gel electrophoresis.

2.3 Exogenous Application of dsRNA on Plants

1. Plants: *Nicotiana tabacum* cv. Xanthi nc and pepper (*Capsicum annuum*) at 3–4 weeks developmental stage.
2. Pepper mild mosaic virus (PMMoV) at 10 μg/ml.
3. Carborundum.
4. Atomizer.
5. PMMoV antisera.

3 Methods

3.1 In Vitro dsRNA Production Methods

Here we describe three approaches for the preparation of DNA templates for in vitro dsRNA production (Fig. 1). The *first approach* (Method 1) is based on cloning the target sequence of interest into a plasmid carrying a T7 RNA polymerase promoter on both sides of the polylinker site in reverse orientations and amplifying the target sequence by PCR using the T7 primer. The *second approach* (Method 2) is based on a two-step PCR to sequentially incorporate a linker sequence, followed by the T7 RNA polymerase promoter sequence, into the 5′ ends of the target sequence. The second approach is quicker as no cloning is required and only the primers required in the first PCR step need to be synthesized for each target sequence. For the second PCR step, a single primer is needed. In both of these approaches the produced target-specific DNA is

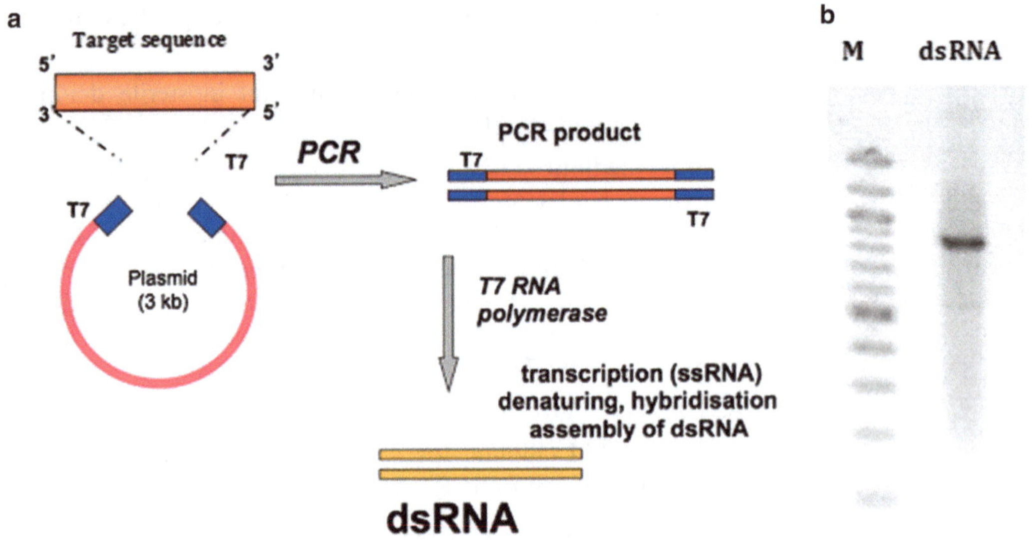

Fig. 1 In vitro dsRNA production (Method 1). (**a**) The target sequence is cloned into a polylinker site of a plasmid vector with opposite T7 promoters. The derived PCR product with the T7 prompter sequence at both ends is used for in vitro transcription. The resulting ssRNAs are hybridized to form dsRNA of the target sequence. (**b**) DsRNA of a target sequence of approx. 850 bp produced using Method 1. M is a 100 bp DNA Ladder (New England BioLabs, Inc.)

subsequently used as a template for the T7 RNA polymerase to produce complementary ssRNA molecules that are hybridized to obtain dsRNA. In the *third approach*, in vitro synthesized dsRNA is obtained in a single reaction that combines the catalytic activities of T7 DdRP and the phi6 RdRP [8]. The DNA template for dsRNA production is amplified by PCR from an appropriate DNA fragment containing the target sequence. The dsRNA is produced using this template in a reaction mixture that combines the transcription and replication reactions (*see* Fig. 3). This protocol does not require any hybridization step that often results in unwanted partially duplexed products, especially if the RNA molecules are long [8]. Instead, the dsRNA molecules are produced by enzymatic synthesis that generates fully complementary dsRNA molecules of high quality.

3.1.1 Method 1: One-Step PCR, Transcription, and Hybridization

1. Prepare cDNA of the target sequence by RT-PCR (*see* **Note 2**). For first strand cDNA synthesis of the target, use the M-MuLV Reverse transcriptase according to manufacturer's instructions in a 20 μl reaction including 90 pmol of the forward target-specific primer PR1 and 0.9 μg template (*see* **Note 3**). For the second strand cDNA synthesis, use 1 μl of the produced cDNA, 0.5 U VentR® DNA Polymerase, and 0.2 μM of forward and reverse target-specific primers PRF/PRR (50 μl reaction). The PCR thermal profile depends on the primer pair and PCR machine, but generally comprise 1 cycle at 94 °C for 3 min,

35 cycles at 94 °C for 1 min, annealing temperature (depending on the primers) for 1 min, 72 °C for 1 min per kb, and 1 cycle at 72 °C for 10 min.

2. Gel-purify the PR1/PR2 PCR product using the QIAquick Gel Extraction Kit (optional step).
3. Clone the purified PRF/PRR PCR product into pCR™ IITopo vector and transform the plasmid construct ("pCR™ IITopo::PRF/PRR-target") into *E. coli* Mach 1 Topo cells. Confirm cloning result by sequencing using M13F/M13R primers.
4. Make a plasmid preparation of pCR™ IITopo::PRF/PRR-target using the QIAprep Spin Miniprep Kit, digest it using the endonucleases corresponding to the restriction sites incorporated in the PRF/PRR primers, and recover the insert from the gel using the QIAquick Gel Extraction Kit. Similarly, digest and gel-purify the LITMUS28i transcription vector. Use a T4 ligase to clone the insert between the T7 promoters of the transcription vector. Transform the plasmid construct ("LITMUS28i::target") into *E. coli* Mach 1 Topo cells and confirm the cloning result by sequencing using M13F/M13R primers.
5. Use the plasmid construct LITMUS28i::target to prepare the linear DNA template for the in vitro transcription by PCR, as follows: set up a PCR reaction containing 1× Phusion HF Buffer including 1.5 mM $MgCl_2$, 0.2 mM of each of dNTPs, 0.4 μM of T7 primer, 1 U of Phusion High-Fidelity DNA Polymerase, and 15–20 ng of LITMUS28i::target as template. The PCR temperature profile comprise an initial denaturation step at 98 °C for 2 min, 35 cycles at 98 °C for 10 s, 50 °C for 30 s, and 72 °C for 1 min per kb, and a final extension at 72 °C for 5 min. The resulting PCR product ("PCR-T7-target") has a T7 RNA polymerase promoter at the 5′ end of both stands. Analyze 1 μl of the PCR product by 1.5 % (w/v) agarose gel electrophoresis in 0.5× TBE buffer at 70 V and visualize under UV light.
6. Purify the PCR-T7-target DNA using the QIAquick Gel Extraction Kit, and check it by agarose gel electrophoresis prior to transcription.
7. Set up the in vitro transcription reactions, using the T7 RiboMAX™ Express RNAi System and approx. 0.7 μg of the PCR-generated DNA template. After transcription, place the reaction tube in a 85 °C water bath and cool down to room temperature overnight. Analyze 1 μl of the transcription reaction by electrophoresis in 1.5 % agarose gel (0.5× TAE buffer). The difference in migration between DNA and dsRNA is small.
8. To confirm quality and quantity of produced dsRNA of the targeted sequence, treat the in vitro produced transcripts with DNase and RNase (provided with the above mentioned kit) in

2× SSC buffer at 37 °C for 1 h. Submit the treated dsRNA to standard phenol–chloroform–isoamyl-alcohol extraction, ethanol precipitation, as follows: If the sample is less than 100 μl, add RNase-free water up to 100 μl. Add an equal volume of phenol–chloroform–isoamyl-alcohol to the reaction solution. Vortex the mixture and then centrifuge for 10 min at 13,000 × *g* at room temperature. Transfer the aqueous phase to a new tube. Add 1/10 volume of 3 M sodium acetate (pH 5.2). Add 2.5 volumes of 100 % ethanol. Leave at −20 °C overnight. Spin at top speed in a standard microcentrifuge at 4 °C for 30 min. Wash with 70 % ethanol followed by a spin for 10 min at 4 °C at top speed. Dry the pellet and resuspended in the same volume of RNase-free water.

9. Analyze the purified dsRNA by 1.5 % agarose gel electrophoresis in 0.5× TAE buffer.

3.1.2 Method 2: Two-Step PCR, Transcription, and Hybridization

Having performed **steps 1–3** as described above in Subheading 3.1.1, a two-step PCR approach can be used to generate the template DNA for in vitro transcription (*see* Fig. 2):

1. Perform the first PCR reaction containing: 1× Phusion HF Buffer including 1.5 mM $MgCl_2$, 0.2 mM of each of dNTPs, 0.2 μM of Linker-PRF and Linker-PRR primers, 1 U of

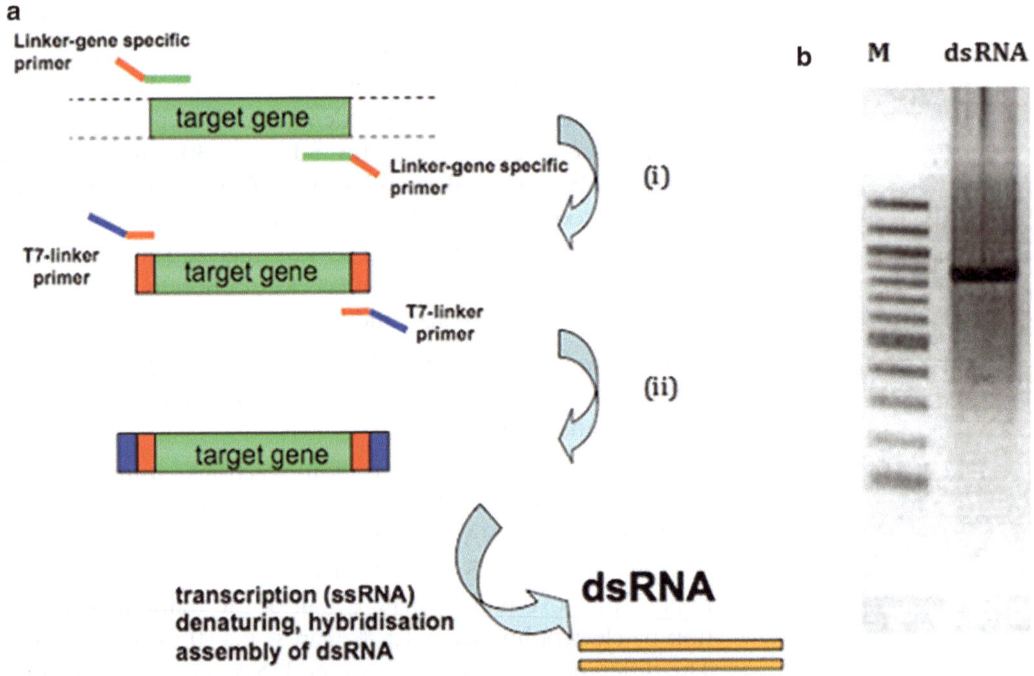

Fig. 2 In vitro dsRNA production (Method 2). (**a**) The target sequence is amplified in a two-step PCR approach, incorporating at both ends sequentially a (i) linker sequence at the first step, and (ii) a T7 promoter sequence at the second step. The derived PCR product with the T7 prompter sequence at both ends is used for in vitro transcription. The resulting ssRNAs are hybridized to form dsRNA of the target sequence. (**b**) DsRNA of a target sequence of approx. 850 bp produced using Method 2. M is a 100 bp DNA Ladder (New England BioLabs, Inc.)

Phusion High Fidelity DNA polymerase, and 15–20 ng of pCR™ IITopo::PRF/PRR-target as DNA template. The PCR temperature profile depends on the primer pair and PCR machine used, but generally comprise an initial denaturation step at 98 °C for 30 s, 35 cycles at 98 °C for 10 s, annealing temperature depending on the primers for 30 s and 72 °C for 1 min per kb, and a final extension at 72 °C for 10 min.

2. Purify the PCR product ("PCR-target-Linker") from gel using the QIAquick Gel Extraction Kit.
3. Perform the second PCR step setting up a reaction containing: 1× Phusion HF Buffer including 1.5 mM $MgCl_2$, 0.2 mM of each of dNTPs, 0.4 μM of primer T7-Linker, 1 U of Phusion High Fidelity DNA polymerase, and 1 μl of 1:100 PCR-target-Linker from the first PCR step as DNA template. The PCR temperature profile comprise an initial denaturation step at 98 °C for 30 s, 35 cycles at 98 °C for 10 s, 55 °C for 30 s, and 72 °C for 1 min per kb, and a final extension at 72 °C for 10 min.
4. Purify the PCR product ("PCR-target-Linker-T7") using the QIAquick PCR Purification Kit and use it as DNA template for in vitro transcription.
5. In vitro transcription is performed as described in Method 1 (**steps** 7 and **8**).

3.1.3 Method 3: One-Step PCR, Transcription, and RNA Replication

1. Amplify the target specific DNA by PCR. Use 1 pg to 200 ng of template DNA (typically 50 ng of plasmid DNA). Mix the template DNA with the components provided in the Replicator RNAi Kit in 50 μl reaction volume according to manufacturer's instructions. The PCR thermal conditions depend on the template and primers used and can be defined based on the information provided by the manufacturer. In the resulting PCR product the selected target specific sequence is flanked by the phage T7 RNA polymerase promoter sequence at one end and by the phage phi6 RNA polymerase promoter sequence at the other end (*see* Fig. 3).
2. Analyze a sample (2 μl) of the PCR reaction product by agarose gel electrophoresis (e.g., 1.0 % agarose gel in TBE buffer) to confirm that the length of the product matches the expected size. Use EtBr- or SYBR Green staining for UV-visualization of the DNA.
3. Purify the remainder of the PCR product using the Wizard SV Gel and PCR Clean-Up System, QIAquick PCR Purification Kit or from gel using the QIAquick Gel Extraction Kit following the protocols provided by the manufacturer. Measure the concentration of the DNA.
4. Produce the target-specific dsRNA from the PCR-amplified DNA template using the RNA polymerases of phage T7 and

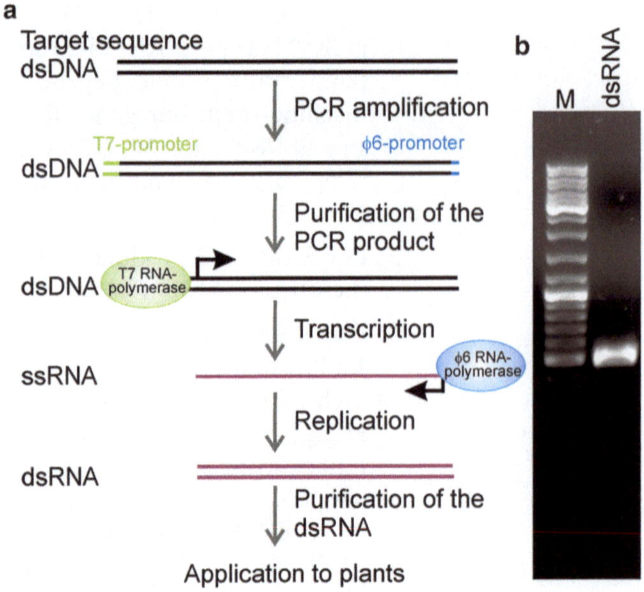

Fig. 3 In vitro dsRNA production (Method 3). (**a**) The target sequence is amplified by PCR to produce templates containing the target sequence flanked by T7 promoter sequence at one end (*green*) and phi6 RdRP promoter sequence (*blue*) at the other end. T7 DdRP directs the synthesis of target-specific ssRNA molecules that are converted to dsRNA by the phi6 RdRP. (**b**) DsRNA of a target sequence of approx. 500 bp produced using Method 3. M is a GeneRuler DNA Ladder mix (Fermentas)

phi6 and buffer components supplied in the Replicator RNAi Kit (*see* Fig. 3) according to the instructions of the manufacturer (*see* **Note 6**). Use 0.5–2 μg of DNA template per 50 μl reaction. Incubate the reaction mixtures at 35 °C for 2–4 h (or overnight).

5. Analyze a sample of the dsRNA product by agarose gel electrophoresis. Standard DNA markers can be used for the rough estimation of the size of the dsRNA. Target specific ssRNA that is not converted to dsRNA can sometimes be detected (*see* **Note 7**).

6. Purify the dsRNA by stepwise precipitation with 2 M and 4 M LiCl using the 8 M LiCl stock provided in the Replicator RNAi Kit according to the instructions of the manufacturer. This procedure selectively removes template DNA and ssRNA as well as most of the NTPs from the reaction mixture. As LiCl may cause unwanted effects in the downstream applications the resulting pellet should be carefully washed with 70 % ethanol. In addition, ammonium acetate precipitation can be applied to improve the purity of the sample. Chromatographic methods may also be used for the purification of dsRNA [21] or the produced dsRNA can be further processed in vitro to

small interfering RNA (siRNA) molecules using Dicer enzyme [8, 22].

7. Measure the concentration of the dsRNA and analyze a sample by agarose gel electrophoresis (for an example *see* Fig. 3). Typically 20–50 µg of dsRNA can be produced in a single 50 µl reaction. However, the yield depends on the target sequence. The Replicator RNAi kit contains reagents for forty 50 µl reactions.

3.2 In Vivo dsRNA Production

Two approaches are described below for the production of dsRNA in genetically modified bacterial cells (in vivo). Both methods utilize the phage T7 DdRP for transcription of target-specific sequences placed under this polymerase's promoter and the transcripts either anneal inside the *E. coli* cells (Method 4) or are amplified by phage phi6 polymerase complex in a phi6 carrier state *Pseudomonas* cell line (Method 5).

3.2.1 Method 4: Production of dsRNA in Escherichia coli Using T7 RNA Polymerase

1. Amplify the target sequence by PCR using specific primers and clone the PCR fragment into *Xba*I and *Hin*dIII sites (restriction sites depend on absence within the target sequence) of L4440. The resulting plasmid construct and the empty L4440 vector are transformed into HT115 (DE3) cells using standard $CaCl_2$ transformation protocols.
2. Inoculate overnight culture of HT115 (DE3) in LB medium containing 100 µg/ml Amp and 12.5 µg/ml Tet and incubate (37 °C with shaking at ca. 220 rpm) overnight. A 30 ml culture is usually enough for the induction of a volume of 2 l of bacterial culture cells for induction.
3. Dilute culture 1:75 (OD_{600} = ca. 0.05) in fresh LB with antibiotics. This is made by centrifuging the required volume of the overnight culture at 3,000×*g* 10 min. Discard the supernatant and resuspend the pellet. Grow the culture to OD_{600} = 0.5 (ca. 3 h).
4. Induce the T7 RNA polymerase gene by the addition of IPTG to the final concentration of 4 µM, and incubate with shaking for 2–4 h at 37 °C.
5. Centrifuge the culture at 13,000×*g* 20 min. Resuspend the pellet into 1/50 vol of 25 mM Tris, 5 mM EDTA, pH 7.5.
6. To estimate the amount of dsRNA produced, lyse 100 µl of the culture by adding 1 M ammonium acetate, phenol–chloroform (1:1) and incubate at 65 °C. Precipitate nucleic acids by adding ethanol and incubate at −20 °C for at least 30 min. Resuspend the pellet in 10 mM Tris, 1 mM EDTA, pH 7.5. Analyze by electrophoresis in 1 % agarose (TBE buffer) and stain with EtBr. Figure 4, panel (a) shows in vitro produced PMMoV replicase gene specific dsRNA, whereas *E. coli* produced dsRNA is shown in Fig. 4, panel (b).

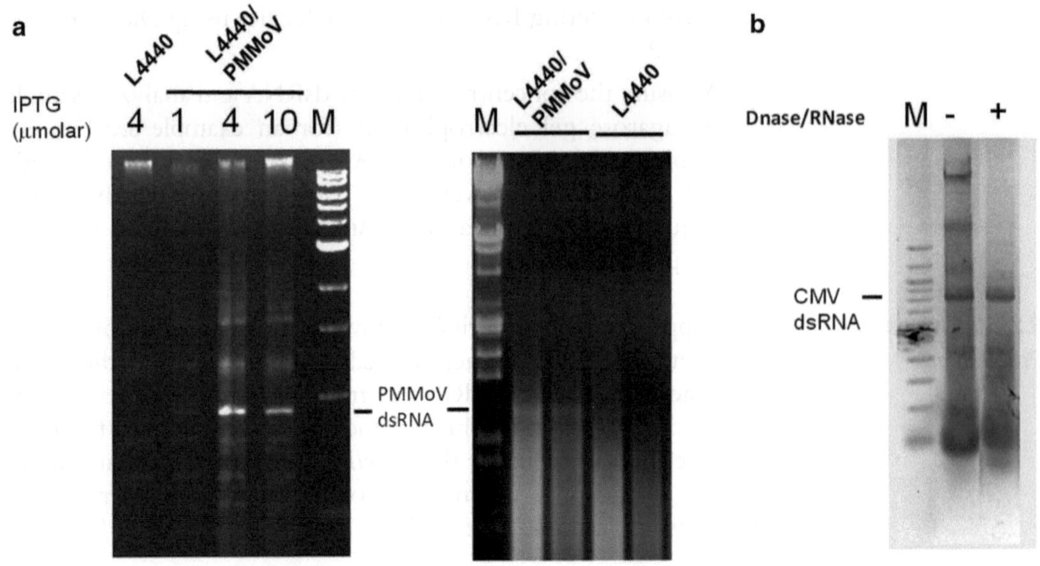

Fig. 4 In vivo dsRNA production (Method 4). (**a**) Production of PMMoV-specific dsRNA in *E. coli* strain HT115 (DE3). Bacteria were transformed with either L4440/PMMoV, a plasmid designed to express a dsRNA molecule consisting of 977 bp of PMMoV-specific sequence, or with the empty plasmid (L4440). Bacterial cultures were induced with IPTG at different concentrations and processed for total nucleic acid content. Samples were resolved by electrophoresis in a 1 % agarose gel before (*left panel*) or after lysis (*right panel*), and visualized by EtBr staining and UV detection. *M* λ molecular weight markers. The position of bacterially expressed PMMoV-specific dsRNA is indicated. (**b**) Production of CMV-specific dsRNA in *E. coli* strain HT115 (DE3). Bacteria were transformed with a cloned target sequence from CMV genome and processed for dsRNA extraction. DsRNA quality was visually assessed on agarose gel, as detailed above

7. Pass the remaining cell culture twice through French Press (1,100 psi), and centrifuge the resulting liquid material at 13,000 × *g* for 20 min.
8. Test the supernatant for its interfering activity on viral infection (*see* **Note 8**).

3.2.2 Method 5: Production of dsRNA in Pseudomonas syringae Using the Phage phi6 Polymerase Complex

1. Prepare electrocompetent *P. syringae* LM2691cells. Use an overnight culture of *P. syringae* LM2691 to inoculate 500 ml LB. Incubate the culture at 28 °C with aeration until the cell density reaches an OD_{550} value of 1.0. Chill the culture for 15 min and collect the cells by centrifugation (2,400 × *g* for 15 min at 4 °C). Re-suspend the pellet in 1/10 vol of cold ultra pure water, centrifuge as above and re-suspend in 1/20 vol of 10 % glycerol, and re-centrifuge as above. Re-suspend the washed cells in cold 10 % glycerol to a final volume of 8 ml. Aliquot the electrocompetent cells in 100 μl batches and store at −80 °C.
2. Perform the electroporation of the electrocompetent *P. syringae* LM2691 cells with plasmids pLM991 and pS_{insert}. Add 0.5–1 μg

of each plasmid to the thawed electrocompetent cells and incubate on ice for 10 min. Transfer the cells to a cuvette and perform electroporation (25 μF, 2.5 kV, 200 Ω). Add immediately 1 ml of SOC medium to the cuvette and transfer the mixture to a 14 ml round-bottomed tube. Incubate at 28 °C for 2 h with aeration (*see* **Note 9**) and plate on LB plates supplemented with 25 μg/ml kanamycin. Incubate the plates at 28 °C until colonies appear (at least 2 days).

3. Isolation of RNA for screening. Inoculate a 5 ml LB (with 25 μg/ml kanamycin) with a single transformant colony and grow overnight at 28 °C with aeration (*see* **Note 10**). Collect the cells by centrifugation (microcentrifuge, 18,900×*g*, 1 min) and re-suspend in 500 μl sterile, ice cold ultra pure water. Add 1 vol of TRIsure (or equivalent acidic phenol agent), mix thoroughly, and incubate on ice for 15 min. Add 1/5 vol of chloroform, vortex thoroughly and centrifuge (microcentrifuge, 18,900×*g*, 15 min, at 4 °C). Collect the aqueous, RNA containing upper phase.
4. Selective precipitation of ssRNA and dsRNA using LiCl. Add 1/3 volume of 8 M LiCl to the RNA sample isolated from the *Pseudomonas* cells to yield a final concentration of 2 M. Incubate at −20 °C for 30 min. Collect the precipitated ssRNA by centrifugation (microcentrifuge, 18,900×*g*, 20 min, 4 °C) and transfer the supernatant to a fresh tube. Add 1/2 volume of 8 M LiCl to the supernatant to yield a final concentration of 4 M. Incubate at −20 °C for 30 min and collect the precipitated dsRNA by centrifugation as above (*see* **Note 11**). Wash the pellet carefully with cold 70 % (v/v) ethanol. Repeat several times to ensure complete removal of LiCl. Air-dry the pellet for 5–10 min at room temperature and resuspend the dsRNA in sterile ultra pure water (*see* **Note 12**).
5. Analyze the RNA samples by agarose gel electrophoresis (1.0 % agarose in TBE buffer) using EtBt or SYBR Green for UV-visualization (*see* Fig. 5, panel (d)). Determine the concentration of dsRNA spectrophotometrically at 260 nm.

3.3 Exogenous Application of dsRNA on Plants for Virus Control: Case Study of PMMoV/Tobacco or Pepper Pathosystems

1. Prepare the mixture (dsRNA, purified virus/virus-infected plant sap) for inoculation. The concentration of the individual components should be optimized for the specific virus–host pathosystem, source of virus inoculum, and dsRNA production method. For example, add 10 μl of bacterial extract containing dsRNA, produced using Method 4, to an equal volume of purified PMMoV (10 μg/ml). To test interfering activity of bacterial extracts on virus infectivity, for the virus–host pathosystem it is advised to test a mixture containing bacterial extract void of dsRNA. Keep the mixtures on ice until use.

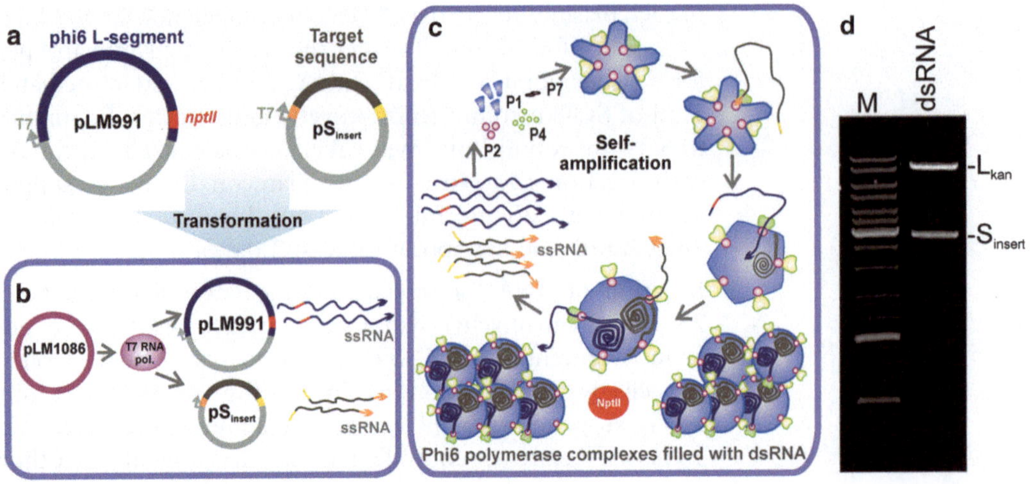

Fig. 5 In vivo dsRNA production in *Pseudomonas syringae* (Method 5). (**a**) Plasmid pLM991 carries a cDNA copy of the φ6L_{kan}-segment that encodes for the phi6 genome segment L (*blue*) with a kanamycin resistance gene *nptII* (*red*) insert. Plasmid pS_{insert} contains the target sequence inserted between phage-specific RNA packaging and replication signals (in *orange* and *yellow*, respectively). Both sequences of interest are placed under the T7 promoter. (**b**) Upon transformation into the host cells, the T7 RNA polymerase (*purple*; expressed from plasmid pLM1086) transiently synthesizes ssRNA from pLM991 and pS_{insert}, which are non-replicative in *P. syringae*. (**c**) The (+)ssRNA of the φ6 L_{kan}-segment directs translation of viral proteins that yield empty polymerase complexes. The polymerase complexes encapsidate the phage-specific ssRNA molecules and upon packaging convert them into dsRNA. These dsRNA segments are used as templates for transcription and lead to amplification of dsRNA-filled polymerase complexes. Stable carrier-state cell lines are obtained by applying a continuous selective pressure, in this case kanamycin resistance mediated by virus replication. (**d**) DsRNA produced using Method 5. The mobility of the L_{kan} (7.6 kb) and S_{insert} (4.1 kb) dsRNA molecules is indicated on the right. M is a GeneRuler DNA ladder (Thermo Scientific)

2.
   (a) *Inoculation by rubbing*: Inoculate the host plant (e.g., two fully expanded leaves of *Nicotiana tabacum* cv. Xanthi nc for the PMMoV-tobacco pathosystem) with the prepared mixture by gently rubbing the leaf surface with the inoculum using carborundum as an abrasive.
   (b) *Inoculation by spraying and rubbing:* Spray the dsRNA solution onto the surface of plant leaves (e.g., leaves of pepper plants) using an atomizer, followed by gently rubbing of the sprayed leaf surface with the virus inoculum (e.g., PMMoV at 10 μg/ml) using carborundum (*see* **Note 13**).
3. Keep the inoculated plants in growth chambers under the conditions suitable for symptom development of the specific virus–host pathosystem (e.g., for the PMMoV-tobacco 16 h light and 8 h dark cycle at 25 °C).
4. Monitor symptom development after inoculation for a period depending of the virus–host pathosystem (*see* Fig. 6, **Note 14**)

Fig. 6 HT115-expressed PMMoV-specific dsRNA interferes with PMMoV infection. (**a**) Opposite half-leaves of *N. tabacum* cv. Xanthi nc plants were inoculated with PMMoV alone or with mixtures of PMMoV and dsRNA containing bacterial extracts prepared using Method 4. The used HT115 extracts harbored either L4440/PMMoV or the empty vector (L4440), as indicated. After 5 days, similar numbers of local lesions were observed in both halves of the leaf shown at the left. No visible local response was observed in the half-leaf inoculated with PMMoV together with the bacterial extracts derived from L4440/PMMoV (right). (**b**) PMMoV was inoculated on pepper plants after spraying with bacterial extract produced using Method 4 harboring L4440/PMMoV or L4440 in the same leaves, as indicated. A healthy, non-inoculated plant is shown as a control. Values at the bottom correspond to virus titer as determined by DAS-ELISA (mean ± standard error) at 25 dpi. Plants were photographed at 40 dpi

5. Assess the virus titer in plants using double antibody sandwich (DAS)-ELISA at 25 days post infection (dpi) (for an example *see* Fig. 6b, bottom).

4 Notes

1. Use of names of chemicals in this chapter implies no approval of them to the exclusion of others. This information is given for the convenience of users, but equivalent products may be used if they can be shown to lead to the same results.
2. For first-strand cDNA synthesis reactions it is possible to use sequence-specific primers, oligo(dT) or random primers. Oligo(dT) and random primers allow many different targets to be studied from the same cDNA pool produced. Each primer option may vary regarding sensitivity, accuracy and yield of the first strand cDNA synthesis.
3. It is recommended to prepare a second dsRNA product of identical size, but with unrelated nucleotide sequence, to be included as a negative control.

4. The phi6 RdRP can convert any given ssRNA to dsRNA regardless of sequence [10]. However, efficiency of the replication reaction is improved if the 3′-end of the template contains several C residues and a stretch of unpaired nucleotides [23, 24].
5. This strain of *P. syringae* is plant pathogenic and has to be treated in accordance with the legislation concerning genetically modified organisms (GMO). Special care should be given to ensure proper disposal. Labware that has been in contact with LM2691 must be sterilized (disposables into GMO bin, glassware decontamination by 0.2 M NaOH).
6. It is recommended that the 10× dsRNA Synthesis Buffer is heated at 35 °C for 10 min before use. Addition of $MnCl_2$ is essential for the catalytic activity of phi6 RdRP [10, 25, 26].
7. The one-step transcription and replication reaction sometimes results in overproduction of ssRNA molecules, implying an imbalance in the activities of the T7 DdRP and phi6 RdRP. Hence, the amount of T7 DdRP in the reaction mixture can be slightly decreased to improve the final ssRNA:dsRNA ratio.
8. The dsRNA-enriched supernatant can be preserved for several weeks at −80 °C by adding an equal volume of isopropanol.
9. Special care should be given to work fast as the cells are weak following electroporation.
10. Some transformants grow very slowly and may require up to 48 h of incubation.
11. The dsRNA pellet is not necessarily visible (pure dsRNA is opaque; a high salt concentration or ssRNA yields a white pellet).
12. Special care should be given not to over-dry the pellet, as it may become difficult to resuspend the dsRNA.
13. Virus infectivity was significantly abolished when plants were sprayed lysed bacteria up to several days before virus inoculation. Extracts derived from bacteria expressing specific viral dsRNA can prevent infection by plant viruses other than PMMoV, such as plum pox virus, potato virus Y and zucchini yellow mosaic virus. Protection conferred by exogenously supplied dsRNA requires that interfering dsRNA sprayed onto leaf surfaces enters the cell together with the virus. Thus, virus inoculation on parts of the plant not covered by the bacterial extracts lead to systemic infection.
14. Plants that are protected against mechanically inoculated viruses may not be protected from virus inoculated by the natural vectors that, in most cases, release the virus particles directly inside the plant cell [27].

Acknowledgment

This work was supported by: (a) a grant to FT from the Spanish Ministry of Science and Technology (BIO2009-10172), (b) the Academy of Finland (grants 250113, 256069, and 272507 to M.M.P. and grants 256197, 255342, and 256518 to D.H.B.), and (c) the COST FA0806 to A.E.V., M.C.H., P.L.S., D.H.B., M.M.P.

References

1. Sanford JC, Johnston SA (1985) The concept of parasite derived resistance-deriving resistance genes from the parasite's own genome. J Theor Biol 113:395–405
2. Abel PP, Nelson RS, De B, Hoffmann N, Rogers SG, Fraley RT, Beachy RN (1986) Delay of disease development in transgenic plants that express the tobacco mosaic virus coat protein gene. Science 232:738–743
3. Padmanabhan C, Zhang X, Jin H (2009) Host small RNAs are big contributors to plant innate immunity. Curr Opin Plant Biol 12:465–472
4. Wang MB, Masuta C, Smith NA, Shimura H (2012) RNA silencing and plant viral diseases. Mol Plant Microbe Interact 25:1275–1285
5. Duan CG, Wang CH, Guo HS (2012) Application of RNA silencing to plant disease resistance. Silence 3:5
6. Brodersen P, Voinnet O (2006) The diversity of RNA silencing pathways in plants. Trends Genet 22:268–280
7. Martínez de Alba AE, Elvira-Matelot E, Vaucheret H (2013) Gene silencing in plants: a diversity of pathways. Biochim Biophys Acta 1829:1300–1308
8. Aalto AP, Sarin LP, van Dijk AA, Saarma M, Poranen MM, Arumäe U, Bamford DH (2007) Large-scale production of dsRNA for siRNA pools for RNA interference utilizing bacteriophage phi6 RNA-dependent RNA polymerase. RNA 13:422–429
9. Timmons L (2006) Construction of plasmids for RNA interference and in vitro transcription of double-stranded RNA. Methods Mol Biol 351:109–117
10. Makeyev EV, Bamford DH (2000) Replicase activity of purified recombinant protein P2 of double-stranded RNA bacteriophage phi6. EMBO J 19:124–133
11. Sun Y, Qiao X, Mindich L (2004) Construction of carrier state viruses with partial genomes of the segmented dsRNA bacteriophages. Virology 319:274–279
12. Timmons L, Court DL, Fire A (2001) Ingestion of bacterially expressed dsRNAs can produce specific and potent genetic interference in *Caenorhabditis elegans*. Gene 263: 103–112
13. Tenllado F, Díaz-Ruíz JR (2001) Double-stranded RNA-mediated interference with plant virus infection. J Virol 75:12288–12297
14. Holeva MC, Sclavounos AP, Kyriakopoulou PE, Voloudakis AE (2006) *In vitro* or *in vivo* produced dsRNA induces tobacco and tomato resistance to *Cucumber mosaic virus* (CMV). 13th Hellenic phytopathological congress of the Greek phytopathological society, Athens, 15–19 Oct 2006
15. Holeva MC, Sclavounos, AP, Kyriakopoulou PE, Voloudakis AE (2007) External application of dsRNA of the capsid protein (CP) or 2b gene of CMV reduces the severity of CMV-infection in tobacco. XIII IS-MPMI congress, Sorrento, Italy, 21–27 July 2007
16. Carbonell A, Martínez de Alba AE, Flores R, Gago S (2008) Double-stranded RNA interferes in a sequence-specific manner with the infection of representative members of the two viroid families. Virology 371:44–53
17. Yin G, Sun Z, Liu N, Zhang L, Song Y, Zhu C, Wen F (2009) Production of double-stranded RNA for interference with TMV infection utilizing a bacterial prokaryotic expression system. Appl Microbiol Biotechnol 84:323–333
18. Tenllado F, Barajas D, Vargas M, Atencio FA, González-Jara P, Díaz-Ruíz JR (2003) Transient expression of homologous hairpin RNA causes interference with plant virus infection and is overcome by a virus encoded suppressor of gene silencing. Mol Plant-Microbe Interact 16:149–158
19. Tenllado F, Martinez-García B, Vargas M, Díaz-Ruíz JR (2003) Crude extracts of bacterially expressed dsRNA can be used to protect plants against virus infections. BMC Biotechnol 3:3
20. Tenllado F, Llave C, Díaz-Ruíz JR (2004) RNA interference as a new biotechnological tool for the control of virus diseases in plants. Virus Res 102:85–96

21. Romanovskaya A, Sarin LP, Bamford DH, Poranen MM (2013) High-throughput purification of double-stranded RNA molecules using convective interaction media monolithic anion exchange columns. J Chromatogr A 1278:54–60
22. Romanovskaya A, Paavilainen H, Nygårdas M, Bamford DH, Hukkanen V, Poranen MM (2012) Enzymatically produced pools of canonical and Dicer-substrate siRNA molecules display comparable gene silencing and antiviral activities against herpes simplex virus. PLoS One 7:e51019
23. Yang H, Makeyev EV, Bamford DH (2001) Comparison of polymerase subunits from double-stranded RNA bacteriophages. J Virol 75:11088–11095
24. Sarin LP, Poranen MM, Lehti NM, Ravantti JJ, Koivunen MR, Aalto AP, van Dijk AA, Stuart DI, Grimes JM, Bamford DH (2009) Insights into the pre-initiation events of bacteriophage phi6 RNA-dependent RNA polymerase: towards the assembly of a productive binary complex. Nucleic Acids Res 37:1182–1192
25. Poranen MM, Salgado PS, Koivunen MRL, Wright S, Bamford DH, Stuart DI, Grimes JM (2008) Structural explanation for the role of Mn^{2+} in the activity of phi6 RNA-dependent RNA polymerase. Nucleic Acids Res 36: 6633–6644
26. Wright S, Poranen MM, Bamford DH, Stuart DI, Grimes JM (2012) Non-catalytic ions direct the RNA-dependent RNA polymerase of bacterial dsRNA virus phi6 from de novo initiation to elongation. J Virol 86:2837–2849
27. Vargas M, Martinez-García B, Díaz-Ruíz JR, Tenllado F (2008) Transient expression of homologous hairpin RNA interferes with PVY transmission by aphids. Virol J 5:42

Chapter 20

Detection of Plant Virus in Meristem by Immunohistochemistry and In Situ Hybridization

Tomofumi Mochizuki and Satoshi T. Ohki

Abstract

Most plant viruses do not infect the shoot apical meristem (SAM) of a host plant, and this virus-free region of meristem tissue has been used to obtain virus-free clones by meristem tip culture. Thus, the validation of viral distribution in meristem tissues is important for ensuring the appropriate excision of virus-free meristem tips. Although immunohistochemical microscopy and in situ hybridization are classical techniques, they allow us to determine the presence or absence of plant viruses in the shoot meristem tissues of a host plant. Briefly, meristem tissues are excised from infected plants, fixed, embedded in paraffin medium, and prepared in semithin sections (10–15 μm). By treating these sections with an antibody against viral protein or with a probe complementary to viral RNA, the viral distribution in the meristem tissue can be clearly observed. Importantly, these procedures are broadly applicable to most virus (and viroid) and host plant combinations.

Key words *Cucumber mosaic virus*, Meristem tissue, Viral distribution, Immunohistochemical microscopy, In situ hybridization, Coat protein, Viral RNA

1 Introduction

The shoot apical meristem (SAM) of a plant contains stem cells that differentiate into leaves or stems [1]. Generally, the SAM has a dome shape of 0.1–0.3 mm in diameter. The cells of the SAM are tiny, with rich cytoplasm and thin cell walls. Histological-morphological observation suggests that SAM cells are divided into two regions (Fig. 1): the tunica (epidermal layer) and the corpus (central zone), the cell sizes and vacuolarization degrees of which are obviously different [2]. Plant viruses are generally considered as not infectious to the SAM, with some exceptions [3]. This absence of viruses in the shoot tissues is of practical importance because virus-free plant clones can be generated from infected plants by culturing excised meristem tips. Because viral contamination in these excised meristem tips affects the production of viral-free

Ichiro Uyeda and Chikara Masuta (eds.), *Plant Virology Protocols*, Methods in Molecular Biology, vol. 1236,
DOI 10.1007/978-1-4939-1743-3_20,

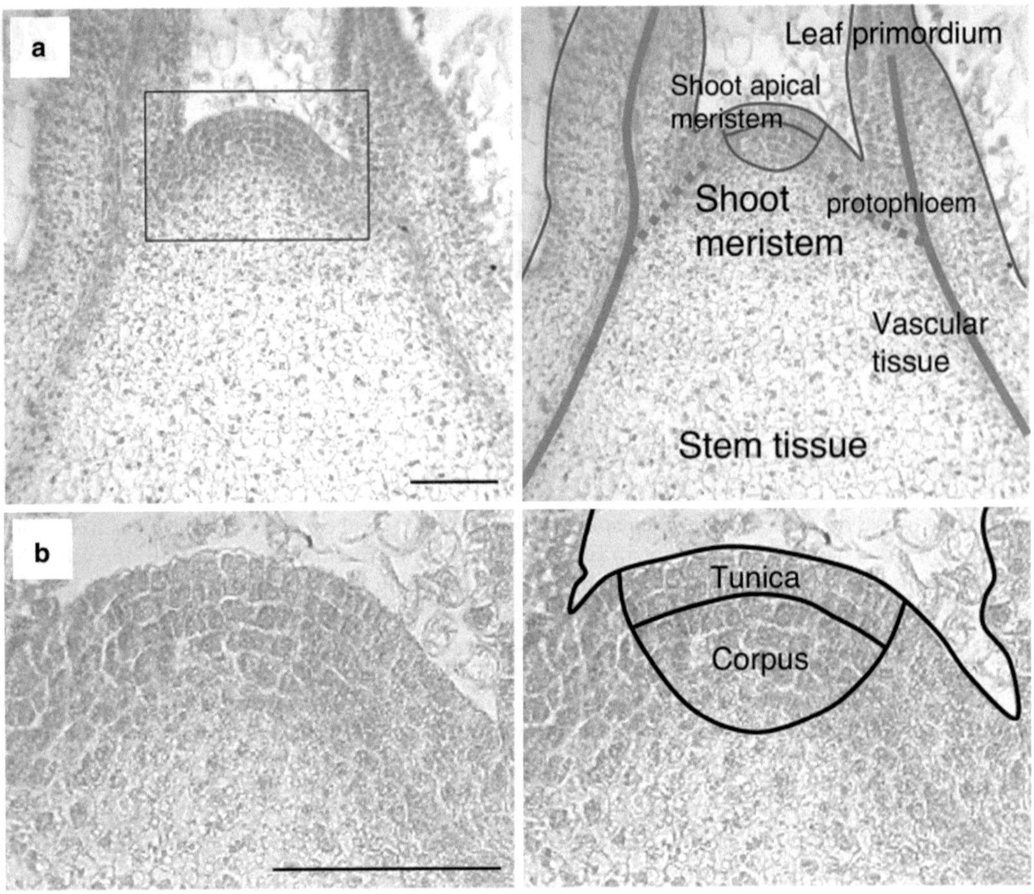

Fig. 1 (**a**) Longitudinal section and diagram of a shoot meristem of a tobacco plant (*Nicotiana tabacum* cv. Xanthi). (**b**) Detailed observation of the *boxed area* of A showing the apical meristem organization with a two-layered tunica and some corpus cells. Bars = 100 μm

clones, understanding viral distribution in the meristem tissue is of particular importance [4].

Observations of viral distribution in the meristem have been conducted using the electron microscope from the early stages of plant viral disease research, and particles of several viruses, including *Tobacco ringspot virus* (TRSV), *Pepper ringspot virus*, *Potato virus X* (PVX), *Odontoglossum ringspot virus*, and *Barley stripe mosaic virus* have been observed in the shoot and/or root apical meristem tissues [5–9]. Conversely, most viruses have not been observed in SAM cells. Mori et al. [10] observed viral distribution in the shoot meristem of several plants using the fluorescent antibody technique and reported that *Cucumber mosaic virus* (CMV), *Tobacco mosaic virus* (TMV), PVX, and *Potato virus Y* (PVY) were not detected in the SAM and leaf primordia; the size of the virus-free areas differed among virus–host plant combinations. We observed the kinetics of viral distribution in the meristem tissue of tobacco infected with several CMV strains using immunohistochemical microscopy (IHM) and in situ hybridization (ISH), finding that the SAM of tobacco, although infected with some strains of CMV at early stages of infection, subsequently recovered from CMV infection [11, 12]. Recently, green fluorescence protein (GFP)-tagged viruses have been used to detect viral distribution kinetics more easily, although the plant viruses that can be studied by the GFP-tagging method are limited [13, 14]. Thus, IHM and ISH remain the most useful and powerful tools for the observation of virus (and viroid) distribution in meristem tissue (e.g., [15–21]). The following sections describe the detailed techniques for IHM and ISH that we have used to analyze CMV distribution in the meristem tissue of inoculated tobacco.

2 Materials

2.1 Virus, Plant, and Inoculation

1. *Nicotiana tabacum* cv. Xanthi or Samsun, seedlings grown on planting soil in a growth chamber or a greenhouse.
2. Purified CMV (>20 μg/mL).
3. Carborundum powder (600 mesh).
4. Commercial cotton swabs: Not sterile.
5. 0.1 M potassium phosphate buffer pH 7.0: 0.1 M KH_2PO_4, 0.1 M K_2HPO_4.

2.2 Preparation of Paraffin and Embedding Meristem Tissue (See Note 1)

1. FAA solution: 50 % ethanol, 5 % formalin, and 5 % acetic acid.
2. Microtubes (2 mL) or small vials.
3. Fine-tipped tweezers: Not sterile.
4. Razor blades: Not sterile.
5. Vacuum aspirator.
6. Graded series of ethanol/distilled water mixtures (50, 70, 90 and 100 %).
7. Xylene.
8. Embedding paraffin: Paraplast Plus (Sigma, USA) or TissuePrep (Fisher Scientific, USA).
9. Paraffin oven.
10. Plastic Petri dishes or embedding cassettes.
11. 70 % glycerol.
12. Surgical knife: Not sterile.
13. Rotary-microtome.
14. MAS-coated hydrophilic glass slide (Matsunami Glass, Japan).
15. Warming plate.

16. Wash dishes and slides racks.
17. Super PAP Pen, a liquid blocker (Daido Sangyo, Japan).

2.3 Immunohistochemical Microscopy (IHM)

1. PBST: 8.1 mM $Na_2HPO_4 \cdot H_2O$, 1.47 mM KH_2PO_4, 2.68 mM KCl, 137 mM NaCl, 0.05 % Tween 20, pH 7.4.
2. Wash dishes and slides racks.
3. Airtight box for moist chamber.
4. Paper towel.
5. Anti-CMV IgG (Japan Plant Protection Association, Japan): 1:1,000 dilution with PBST containing 4 % skimmed milk.
6. Alkaline phosphatase-conjugated anti-rabbit IgG (Sigma): 1:1,000 dilution with PBST containing 4 % skimmed milk.
7. Coloring buffer for VECTOR blue: 100 mM Tris–HCl, pH 8.2.
8. Alkaline phosphatase substrate kit III (VECTOR blue) (Vector Lab, USA).
9. 70 % glycerol or commercial mounting medium.
10. Light microscope (e.g. Olympus RX-50, Japan).

2.4 In Situ Hybridization (ISH)

1. Prehybridization solution: 50 % formamide, 300 mM NaCl, 10 mM Tris–HCl pH 8.0, 1 mM ethylenediaminetetraacetic acid, 1× Denhardt's solution, 0.25 % sodium lauryl sulfate, 1 % dextran sulfate.
2. Digoxygenin (DIG)-labeled RNA probe, transcribed from the plasmid clone using the DIG RNA Labeling Kit (Roche Diagnostics, Schweiz) (*see* **Note 2**).
3. 50 % formamide solution.
4. 2× SSC: 0.3 M NaCl, 0.03 M sodium citrate (*see* **Note 3**).
5. Buffer 1: 0.1 M maleic acid, 0.15 M NaCl, pH 7.5.
6. Alkaline phosphatase-conjugated anti-DIG antibody (Roche Diagnostics): 1:1,000 dilution with buffer 1 containing 1 % blocking solution.
7. AP buffer: 100 mM Tris, 100 mM NaCl, 5 mM $MgCl_2 \cdot 6H_2O$, pH 9.5.
8. BCIP/NBT Liquid Substrate System (Sigma).
9. 70 % glycerol or commercial mounting medium.

3 Methods

For IHM and ISH, the meristem tissues (approximately 5 mm) excised from the infected plants are fixed and embedded in paraffin medium, and semithin sections (10–15 μm) of the meristem tissues are then obtained. By using an antibody against viral protein or a probe complementary to viral or viroid RNA, the viral and viroid

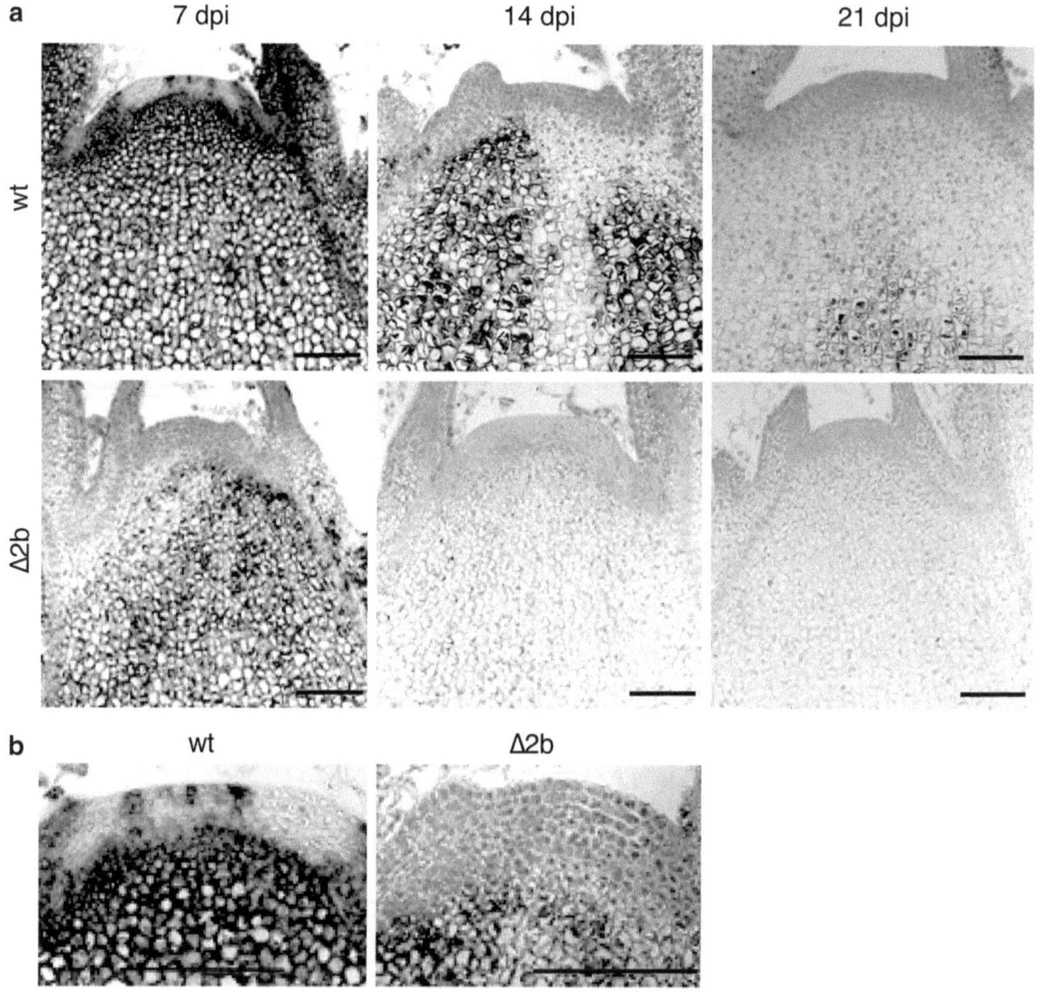

Fig. 2 Representative images from immunohistochemical microscopy that show the distribution of *Cucumber mosaic virus* (CMV) with (wt) and without (Δ2b) the silencing suppressor 2b protein in the SAM of inoculated tobacco plants. Longitudinal sections of meristem tissues of the tobacco plants at 7, 14, and 21 days postinoculation (dpi) were treated with an antibody to the CMV coat protein (CP). CMV-infected cells are stained darkly. (**a**) Low-magnification (100×) images show that both wt and Δ2b CMVs infect most meristem cells near the SAM at 7 dpi. However, at 14 dpi, Δ2b infection is not observed in the meristem tissue, whereas the wt is present. (**b**) High magnification (200×) images show the detailed distribution of CMV in the SAM at 7 dpi, when the viruses are at their maximum spread in meristem tissues. Wt CMV, but not Δ2b, infected SAM cells. Bars = 100 μm

distribution in each section of meristem tissue can be clearly detected. As a demonstration, we sequentially observed the distribution kinetics of CMV with or without a silencing suppressor in tobacco meristem tissues. Representative results are shown in Fig. 2 and Table 1. CMV with a silencing suppressor (wt) invaded SAM at 7 dpi, and levels of the virus gradually reduced at 14 and 21 days postinoculation (dpi). Conversely, CMV without a silencing suppressor (Δ2b) was not detected in SAM at 7 dpi and had disappeared from the meristem tissues at 14 and 21 dpi.

Table 1
Distribution of *Cucumber mosaic virus* with (wt) and without (Δ2b) the silencing suppressor in the shoot meristem tissues of infected tobacco plants

Virus	dpi	Number of sections observed[a]		
		A	B	C
wt	7	14	1	0
	14	0	15	0
	21	0	15	0
Δ2b	7	0	12	3
	14	0	7	8
	21	0	2	13

[a]The kinetics of CMV infections in the shoot meristem were assayed using immunohistochemical microscopy at 7, 14, and 21 days postinoculation (dpi). Fifteen individual plants were observed

A signals detected in the SAM cells (tunica-corpus), *B* signals detected in the meristem but not in the SAM, *C* no signals in the meristem tissue sections

3.1 Sampling, Fixing, and Embedding the Meristem Tissue

1. Because viral distribution in the meristem tissue varies according to the growth environment, growth stage of inoculated plant, and days after virus inoculation, it is preferable to use a large number of samples and multiple sampling periods. For the CMV–tobacco combination, we usually observe 15 or more individual samples at 7, 14, and 21 dpi [11, 12, 21]. In this case, at least 50 individual tobacco plants per experimental group are necessary. The largest leaf of tobacco plants in the five-to-six-leaf stage is mechanically inoculated with purified CMV at 20 μg/ml, and the inoculated tobacco plants are then grown in a growth chamber at 24 °C. The preparation of mock-inoculated healthy plants is necessary to find a nonspecific stain.
2. In sampling, small leaves (>approximately 2 mm) around the meristem are carefully removed using tweezers and a razor blade (Fig. 3b, c). Approximately 5 mm of the longitudinal length of meristem tissue is excised from the shoot (Fig. 3d). The sampled meristem tissues are immediately soaked into 5–10 times volume (w/v) of FAA solution in 2 mL microtubes or small vials on ice (Fig. 3e) (*see* **Note 4**). To remove any air within the tissue that would interfere with fine fixation and to promote the permeation of FAA solution into the meristem tissue, vacuum deairing is conducted for 10 min. This process is repeated several times until the bubbles derived from the meristem tissue had nearly vanished. The meristem samples are then fixed in the FAA solution at 4 °C overnight. Meristem tissues can be stored in FAA for long periods.

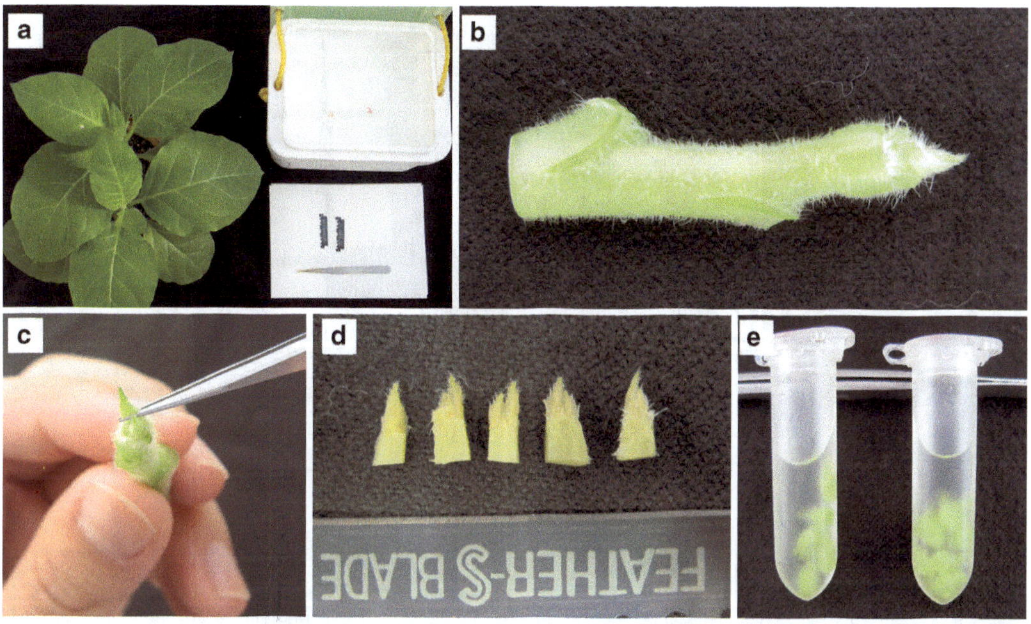

Fig. 3 Photographs of sampling procedures for tobacco meristem tissue. (**a**) The instruments used in the excision of meristem tissue (fine-tipped tweezers, razor blades, and microtubes filled with FAA solution in the ice box). (**b**, **c**) unwanted small leaves (>approximately 2 mm) on the shoot top are removed using tweezers and a razor blade. (**d**) Excised meristem tissues approximately 5 mm in longitudinal length. (**e**) Excised meristem tissues are immediately soaked in FAA solution in the 2 mL microtubes

3. The fixed samples are dehydrated by a graded series of ethanol/water mixtures (50 and 70 % for 10 min on ice, then 90 % for 10 min and 100 % for two periods of 30 min at room temperature). Following treatment with an interim solution (100 % ethanol:xylene = 1:1) for 30 min, the 100 % ethanol is replaced by 100 % xylene for two periods of 30 min (*see* **Note 5**)
4. Paraffin pieces are melted in a beaker using a paraffin oven at 55–65 °C (Fig. 4a) (*see* **Note 6**).
5. An amount of melted paraffin equivalent to the xylene is added to the microtubes or vials, and the samples are incubated in an oven at 55–65 °C until the xylene had completely evaporated (2–3 days).
6. A solution of 70 % glycerol is swabbed in a Petri dish before the melted paraffin is added to ensure the easy removal of the caked paraffin block from the dish. The meristem tissues of each experimental group in the melted paraffin are transferred with warmed tweezers into the Petri dish until it is filled with paraffin and aligned to leave enough space between the meristem tissues (Fig. 4a). Alternatively, each individual meristem tissue can be separately transferred into an embedding cassette filled with melted paraffin. The paraffin is then solidified at

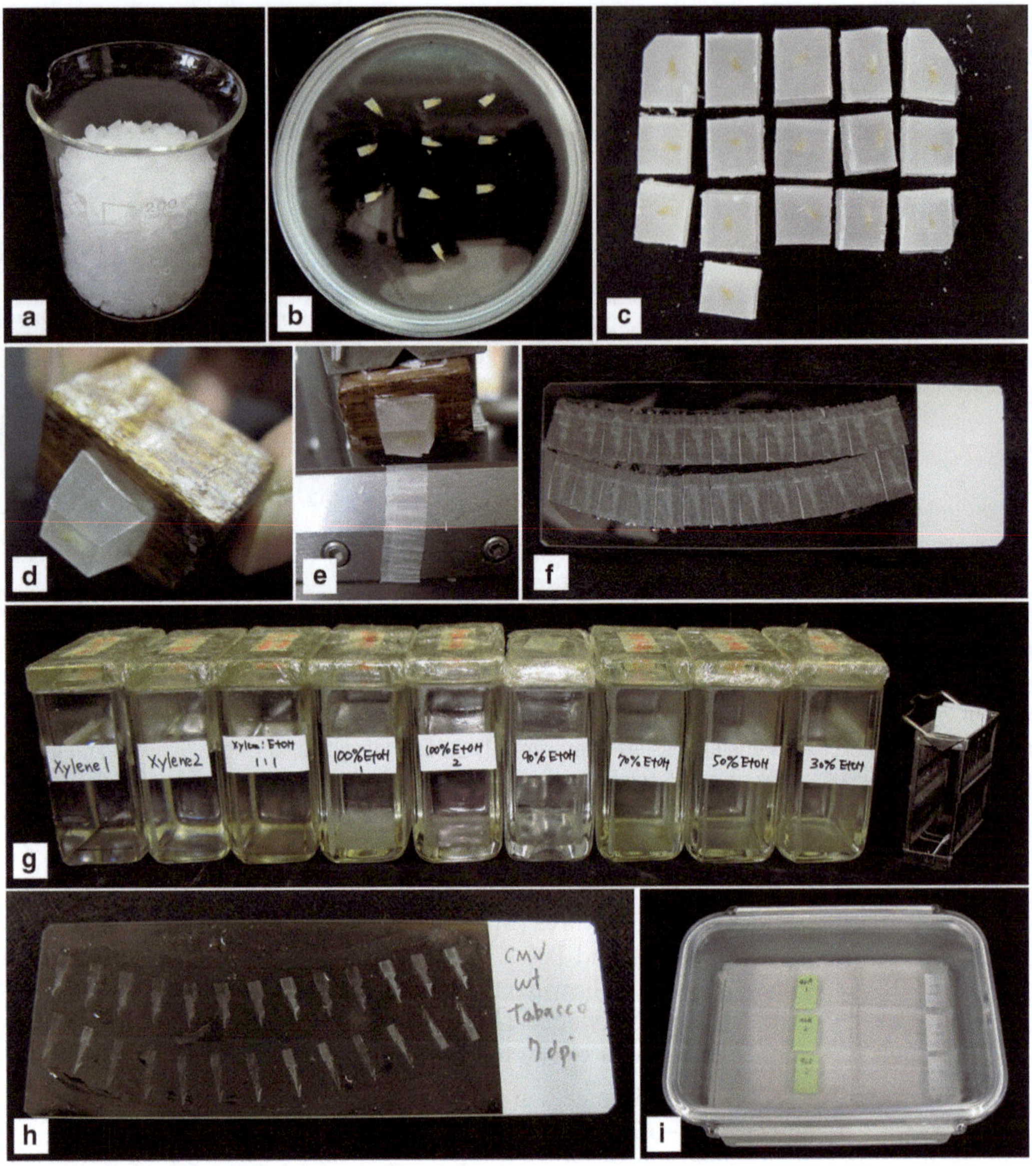

Fig. 4 Representative photographs of paraffin embedding, tissue processing for serial semithin sections, and histochemical procedures. (**a**) Paraffin pieces are melted in a beaker. (**b**) Meristem tissues of the same experimental group are transferred into a Petri dish filled with melted paraffin and aligned. (**c**) Paraffin blocks are carved into individual samples. (**d**) Excess paraffin around each sample is trimmed using a surgical knife. Note that the top and the bottom of the paraffin block are parallel. (**e**) Semithin serial sections are cut with a rotary-microtome. (**f**) The serial sections including SAM cells are placed in distilled water on a slide glass. (**g**) The paraffin sections on the slide glass are dewaxed and hydrated using a staining dish and slide rack. (**h**) To prevent outflow of the reaction solution of the antigen or probe from the slides, the deparaffinized tissue sections are surrounded using a Super PAP Pen. (**i**) The slides are placed in an airtight box as the meristem tissue sections are treated with antibody solution, hybridization solution, or a color substrate dropped onto the sections

room temperature. After caking, the paraffin blocks are kept at 4 °C until use (*see* **Note 7**)

7. Paraffin blocks are removed from the Petri dish or the embedding cassettes and carved into individual samples (Fig. 4c). Excess paraffin around the samples is trimmed using a surgical knife (*see* **Note 8**). To obtain good serial sections, the top and bottom of the paraffin block are trimmed in parallel (Fig. 4d). Semithin (10–15 μm) serial sections (paraffin ribbons) are cut with a rotary-microtome (Fig. 4e) (*see* **Note 9**), placed in distilled water on a slide glass (Fig. 4f) (*see* **Note 10**), and then thoroughly dried on a warming plate at 37 °C overnight.
8. The paraffin sections on the slide glass are dewaxed in xylene for two periods of 15 min, washed in 100 % ethanol for two periods of 5 min, and hydrated in a graded series of ethanol/water mixtures (70, 50, 30 %, and distilled water for 5 min each) using staining dishes and slide racks (Fig. 4g). The sections were then used for IHM or ISH. To prevent the outflow of the reaction solution of antigen or probe from the slides, the deparaffinized tissue sections are surrounded using a Super PAP Pen (Fig. 4h). The sections are then air-dried.

3.2 Immunohistochemical Microscopy (IHM)

1. The sections are blocked in PBST containing 4 % skimmed milk for 30 min using staining dishes and slide racks.
2. The slides are placed on a wet paper towel in an airtight box (Fig. 4i), and the antibody solution (1,000×CMV CP IgG in PBST containing 4 % skimmed milk) is dropped onto the sections. The sections are incubated with the antibody for 2 h at 37 °C (or overnight at 4 °C) and are then washed three times using PBST for 5 min in the staining dishes.
3. The sections are treated with an alkaline phosphatase-conjugated secondary antibody for 2 h at 37 °C in an airtight box and are then washed three times using PBST for 5 min in the staining dishes.
4. The sections are incubated with a color substrate kit (VECTOR blue) in the dark until the positive control section is well stained. CMV-infected cells are stained dark blue (*see* **Note 11**).
5. The stained sections are air-dried and covered with a coverslip with 70 % glycerol or a commercial mounting medium.
6. The stained sections are examined under light microscopy and photographed (*see* **Note 12**).

3.3 In Situ Hybridization (ISH)

1. The sections are washed with 2× SSC for 5 min using staining dishes, and the slides are then placed on a wet paper towel with 50 % formamide solution in an airtight box. A prehybridization solution is dropped onto the sections on the slides, and the sections are prehybridized for 1 h at 42 °C in a hybridization oven.

2. After removal of the prehybridization solution, the sections in the airtight box are incubated in a hybridization solution containing digoxygenin (Roche Diagnostics) (DIG)-labeled RNA probe complementary to the conserved 3′-UTR sequence of the CMV genomic RNAs (approximately 300 nt) for 12 h at 42 °C in a hybridization oven.
3. The slides are washed twice with low-stringency wash buffer (2× SSC) for 15 min at room temperature, then washed twice at 50 °C with high-stringency wash buffer (first with 0.2× SSC and then with 0.1× SSC) to reduce background and nonspecific binding of the probe.
4. The sections are washed in buffer 1 (0.1 M maleic acid and 0.15 M NaCl, pH 7.5) for 5 min, followed by incubation in a blocking solution for 30 min in the wash dishes. The sections are then incubated in a 1:1,000 dilution of alkaline phosphatase-conjugated anti-DIG antibody (Roche Diagnostics) for 1 h at room temperature in an airtight box. The sections are washed twice in buffer 1 containing 0.05 % Tween 20 for 15 min.
5. Following equilibration with AP buffer for 5 min, the sections are incubated with the color substrate solution (BCIP/NBT Liquid Substrate System) in the dark (*see* **Note 13**).
6. The stained sections are air-dried and covered by a coverslip with 70 % glycerol or a commercial mounting medium.
7. The stained sections are examined with light microscopy and photographed.

4 Notes

1. Because viral and viroid RNAs are abundant in the infected cells and are very stable in the fixed tissues, the use of RNase-free reagents is unnecessary in all procedures. If the in situ hybridization technique is applied to detect the mRNA distribution in the tissue, all reagents and instruments should be treated with diethyl pyrocarbonate.
2. We use an approximately 300-nt DIG-labeled RNA probe complementary to the conserved 3′-UTR sequence of the CMV genomic RNAs. When the DIG-labeled RNA probe is too long (>1 kb), hydrolysis of the labeled RNA probe is required to improve probe penetration in the tissue sections. Generally, a 100–150 nt RNA fragment causes optimal penetration. Hydrolysis protocol: 30 μl 200 mM Na_2CO_3 and 20 μl 200 mM $NaHCO_3$ are added to 50 μl labeled RNA probe and then incubated at 60 °C for *t* min. $t = (L_0 - L_f)/(0.11 \times L_0 \times L_f)$. L_0 = initial size of probe RNA (kb), L_f = expected final size of probe (kb). After hydrolysis, the probe is purified by ethanol

precipitation. Note that RNase-free reagents are necessary for RNA probe preparation. DIG-labeled synthetic oligo-DNA is also useful for in situ hybridization that distinguishes between CMV strains from subgroup I and subgroup II [22].

3. When SSC is used for high-stringency wash buffer (0.2× SSC and 0.1× SSC), 2× SSC is diluted by distilled water.
4. The advantages of using FAA fixative are that (a) FAA can rapidly penetrate meristem tissues, (b) FAA is low-cost, (c) FAA can be preserved for long periods at 4 °C, and (d) plant tissues can be stored in FAA for long periods. If the size of the excised meristem tissues is too large, the penetration of FAA fixative into the tissue becomes insufficient, and good serial sections are not obtained.
5. During the replacement of the graded series of ethanol/water mixtures or xylene, the meristem tissue samples should not become dry. Some liquid from the previous series is left in the tube so that the samples do not dry out. The chlorophyll pigment in the cells is decolorized, and the tissues become transparent during the ethanol treatment.
6. The melting temperature varies for different paraffin products.
7. The paraffin-embedded tissues can be preserved for long periods (several years) at 4 °C in a refrigerator.
8. The removed excess paraffin pieces can be recycled by melting in a paraffin oven.
9. Cutting parallel to the longitudinal axis of the stem is very important for obtaining distinct and precise SAM cell images. The cut sections are often rolled, which prevents the acquisition of a good paraffin ribbon. In most cases, this problem is due to static electricity on the paraffin sections during processing. To alleviate this issue, the room humidity is raised, the thickness of the paraffin sections is further reduced, cuts are made slowly, or the used blade of the rotary-microtome is replaced with a new blade.
10. As serial sections are obtained from one end of the meristem section to the other, only some sections contained SAM cells. Because the SAM of tobacco has a clear dome shape, visual recognition and selection of the serial sections including SAM is possible. If both CP and viral RNA are detected from an individual meristem sample, two to four sections from an individual paraffin ribbon are separately placed onto two slide glasses; one is used for IHM, and the other is used for ISH.
11. Good staining intensity is generally obtained with a 20–30 min incubation time. Staining intensity may be increased by extending the incubation time up to 4 h, although background staining may also increase. Color development should be stopped before the negative control healthy tissue is stained.

12. We recommend that photographs be taken of all samples at both high (200×) and low (100×) magnification. The degree of CMV distribution in the meristem was defined in three categories: A, signals detected in the SAM cells (tunica-corpus), B, signals detected in the meristem but not in the SAM; C, no signals in the meristem (Fig. 2a and Table 1).
13. Using the BCIP/NBT Liquid Substrate System (Sigma), staining incubation time can be increased up to overnight at room temperature in the dark, but the color development reaction must be stopped before the negative control healthy tissue is stained. If a self-prepared reagent of BCIP/NBT solution is used, background staining increases during overnight incubation.

Acknowledgments

The authors would like to thank Dr. Yosuke Matsushita (NARO Institution of Floricultural Science, Japan) for his kind comments on the manuscript.

References

1. Barton MK (2010) Twenty years on: the inner workings of the shoot apical meristem, a developmental dynamo. Dev Biol 341:95–113
2. Evert RF (2006) Apical meristems. In: Esau's plant anatomy: meristems, cells, and tissues of the plant body – their structure, function and development, 3rd edn. Wiley, Hoboken, NJ, pp 133–174
3. Hull R (2002) Matthews' plant virology, 4th edn. Academic, San Diego, CA, pp 407–408
4. Faccioli G, Marani F (1998) Virus elimination by meristem tip culture and tip micrografting. In: Hadidi A, Khetarpal RK, Koganezawa H (eds) Plant virus disease control. APS Press, St. Paul, MN, pp 346–380
5. Appiano A, Pennazio S (1972) Electron microscopy of potato meristem tips infected with potato virus X. J Gen Virol 14:273–276
6. Kitajima EW, Costa AS (1969) Association of pepper ringspot virus (Brazilian tobacco rattle virus) and host cell mitochondria. J Gen Virol 4:177–181
7. Lin NS, Langenberg WG (1984) Distribution of barley stripe mosaic-virus protein in infected wheat root and shoot tips. J Gen Virol 65: 2217–2224
8. Roberts DA, Christie RG, Archer MC Jr (1970) Infection of apical initials in tobacco shoot meristems by tobacco ringspot virus. Virology 42:217–220
9. Toussaint A, Dekegel D, Vanheule G (1984) Distribution of *Odontoglossum ringspot virus* in apical meristems of infected *Cymbidium* cultivars. Physiol Plant Pathol 25:297–305
10. Mori K, Hosokawa D, Watanabe M (1982) Studies on multiplication and distribution of viruses in plants by use of fluorescent antibody technique I. Multiplication and distribution of viruses in shoot apices. Ann Phytopathol Soc Jpn 48:433–443
11. Mochizuki T, Ohki ST (2004) Shoot meristem tissue of tobacco inoculated with *Cucumber mosaic virus* is infected with the virus and subsequently recovers from infection by RNA silencing. J Gen Plant Pathol 70:363–366
12. Mochizuki T, Ohki ST (2005) Amino acid 129 in the coat protein of *Cucumber mosaic virus* primarily determines invasion of the shoot apical meristem of tobacco plants. J Gen Plant Pathol 71:326–332
13. Qu F, Ye X, Hou G, Sato S, Clemente TE, Morris TJ (2005) RDR6 has a broad-spectrum but temperature-dependent antiviral defense role in *Nicotiana benthamiana*. J Virol 79: 15209–15217
14. Schwach F, Vaistij FE, Jones L, Baulcombe DC (2005) An RNA-dependent RNA polymerase prevents meristem invasion by *Potato virus X* and is required for the activity but not the

production of a systemic silencing signal. Plant Physiol 138:1842–1852

15. Di Serio F, Martinez de Alba AE, Navarro B, Gisel A, Flores R (2010) RNA-dependent RNA polymerase 6 delays accumulation and precludes meristem invasion of a viroid that replicates in the nucleus. J Virol 84: 2477–2489
16. Dong F, Mochizuki T, Ohki ST (2010) *Tobacco ringspot virus* persists in the shoot apical meristem but not in the root apical meristem of infected tobacco. Eur J Plant Pathol 126: 117–122
17. Gosalvez-Bernal B, Garcia-Castillo S, Pallas V, Sanchez-Pina MA (2006) Distribution of carnation viruses in the shoot tip: exclusion from the shoot apical meristem. Physiol Mol Plant Pathol 69:43–51
18. Gosalvez-Bernal B, Genoves A, Navarro JA, Pallas V, Sanchez-Pina MA (2008) Distribution and pathway for phloem-dependent movement of *Melon necrotic spot virus* in melon plants. Mol Plant Pathol 9: 447–461
19. Martín-Hernández AM, Baulcombe DC (2008) Tobacco rattle virus 16-kilodalton protein encodes a suppressor of RNA silencing that allows transient viral entry in meristems. J Virol 82:4064–4071
20. Matsushita Y, Usugi T, Tsuda S (2011) Distribution of *tomato chlorotic dwarf viroid* in floral organs of tomato. Eur J Plant Pathol 130:441–447
21. Sunpapao A, Nakai T, Dong F, Mochizuki T, Ohki ST (2009) The 2b protein of *Cucumber mosaic virus* is essential for viral infection of the shoot apical meristem and for efficient invasion of leaf primordia in infected tobacco plants. J Gen Virol 90:3015–3021
22. Takeshita M, Shigemune N, Kikuhara K, Furuya N, Takanami Y (2004) Spatial analysis for exclusive interactions between subgroups I and II of *Cucumber mosaic virus* in cowpea. Virology 328:45–51

Index

Ichiro Uyeda and Chikara Masuta (eds.), *Plant Virology Protocols*, Methods in Molecular Biology, vol. 1236, DOI 10.1007/978-1-4939-1743-3, © Springer Science+Business Media New York 2015

F

G

H

I

L

M

N

O

P

W

Y

MIX
Papier aus verantwortungsvollen Quellen
Paper from responsible sources
FSC® C105338

If you have any concerns about our products,
you can contact us on
ProductSafety@springernature.com

In case Publisher is established outside the EU,
the EU authorized representative is:
Springer Nature Customer Service Center GmbH
Europaplatz 3, 69115 Heidelberg, Germany

Printed by Libri Plureos GmbH
in Hamburg, Germany